Thermomechanical Behavior of Dissipative Composite Materials

Series Editor
Yves Rémond

Thermomechanical Behavior of Dissipative Composite Materials

George Chatzigeorgiou
Nicolas Charalambakis
Yves Chemisky
Fodil Meraghni

ISTE PRESS
ELSEVIER

First published 2017 in Great Britain and the United States by ISTE Press Ltd and Elsevier Ltd

ISTE Press Ltd
27-37 St George's Road
London SW19 4EU
UK

www.iste.co.uk

Elsevier Ltd
The Boulevard, Langford Lane
Kidlington, Oxford, OX5 1GB
UK

www.elsevier.com

Notices
Knowledge and best practice in this field are constantly changing. As new research and experience broaden our understanding, changes in research methods, professional practices, or medical treatment may become necessary.

Practitioners and researchers must always rely on their own experience and knowledge in evaluating and using any information, methods, compounds, or experiments described herein. In using such information or methods they should be mindful of their own safety and the safety of others, including parties for whom they have a professional responsibility.

To the fullest extent of the law, neither the Publisher nor the authors, contributors, or editors, assume any liability for any injury and/or damage to persons or property as a matter of products liability, negligence or otherwise, or from any use or operation of any methods, products, instructions, or ideas contained in the material herein.

British Library Cataloguing-in-Publication Data
A CIP record for this book is available from the British Library
Library of Congress Cataloging in Publication Data
A catalog record for this book is available from the Library of Congress
ISBN 978-1-78548-279-3

Printed and bound in the UK and US

Contents

Foreword 1

This book, by Dr. George Chatzigeorgiou, Dr. Nicolas Charalambakis, Dr. Yves Chemisky and Prof. Fodil Meraghni, deals with composite materials and their complex mechanical behavior. It takes an important new position among the current literature, often very specialized, existing in this domain.

The authors are well-recognized scientists in the materials community and experienced mechanicians, researchers and teaching researchers. They have worked for many years on these materials, used in all domains of industry, particularly in the automotive and aeronautical industry. We can additionally mention their integration within the LEM3 Laboratory, University of Lorraine, Arts et Métiers and CNRS, a Laboratory of Excellence within the Materials Community with Worldwide recognition.

After a general introductory chapter on the notations and continuum mechanics theory they will use, the authors propose a summary of the numerical methods useful for engineers. They then show how composite materials, integrating useful but complex heterogeneities, must be dealt with in mechanics theory. They then focus on periodic and random reinforcements composite behavior, for which the homogenization techniques enable changing scales in a rational manner. This provides the knowledge of the desired macroscopic mechanical properties of a given microstructure, but also the analysis of the local phenomena induced by an external mechanical load on the structure.

This book is rigorously written and includes the latest scientific developments. It proposes a simple and complete view of the existing knowledge of the mechanical behavior of these materials, and is suitable for students and engineers seeking to familiarize themselves with the mechanics of composite materials, and develop complex, optimized and light-weighted structures.

Prof. YVES RÉMOND
University of Strasbourg – CNRS

Foreword 2

The art of mixing heterogeneous materials to improve the bulk properties of mixtures, such as stiffness and strength, has been refined over thousands of years, especially for the civil and military needs of civilizations. During the last two centuries and since the industrial revolution, composite materials, synthetic and natural composites, have been widely used in transportation electronics and biomedical application, among others. The need to develop mathematical models to describe the macroscopic response of heterogeneous materials arose alongside advances in imaging methods that revealed microstructural characteristics contributing to their macroscopic response. In the case of metals, specialized heat treatments combined with other processing techniques led to the creation of desirable microstructures for optimal thermomechanical properties, while in polymeric, metallic and ceramic matrix composite materials, fiber volume fractions and their placement together with the selection of other processing parameters resulted in designing desirable macroscopic or effective bulk properties.

Scientists and engineers have tried to address two key questions related to macroscopic properties of heterogeneous materials: Given sufficient information about their microstructure, how can macroscopic or effective bulk properties of heterogeneous materials be accurately modeled; and, given desired macroscopic properties, how can optimal microstructures be chosen which result in such macroscopic properties. The upscaling process from the micro length scales, where microstructures are realizable to the macroscale, formed the forward homogenization problem that resulted in the development of various analytical and computational techniques over the last few decades. Averaging methods were developed to estimate the macroscopic thermomechanical properties of composites and other heterogeneous media with input of their constituents properties, their morphology and volume fractions. This effort initially focused on polycrystalline metals with heterogeneous precipitates during the middle of the 20th century, shifting to polymeric fibrous composites during the latter part of the century, resulting in what is widely accepted now as the field of micromechanics.

A key contribution that formed the mathematical basis for the development of micromechanics is the Eshelby solution, which is the fundamental elasticity solution of the elastic field surrounding an ellipsoidal inclusion, with a prescribed eigenstrain, embedded into an infinite elastic medium. Since its publication in 1957, and especially its application to heterogeneous media with an extension to what is now known as the Eshelby equivalence principle, many of the averaging upscaling methodologies for composites and heterogeneous microstructures are based on the Eshelby solution. In addition to averaging techniques, like the self-consistent and Mori-Tanaka methods, based on extensions of the Eshelby solution in order to account for particle interactions, computational methods have been developed for heterogeneous media with periodic microstructures, following analytical upscaling or homogenization methods, especially in estimating bounds on macroscopic properties. The use of multiple scales and asymptotic series over a representative volume element allowed for the development of rigorous formulations to be derived and in most cases numerical methods like finite elements to be effectively utilized.

The field of micromechanics is still very active today as new challenges arise from multifunctional materials with coupled properties, for example piezoelectric, electrostrictive, piezomagnetic and magnetostrictive, and especially when microstructure plays a critical role for multifunctional behavior, as is the case for most nanocomposites. An additional challenge is that field interactions in multifunctional heterogeneous materials, such as electromagnetic and thermomechanical, are nonlinear. Nonlinear dissipative heterogeneous materials bring an additional challenge due to path dependence of their macroscopic behavior and in some cases explicit time dependence. This necessitates the development of incremental formulations for the macroscopic response of history dependent heterogeneous materials. A serious effort to include some of the current methodologies and trends, especially the nonlinear behavior of heterogeneous dissipative materials and the necessary analytical and computational tools, is the focus of this book.

The authors follow a rigorous methodology that first introduces the conservation principles and the constitutive laws based on thermodynamics principles. The thermodynamic framework is structured in such a way that allows its direct implementation in homogenization techniques proposed for composites. The book also introduces the computational tools originally developed for homogeneous inelastic dissipative media, such as the return mapping algorithms, with an adaptation for the micromechanics techniques to be used for heterogeneous media. The authors address rigorously periodic homogenization, based on the asymptotic expansion method, for the fully coupled thermomechanical homogenization framework for dissipative materials. Later in the book, they also address the Eshelby-based micromechanical approaches for random media and their extension to fully coupled thermomechanical processes in dissipative composites.

Overall, the authors have developed a consistent formulation of the thermomechanical response of inelastic, history dependent constituents and they have incorporated it, using a multiscale homogenization formulation to estimate the macroscopic history dependent response of heterogeneous materials. They have also provided, along with the formulation, the necessary computational tools for the actual implementation. Graduate students and researchers in engineering sciences, applied mathematics and mechanics of materials, as well as many professionals developing new heterogeneous materials with desirable thermomechanical responses and energy dissipation capacity, will find this book a very useful companion, written by experts in the field.

Dimitris C. LAGOUDAS
Texas A&M University, USA

Preface

The objective of this book is to present a consistent methodology for the study and numerical computation of the thermomechanical response of materials and structures, with a particular focus on composites. We recognize that there are plenty of studies in the literature that successfully treat, both theoretically and computationally, the thermomechanical response of various categories of dissipative materials or composites and that are not included in this manuscript. Our aim though is not to present an extensive review of these studies, but to propose to the scientific community a general framework for studying the vast majority of dissipative materials and composites under fully coupled thermomechanical loading conditions. We cover many aspects of the modeling process, so that the reader is able to find how to: (i) identify the conservation laws and thermodynamic principles that need to be respected by any solid material, (ii) construct proper constitutive laws for various types of dissipative processes, both rate-independent and rate-dependent, by utilizing an appropriate thermodynamic framework, (iii) design robust numerical algorithms that permit accuracy and efficiency in the calculations of very complicated constitutive laws, (iv) extend all the previous points to the study of composites, utilizing rigorous homogenization theories for materials and structures with both periodic and random microstructure. For the last point, the book explores the concepts of periodic homogenization, namely the asymptotic expansion homogenization method, as well as various micromechanics theories based on the Eshelby approach. We believe that our book, with the topics it covers, will be useful to both young and advanced researchers that want to obtain a general guide to properly studying the thermomechanical response of dissipative materials and composites, and to identifying robust and accurate computational schemes.

Chapter 1 is devoted to a quick presentation of tensor calculus, both in Cartesian and curvilinear coordinate systems, as well as the various symbols denoting tensor operations that are utilized throughout the book. While the indicial notation with the Einstein summation rule is very helpful in many situations, the large number of indices, mainly introduced in Chapter 3 on computational methods and Chapters 5 and 6 on homogenization theories, requires a more elegant representation of tensors. Thus, the tensorial notation with bold fonts for vectors and higher order tensors is chosen in the book. The first chapter also discusses the Voigt notation, which is particularly useful for representing second and fourth order tensors as vectors and matrices respectively, simplifying the computational procedures. In addition, a special notation for isotropic fourth order tensors is included in section 1.1.3.

Chapter 2 is a short summary of the continuum mechanics theory and the identification of constitutive laws based on thermodynamic principles. The first four sections discuss the general principles of continuum mechanics (kinetics, kinematics, conservation laws, thermodynamics) and their reduction when small deformation procedures are considered. The fifth section of the chapter focuses on the description of constitutive laws for dissipative materials using a proper thermodynamic framework. The presented framework is quite general in order to include many types of mechanisms: viscoelasticity, plasticity, viscoplasticity and continuum damage constitutive laws are discussed in this section. Even though not mentioned explicitly, the framework is also capable of describing phase transformation mechanisms, such as martensitic transformation occurring, for example, in shape memory alloys. The chapter closes with the presentation of a thermomechanical parameters identification strategy through appropriate experimental protocol for an elastoplastic material.

Chapter 3 presents rigorous integration methods that make it possible to identify and computationally simulate the response of materials and structures under quasi-static loading conditions when nonlinear mechanisms appear. The chapter introduces the general iterative scheme that can be applied for solving a boundary value problem using the finite element method (though without entering into details on finite element computations) and focuses mainly on the numerical implementation of the constitutive law for a homogeneous material. The presented methodologies are based on the well known return mapping algorithm scheme. Analytical description on the methods is given for the case of rate independent plasticity. Numerical applications in plasticity and viscoelasticity are also included in the chapter. The numerical procedures discussed here, even though applied to homogeneous materials, can be considered as the basis for the design of numerical algorithms applied to composites.

Chapter 4 introduces the notion of homogenization, which is presented in two different frameworks: one that focuses on homogenization from an engineering point of view, and one that describes the principles of mathematical homogenization. Both frameworks eventually lead to the same conclusions, even though they start from a different theoretical background. The engineering oriented homogenization accounts for composites with periodic or random microstructure and its concepts are mainly utilized in Chapter 6. On the other hand, the mathematical homogenization is the basis of the asymptotic expansion homogenization method that is utilized in Chapter 5 to describe the homogenization of composites with periodic microstructure.

Chapter 5 focuses on composite materials and structures with periodic microstructure. The asymptotic expansion homogenization method is employed for identifying the microscopic and macroscopic principles that the actual composite obeys (kinetics, kinematics, conservation laws, thermodynamics). For quasi-static processes and composites with dissipative material constituents a numerical scheme is proposed for solving the fully coupled thermomechanical problem iteratively and simultaneously in the two scales (microscopic and macroscopic). An illustrative example of multilayered composite material is discussed and various numerical applications with elastic, plastic and viscoplastic responses are presented.

Chapter 6 discusses homogenization techniques for composite materials and structures with random microstructure. The described methodologies belong to the mean field or Eshelby based theories. After presenting the classical Eshelby problems the chapter introduces three homogenization techniques: the Mori-Tanaka, the self consistent and the Ponte-Castañeda and Willis method. For quasi-static processes and composites with dissipative material constituents an iterative numerical scheme is proposed for solving the fully coupled thermomechanical homogenization problem by using the concentration tensors calculated with an Eshelby-based approach. An illustrative example of a composite material consisting of a matrix phase and spherical particles is presented and various numerical applications with elastic and plastic responses are examined.

Appendices 1 and 2 are devoted to large deformation processes and their connection with the homogenization theories. Appendix 1 discusses the average theorems in the reference configuration of a deformable body, i.e. the stress is measured through the Piola stress tensor and the deformation is identified through the deformation gradient tensor. Appendix 2 presents the extension of the asymptotic expansion homogenization method for composites with periodic microstructure, using the same stress and deformation measurements (though without entering into details of proper microscopic and macroscopic thermodynamic potentials). While these two appendices do not provide the global picture of a general homogenization framework in large deformation processes, they can be considered as important contributions towards such a target.

We would like to acknowledge our colleague, Professor Etienne Patoor, for his assistance and encouragement and express our sincere gratitude to Professor Yves Rémond and Dimitris Lagoudas for honoring us with their respective forewords. We also thank Professor André Chrysochoos and Dr. Gilles Robert for our fruitful discussions about thermomechanics, as well as Professor François Murat for his remarks and comments about the periodic homogenization. Additionally, we highly appreciate the scientific discussions we had with our international partners from Texas A&M University (Assistant Professor Darren Hartl), VirginiaTech (Associate Professor Gary Don Seidel), FAU Erlangen-Nürnberg (Professor Paul Steinmann), Bilkent University (Assistant Professor Ali Javili), Swansea University (Senior Lecturer Mokarram Hossain), TU Bergakademie Freiberg (Professor Björn Kiefer), University of Virginia (Professor Marek-Jerzy Pindera) and TU Dortmund (Professor Andreas Menzel). Last but not least, we would like to warmly thank all the members of the SMART team of LEM3 laboratory for their support. Special thanks belong to our current and former Ph.D. students and postdocs Francis Praud, El-Hadi Tikarrouchine, Nadia Achour-Renault, Clément Nony-Davadie, Dr. Nicolas Despringre, Dr. Dimitris Chatziathanasiou, Dr. Kevin Bonnay and Dr. Dimitris Anagnostou for their valuable comments and remarks, as well as the Experimental Engineer Laurent Peltier, the Research Engineer Dr. Pierre Charbonnier and Mr. Patrick Moll for their assistance with the experimental protocol in the parameter identification procedure described in Chapter 2.

George CHATZIGEORGIOU
Nicolas CHARALAMBAKIS
Yves CHEMISKY
Fodil MERAGHNI
July 2017

Nomenclature

$\boldsymbol{I}$	Second order identity tensor
$\boldsymbol{\mathcal{I}}$	Symmetric fourth order identity tensor
ϵ	Third order permutation tensor
$\boldsymbol{X}$	Position vector in reference configuration
$\boldsymbol{x}$	Position vector in current configuration/under small deformations assumption
t	Time
ρ	Density
$\boldsymbol{u}$	Displacement vector
$\boldsymbol{F}$	Second order deformation gradient tensor
$\boldsymbol{E}$	Second order Green-Lagrange strain tensor
$\boldsymbol{\varepsilon}$	Second order strain tensor under small deformations assumption
$\boldsymbol{t}$	Traction vector
$\boldsymbol{\sigma}$	Second order Cauchy's stress tensor
$\boldsymbol{\Sigma}$	Second order Piola stress tensor
$\boldsymbol{b}$	Body forces vector per unit mass

θ	Absolute temperature
$\boldsymbol{\nabla\theta}$	Temperature gradient vector
$\boldsymbol{q}$	Heat flux vector
q^s	Prescribed heat flux on boundary
e	Internal energy per unit mass
$\mathcal{E}$	Internal energy per unit volume
$\mathcal{R}$	Heat sources per unit mass
ς	Specific entropy per unit mass
η	Specific entropy per unit volume
r	Difference between the rates of the mechanical work and the internal energy
$\mathcal{Q}$	Thermal energy rate per unit volume entering or leaving a material point
Ψ	Helmholtz free energy potential
G	Gibbs free energy potential
$\boldsymbol{\xi}$	Internal state variables
$\boldsymbol{\Xi}_\Psi$	Thermodynamic forces derived from Ψ
$\boldsymbol{\Xi}_G$	Thermodynamic forces derived from G
γ	Internal entropy production per unit volume
γ_{loc}	Local entropy production (intrinsic dissipation)
γ_{con}	Entropy production by heat conduction
c_v	Specific heat capacity at constant volume
c_p	Specific heat capacity at constant pressure
E	Young's modulus

$\boldsymbol{\omega}_{\varepsilon}$ Test functions vector for displacements in weak formulation

ω_{θ} Test functions scalar for temperature in weak formulation

$\boldsymbol{\kappa}$ Second order thermal conductivity tensor

$\boldsymbol{D}^{\varepsilon}$ Fourth order tangent modulus tensor: variation of stress with strain

$\boldsymbol{D}^{\theta}$ Second order tangent modulus tensor: variation of stress with temperature

$\boldsymbol{R}^{\varepsilon}$ Second order tangent modulus tensor: variation of r with strain

R^{θ} Scalar tangent modulus: variation of r with temperature

$\boldsymbol{\Gamma}^{\varepsilon}$ Second order tangent modulus tensor: variation of γ_{loc} with strain

Γ^{θ} Scalar tangent modulus: variation of γ_{loc} with temperature

$\boldsymbol{H}^{\varepsilon}$ Second order tangent modulus tensor: variation of η with strain

H^{θ} Scalar tangent modulus: variation of η with temperature

$\boldsymbol{C}$ Elasticity fourth order tensor

$\boldsymbol{\alpha}$ Thermal expansion coefficients second order tensor

$\boldsymbol{\kappa}$ Thermal conductivity second order tensor

$\boldsymbol{S}$ Eshelby tensor

$\boldsymbol{S}^{\kappa}$ Eshelby-like tensor for thermal conductivity

$\boldsymbol{T}^{\varepsilon}$ Strain – strain interaction tensor

$\boldsymbol{T}^{\theta}$ Strain – temperature interaction tensor

$\boldsymbol{T}^{\kappa}$ Temperature gradient interaction tensor

$\boldsymbol{A}^{\varepsilon}$ Strain – strain concentration tensor

$\boldsymbol{A}^{\theta}$ Strain – temperature concentration tensor

$\boldsymbol{A}^{\kappa}$ Temperature gradient concentration tensor

ϵ Characteristic length scale of composite with periodic microstructure

f_r Volume fraction of the r_{th} phase/inhomogeneity in composite media

1

Mathematical Concepts

The scope of this chapter is to provide a summary of useful identities and properties from tensor analysis and tensor calculus both in Cartesian and curvilinear coordinate systems. It can be seen as a helpful introduction to the various notations that appear throughout the book, especially the "matrix notation" which substitutes the usual indicial notation. In-depth analyses of tensors, tensor operations and differential geometry as applied to continuum mechanics have been presented by Lai *et al.* [LAI 10] and Ciarlet [CIA 05].

In addition to the tensor analysis, the Voigt notation for second- and fourth-order tensors is presented, along with some useful properties that permit us to substitute tensor with matrix operations. These substitutions are very helpful for the numerical implementation of problems that involve fourth-order tensors with minor symmetries.

A small discussion about the Hill notation for isotropic tensors is also included. This tensor representation is utilized in the examples in Chapter 6.

1.1. Tensors in Cartesian coordinates

A tensor is defined as a geometric object that describes linear relations between scalars, vectors or other tensors and thus it can be seen as a multilinear map. In mathematics and physics, a tensor is a very general object, intrinsically defined from a vector space, that is independent of coordinate systems. This section focuses on tensors represented in Cartesian coordinate systems, while later in this chapter tensors in curvilinear coordinates are discussed.

Direct notation is generally adopted in this chapter. Occasional use is made of index notation, the Einstein summation convention for repeated indices being implied. Unless stated otherwise, the indices in tensors take values from 1 to 3. Bold letters are used to denote vectors and tensors in order to distinguish them from scalars. When the vectors and tensors appear in their indicial notation, they have normal form (no bold font). $\boldsymbol{R}$ denotes a second-order rotation tensor with $\boldsymbol{R}^T = \boldsymbol{R}^{-1}$ and $\det\boldsymbol{R} = 1$. $\boldsymbol{I}$ denotes the second-order identity tensor whose components are given by the Kronecker delta δ_{ij},

$$I_{ij} = \delta_{ij}, \quad \delta_{ij} = \begin{cases} 1\,, & i = j, \\ 0\,, & i \neq j. \end{cases}$$

The third-order Cartesian permutation tensor ϵ is defined as

$$\epsilon_{ijk} = \begin{cases} 1\,, & \text{if } (i,j,k) \text{ is } (1,2,3),\ (2,3,1) \text{ or } (3,1,2), \\ -1\,, & \text{if } (i,j,k) \text{ is } (1,3,2),\ (2,1,3) \text{ or } (3,2,1), \\ 0\,, & \text{if } i = j,\ j = k \text{ or } k = i. \end{cases}$$

The symmetric fourth-order identity tensor $\boldsymbol{\mathcal{I}}$ is defined as

$$\mathcal{I}_{ijkl} = \frac{1}{2}[\delta_{ik}\delta_{jl} + \delta_{il}\delta_{jk}].$$

The vector $\boldsymbol{x}$ with components x_q represents the Cartesian coordinates. Moreover, a dot above a tensor denotes the total derivative with respect to time, i.e.

$$\dot{\{\bullet\}} = \frac{D\{\bullet\}}{Dt}.$$

1.1.1. *Tensor operations*

A tensor $\boldsymbol{A}$ of order n is a quantity that can be transformed from one coordinate system $\boldsymbol{x}$ to another coordinate system $\boldsymbol{x}'$ according to the formulas

$$A'_{i_1 i_2 \dots i_n} = R_{j_1 i_1} R_{j_2 i_2} \dots R_{j_n i_n} A_{j_1 j_2 \dots j_n},$$
$$A_{i_1 i_2 \dots i_n} = R_{i_1 j_1} R_{i_2 j_2} \dots R_{i_n j_n} A'_{j_1 j_2 \dots j_n}.$$

A zeroth-order tensor is a scalar and a first-order tensor is a vector. The following tensor operations are introduced:

– A tensor $\boldsymbol{A}$ of order n, a tensor $\boldsymbol{B}$ of order n and a scalar c produce the n_{th} order tensors

$$\boldsymbol{C} = \boldsymbol{A} + \boldsymbol{B} \quad \text{with } C_{i_1 i_2 \dots i_n} = A_{i_1 i_2 \dots i_n} + B_{i_1 i_2 \dots i_n},$$
$$\boldsymbol{C} = c\boldsymbol{A} \quad \text{with } C_{i_1 i_2 \dots i_n} = cA_{i_1 i_2 \dots i_n}.$$

– A tensor $\boldsymbol{A}$ of order $n+1$ and a tensor $\boldsymbol{B}$ of order $m+1$ produce the $[n+m]_{\text{th}}$ order tensors

$$\boldsymbol{C} = \boldsymbol{A} \cdot \boldsymbol{B} \quad \text{with } C_{i_1 i_2 \dots i_{n+m}} = A_{i_1 i_2 \dots i_n q} B_{q i_{n+1} i_{n+2} \dots i_{n+m}},$$
$$\boldsymbol{C} = \boldsymbol{A} \tilde{\cdot} \boldsymbol{B} \quad \text{with } C_{i_1 i_2 \dots i_{n+m}} = A_{i_1 i_2 \dots i_n q} B_{i_{n+1} i_{n+2} \dots i_{n+m} q},$$
$$\boldsymbol{C} = \boldsymbol{A} \underset{\sim}{\cdot} \boldsymbol{B} \quad \text{with } C_{i_1 i_2 \dots i_{n+m}} = A_{q i_1 i_2 \dots i_n} B_{q i_{n+1} i_{n+2} \dots i_{n+m}}.$$

The first operation is called a single contraction product.

– A tensor $\boldsymbol{A}$ of order $n+2$ and a tensor $\boldsymbol{B}$ of order $m+2$ produce the $[n+m]_{\text{th}}$ order tensors

$$\boldsymbol{C} = \boldsymbol{A} : \boldsymbol{B} \quad \text{with } C_{i_1 i_2 \dots i_{n+m}} = A_{i_1 i_2 \dots i_n pq} B_{pq i_{n+1} i_{n+2} \dots i_{n+m}},$$
$$\boldsymbol{C} = \boldsymbol{A} \tilde{:} \boldsymbol{B} \quad \text{with } C_{i_1 i_2 \dots i_{n+m}} = A_{i_1 i_2 \dots i_n pq} B_{i_{n+1} i_{n+2} \dots i_{n+m} pq},$$
$$\boldsymbol{C} = \boldsymbol{A} \underset{\sim}{:} \boldsymbol{B} \quad \text{with } C_{i_1 i_2 \dots i_{n+m}} = A_{pq i_1 i_2 \dots i_n} B_{pq i_{n+1} i_{n+2} \dots i_{n+m}}.$$

The first operation is called a double contraction product. When $\boldsymbol{A}$ and $\boldsymbol{B}$ are second-order tensors, the first operation is called a scalar tensorial product, since the result is a scalar quantity.

– A tensor $\boldsymbol{A}$ of order n and a tensor $\boldsymbol{B}$ of order m produce the $[n+m]_{\text{th}}$ order tensor

$$\boldsymbol{C} = \boldsymbol{A} \otimes \boldsymbol{B} \quad \text{with } C_{i_1 i_2 \ldots i_{n+m}} = A_{i_1 i_2 \ldots i_n} B_{i_{n+1} i_{n+2} \ldots i_{n+m}}.$$

This operation is called a dyadic product.

– A tensor $\boldsymbol{A}$ of order n and a tensor $\boldsymbol{B}$ of order 1 produce the n_{th} order tensors

$$\begin{array}{ll} \boldsymbol{C} = \boldsymbol{A} \times \boldsymbol{B} & \text{with } \boldsymbol{A} \times \boldsymbol{B} = [\boldsymbol{A} \otimes \boldsymbol{B}]\!:\!\boldsymbol{\epsilon}, \\ \boldsymbol{C} = \boldsymbol{B} \times \boldsymbol{A} & \text{with } \boldsymbol{B} \times \boldsymbol{A} = \boldsymbol{\epsilon}\!:\![\boldsymbol{B} \otimes \boldsymbol{A}]. \end{array}$$

These two operations are called a cross product.

1.1.1.1. *Special operations for tensors up to the fourth-order*

For tensors up to the fourth-order, the following special type of tensor operations are introduced:

– A tensor $\boldsymbol{A}$ of order 2 and a tensor $\boldsymbol{B}$ of order 2 produce the fourth-order tensors

$$\boldsymbol{C} = \boldsymbol{A} \,\overline{\otimes}\, \boldsymbol{B} \quad \text{with } C_{ijkl} = A_{ik} B_{jl},$$

$$\boldsymbol{C} = \boldsymbol{A} \,\underline{\otimes}\, \boldsymbol{B} \quad \text{with } C_{ijkl} = A_{il} B_{jk}.$$

– A tensor $\boldsymbol{A}$ of order 3 and a tensor $\boldsymbol{B}$ of order 1 produce the second-order tensor

$$\boldsymbol{C} = \boldsymbol{A} \,\bar{\cdot}\, \boldsymbol{B} \quad \text{with } C_{ij} = A_{ikj} B_k.$$

– A tensor $\boldsymbol{A}$ of order 3 and a tensor $\boldsymbol{B}$ of order 3 produce the first-order tensor (i.e. the vector)

$$\boldsymbol{C} = \boldsymbol{A} \,\bar{:}\, \boldsymbol{B} \quad \text{with } C_i = A_{jik} B_{jk}.$$

– A tensor $\boldsymbol{A}$ of order 4 and a tensor $\boldsymbol{B}$ of order 2 produce the second-order tensor

$$\boldsymbol{C} = \boldsymbol{A} \,\bar{:}\, \boldsymbol{B} \quad \text{with } C_{ij} = A_{ikjl} B_{kl}.$$

– For a second-order tensor $\boldsymbol{A}$, we can identify:

$$\begin{aligned}
\text{its transpose} \quad & \boldsymbol{A}^T \quad \text{with} \quad A^T_{ij} = A_{ji}, \\
\text{its inverse} \quad & \boldsymbol{A}^{-1} \quad \text{with} \quad \boldsymbol{A} \cdot \boldsymbol{A}^{-1} = \boldsymbol{I}, \\
\text{its inverse transpose} \quad & \boldsymbol{A}^{-T} \quad \text{with} \quad A^{-T}_{ij} = A^{-1}_{ji}, \\
\text{its trace} \quad & \text{tr}\boldsymbol{A} \quad \text{with} \quad \text{tr}\boldsymbol{A} = A_{ii} = \boldsymbol{A} : \boldsymbol{I}, \\
\text{its determinant} \quad & \det\boldsymbol{A} \quad \text{with} \quad \det\boldsymbol{A} = \epsilon_{ijk} A_{1i} A_{2j} A_{3k}.
\end{aligned}$$

The second and the third expressions hold as long as $\boldsymbol{A}$ is invertible. When $\boldsymbol{A} = \boldsymbol{A}^T$, then $\boldsymbol{A}$ is called symmetric.

– For a fourth-order tensor $\boldsymbol{A}$, we can identify its transpose $\boldsymbol{A}^T$ with the property $A^T_{ijkl} = A_{klij}$.

– A fourth-order tensor $\boldsymbol{A}$ is considered to have minor symmetries if $A_{ijkl} = A_{jikl} = A_{ijlk}$. This tensor is considered to have major and minor symmetries if, in addition, $A_{ijkl} = A_{klij}$. When $\boldsymbol{A}$ has minor symmetries, its inverse tensor $\boldsymbol{A}^{-1}$ may exist with the property

$$\boldsymbol{A} : \boldsymbol{A}^{-1} = \boldsymbol{A}^{-1} : \boldsymbol{A} = \frac{1}{2} [\boldsymbol{I} \,\overline{\otimes}\, \boldsymbol{I} + \boldsymbol{I} \,\underline{\otimes}\, \boldsymbol{I}] = \boldsymbol{\mathcal{I}}.$$

1.1.2. *Tensor derivatives*

From differential geometry we have the following definitions for an n_{th} order tensor in Cartesian coordinates [STE 08, STE 15]:

1) The gradient $\text{grad}\boldsymbol{A}$ is a tensor of order $n + 1$, defined by the relation

$$[\text{grad}A]_{i_1 i_2 \ldots i_n i_{n+1}} = \frac{\partial A_{i_1 i_2 \ldots i_n}}{\partial x_{i_{n+1}}}.$$

2) The symmetric gradient $\text{grad}_{\text{sym}}\boldsymbol{A}$ is a tensor of order $n+1$, defined by the relation

$$\left[\text{grad}_{\text{sym}}A\right]_{i_1 i_2 \dots i_n i_{n+1}} = \frac{1}{2}\left[\frac{\partial A_{i_1 i_2 \dots i_{n-1} i_n}}{\partial x_{i_{n+1}}} + \frac{\partial A_{i_1 i_2 \dots i_{n-1} i_{n+1}}}{\partial x_{i_n}}\right].$$

3) The divergence $\text{div}\boldsymbol{A}$ is a tensor of order $n-1$, defined by the relation

$$\text{div}\boldsymbol{A} = \text{grad}\boldsymbol{A}\colon\boldsymbol{I}, \quad \text{or} \quad [\text{div}A]_{i_1 i_2 \dots i_{n-1}} = \frac{\partial A_{i_1 i_2 \dots i_{n-1} i_n}}{\partial x_{i_n}}.$$

4) The curl $\text{curl}\boldsymbol{A}$ is a tensor of order n, defined by the relation

$$\text{curl}\boldsymbol{A} = -\text{grad}\boldsymbol{A}\colon\boldsymbol{\epsilon}, \quad \text{or} \quad [\text{curl}A]_{i_1 i_2 \dots i_n} = -\frac{\partial A_{i_1 i_2 \dots i_{n-1} j}}{\partial x_k}\epsilon_{jki_n}.$$

Moreover, for a tensor $\boldsymbol{A}$ of order $n > 0$ and a tensor $\boldsymbol{B}$ of order $m > 0$, the tensor $\dfrac{\partial \boldsymbol{A}}{\partial \boldsymbol{B}}$ of order $n+m$ is defined as

$$\left[\frac{\partial A}{\partial B}\right]_{i_1, i_2, \dots, i_{m+n}} = \frac{\partial A_{i_1, i_2, \dots, i_n}}{\partial B_{i_{n+1}, i_{n+2}, \dots, i_{n+m}}}.$$

Using these definitions and operations, certain relations for the derivatives of products between two tensors $\boldsymbol{A}$ and $\boldsymbol{B}$ can be obtained:

For an n_{th} order tensor $\boldsymbol{A}$ and an m_{th} order tensor $\boldsymbol{B}$,

$$\begin{aligned}
\text{grad}(\boldsymbol{A}\cdot\boldsymbol{B}) &= \text{grad}\boldsymbol{A}\cdot\boldsymbol{B} + \boldsymbol{A}\cdot\text{grad}\boldsymbol{B}, \\
\text{grad}(\boldsymbol{A}\colon\boldsymbol{B}) &= \text{grad}\boldsymbol{A}\colon\boldsymbol{B} + \boldsymbol{A}\colon\text{grad}\boldsymbol{B}, \\
\text{grad}(\boldsymbol{A}\otimes\boldsymbol{B}) &= \text{grad}\boldsymbol{A}\otimes\boldsymbol{B} + \boldsymbol{A}\otimes\text{grad}\boldsymbol{B}.
\end{aligned}$$

For an n_{th} order tensor $\boldsymbol{A}$ and a vector $\boldsymbol{B}$,

$$\begin{aligned}
\text{grad}(\boldsymbol{A}\times\boldsymbol{B}) &= \text{grad}\boldsymbol{A}\times\boldsymbol{B} + \boldsymbol{A}\times\text{grad}\boldsymbol{B}, \\
\text{grad}(\boldsymbol{B}\times\boldsymbol{A}) &= \text{grad}\boldsymbol{B}\times\boldsymbol{A} + \boldsymbol{B}\times\text{grad}\boldsymbol{A},
\end{aligned}$$

$$\begin{aligned}
\text{div}(\boldsymbol{A}\otimes\boldsymbol{B}) &= \text{grad}\boldsymbol{A}\cdot\boldsymbol{B}+\boldsymbol{A}\,\text{div}\boldsymbol{B},\\
\text{div}(\boldsymbol{B}\otimes\boldsymbol{A}) &= \text{grad}\boldsymbol{B}\,\tilde{\cdot}\,\boldsymbol{A}+\boldsymbol{B}\otimes\text{div}\boldsymbol{A},\\
\text{div}(\boldsymbol{B}\times\boldsymbol{A}) &= \boldsymbol{\epsilon}\!:\![\text{grad}\boldsymbol{B}\,\tilde{\cdot}\,\boldsymbol{A}]+\boldsymbol{B}\times\text{div}\boldsymbol{A}.
\end{aligned}$$

For a scalar A and an n_{th} order tensor $\boldsymbol{B}$,

$$\text{div}(A\boldsymbol{B}) = \boldsymbol{B}\cdot\text{grad}A+A\,\text{div}\boldsymbol{B}.$$

For two vectors $\boldsymbol{A}$ and $\boldsymbol{B}$,

$$\begin{aligned}
\text{div}(\boldsymbol{A}\otimes\boldsymbol{B}) &= \text{grad}\boldsymbol{A}\cdot\boldsymbol{B}+\boldsymbol{A}\otimes\text{div}\boldsymbol{B},\\
\text{div}(\boldsymbol{A}\times\boldsymbol{B}) &= \boldsymbol{B}\cdot\text{curl}\boldsymbol{A}-\boldsymbol{A}\cdot\text{curl}\boldsymbol{B}.
\end{aligned}$$

For a vector $\boldsymbol{A}$ and a second-order tensor $\boldsymbol{B}$,

$$\begin{aligned}
\text{div}(\boldsymbol{A}\cdot\boldsymbol{B}) &= \boldsymbol{B}\!:\!\text{grad}\boldsymbol{A}+\boldsymbol{A}\cdot\text{div}\boldsymbol{B},\\
\text{div}(\boldsymbol{A}\otimes\boldsymbol{B}) &= \text{grad}\boldsymbol{A}\,\tilde{\cdot}\,\boldsymbol{B}+\boldsymbol{A}\otimes\text{div}\boldsymbol{B},\\
[\text{div}(\boldsymbol{A}\otimes\boldsymbol{B})]^T &= \boldsymbol{B}\,\tilde{\cdot}\,\text{grad}\boldsymbol{A}+\text{div}\boldsymbol{B}\otimes\boldsymbol{A},\\
\text{div}(\boldsymbol{B}\cdot\boldsymbol{A}) &= \boldsymbol{A}\cdot\text{div}\boldsymbol{B}^T+\boldsymbol{B}^T\!:\!\text{grad}\boldsymbol{A}.
\end{aligned}$$

1.1.3. *Vector and second-order tensor properties*

The following properties hold for two vectors $\boldsymbol{a}$, $\boldsymbol{b}$ and a scalar c:

$$\begin{aligned}
\boldsymbol{a}\cdot\boldsymbol{b}=\boldsymbol{b}\cdot\boldsymbol{a},&\quad \boldsymbol{a}\times\boldsymbol{b}=-\boldsymbol{b}\times\boldsymbol{a},\\
\boldsymbol{a}\times\boldsymbol{a}=\boldsymbol{0},&\quad [\boldsymbol{a}\otimes\boldsymbol{b}]^T=\boldsymbol{b}\otimes\boldsymbol{a},\\
\text{div}(\text{curl}\boldsymbol{a})=0,&\quad \text{curl}(\text{grad}c)=\boldsymbol{0}.
\end{aligned}$$

The following properties hold for two second-order tensors $\boldsymbol{A}$ and $\boldsymbol{B}$ (the inverse operations are valid only if the considered tensors are invertible):

$$[\boldsymbol{A}+\boldsymbol{B}]^T=\boldsymbol{A}^T+\boldsymbol{B}^T,\qquad [\boldsymbol{A}\cdot\boldsymbol{B}]^T=\boldsymbol{B}^T\cdot\boldsymbol{A}^T,$$

$$
\begin{aligned}
[\boldsymbol{A}\cdot\boldsymbol{B}]^{-1} &= \boldsymbol{B}^{-1}\cdot\boldsymbol{A}^{-1}, & \boldsymbol{A}:\boldsymbol{B} &= \boldsymbol{B}:\boldsymbol{A},\\
[\boldsymbol{A}\otimes\boldsymbol{B}]^{T} &= \boldsymbol{B}\otimes\boldsymbol{A}, & \mathrm{tr}(\boldsymbol{A}^{T}) &= \mathrm{tr}\boldsymbol{A},\\
\det(\boldsymbol{A}^{T}) &= \det\boldsymbol{A}, & \det(\boldsymbol{A}^{-1}) &= [\det\boldsymbol{A}]^{-1},\\
\frac{\partial\det\boldsymbol{A}}{\partial\boldsymbol{A}} &= \boldsymbol{A}^{-T}\det\boldsymbol{A}, & \det(\boldsymbol{A}\cdot\boldsymbol{B}) &= \det(\boldsymbol{A})\det(\boldsymbol{B}).
\end{aligned}
$$

1.1.3.1. *A useful identity for continuum mechanics*

For a second-order symmetric tensor $\boldsymbol{n}$, we can identify the deviatoric part $\boldsymbol{n}'$ and the deviatoric norm n^{dev} as

$$
\boldsymbol{n}' = \boldsymbol{n} - \frac{1}{3}[\mathrm{tr}\boldsymbol{n}]\boldsymbol{I},\ n^{\mathrm{dev}} = \sqrt{c\,\boldsymbol{n}':\boldsymbol{n}'},
$$

where c is a known scalar. Then, the following property holds:

$$
\frac{\partial n^{\mathrm{dev}}}{\partial\boldsymbol{n}} = \frac{c\,\boldsymbol{n}'}{n^{\mathrm{dev}}}.
$$

Indeed, using the identity $\boldsymbol{n}':\boldsymbol{I} = \mathrm{tr}\boldsymbol{n}' = 0$, we can write

$$
\begin{aligned}
\frac{\partial n^{\mathrm{dev}}}{\partial\boldsymbol{n}} &= \frac{c\,\boldsymbol{n}'}{n^{\mathrm{dev}}}:\frac{\partial\boldsymbol{n}'}{\partial\boldsymbol{n}} = \frac{c\,\boldsymbol{n}'}{n^{\mathrm{dev}}}:\left[\frac{\partial\boldsymbol{n}}{\partial\boldsymbol{n}} - \frac{1}{3}\boldsymbol{I}\otimes\frac{\partial(\mathrm{tr}\boldsymbol{n})}{\partial\boldsymbol{n}}\right]\\
&= \frac{c\,\boldsymbol{n}':\mathcal{I}}{n^{\mathrm{dev}}} - \frac{\boldsymbol{n}':\boldsymbol{I}}{n^{\mathrm{dev}}}\frac{\partial(\mathrm{tr}\boldsymbol{n})}{\partial\boldsymbol{n}} = \frac{c\,\boldsymbol{n}'}{n^{\mathrm{dev}}}.
\end{aligned}
$$

1.1.4. *Voigt notation for second- and fourth-order tensors*

Voigt notation is a practical way to reduce the order of symmetric tensors for simplifying the calculations. The Voigt representation follows the "interchange rule" 11→1, 22→2, 33→3, 12→4, 13→5, 23→6. Specifically:

– A symmetric second-order tensor has only six independent components and can be described with the help of a 6×1 vector. There are two types of vectors that are considered in this method, the "stress-type":

$$
\boldsymbol{\sigma} = \begin{bmatrix}\sigma_{11} & \sigma_{22} & \sigma_{33} & \sigma_{12} & \sigma_{13} & \sigma_{23}\end{bmatrix}^{T}
$$

$$= \begin{bmatrix} \sigma_1 & \sigma_2 & \sigma_3 & \sigma_4 & \sigma_5 & \sigma_6 \end{bmatrix}^T,$$

and the "strain type":

$$\widetilde{\varepsilon} = \begin{bmatrix} \varepsilon_{11} & \varepsilon_{22} & \varepsilon_{33} & 2\varepsilon_{12} & 2\varepsilon_{13} & 2\varepsilon_{23} \end{bmatrix}^T$$
$$= \begin{bmatrix} \varepsilon_1 & \varepsilon_2 & \varepsilon_3 & 2\varepsilon_4 & 2\varepsilon_5 & 2\varepsilon_6 \end{bmatrix}^T.$$

In this context, the superscript T denotes the usual transpose operator, i.e. it transforms the 1×6 matrix to a 6×1 vector. It should also be noted that this Voigt notation slightly deviates from what is commonly used in the literature: it considers the 12 shear component as the fourth element and the 23 component as the sixth element. This slight deviation is helpful in numerical implementations, since in plane strain or plane stress problems we can work on the 1-2 plane and reduce directly the stress or strain tensor to 4×1 vectors.

Using these forms, the scalar tensorial product $W = \sigma_{ij}\varepsilon_{ij}$ can be written with the help of matrix operations as

$$W = \boldsymbol{\sigma}^T \cdot \widetilde{\varepsilon}, \quad \text{or} \quad W = \widetilde{\varepsilon}^T \cdot \boldsymbol{\sigma},$$

where the symbol $(\,\cdot\,)$ in the Voigt notation denotes the classical matrix multiplication.

– A fourth-order tensor that has minor symmetries (i.e. $A_{ijkl} = A_{jikl} = A_{ijlk}$) represents a linear relation between symmetric second-order tensors, it has only 36 independent components and can be represented as a 6×6 matrix. The two Voigt representations of second-order tensors necessitate the identification of four Voigt representations of fourth-order tensors in order to express in matrix notation the following tensorial products:

$$\sigma_{ij} = L_{ijkl}\varepsilon_{kl}, \quad \varepsilon_{ij} = M_{ijkl}\sigma_{kl}, \quad \varepsilon^r_{ij} = T_{ijkl}\varepsilon^0_{kl}, \quad \sigma^r_{ij} = H_{ijkl}\sigma^0_{kl},$$
$$\sigma_{kl} = \varepsilon_{ij}P_{ijkl}, \quad \varepsilon_{kl} = \sigma_{ij}Q_{ijkl}, \quad \varepsilon^r_{kl} = \varepsilon^0_{ij}F_{ijkl}, \quad \sigma^r_{kl} = \sigma^0_{ij}G_{ijkl}.$$

Thus, except from the standard 6×6 matrix form

$$\boldsymbol{A} = \begin{bmatrix} A_{1111} & A_{1122} & A_{1133} & A_{1112} & A_{1113} & A_{1123} \\ A_{2211} & A_{2222} & A_{2233} & A_{2212} & A_{2213} & A_{2223} \\ A_{3311} & A_{3322} & A_{3333} & A_{3312} & A_{3313} & A_{3323} \\ A_{1211} & A_{1222} & A_{1233} & A_{1212} & A_{1213} & A_{1223} \\ A_{1311} & A_{1322} & A_{1333} & A_{1312} & A_{1313} & A_{1323} \\ A_{2311} & A_{2322} & A_{2333} & A_{2312} & A_{2313} & A_{2323} \end{bmatrix},$$

written for simplicity as

$$\boldsymbol{A} = \begin{bmatrix} A_{11} & A_{12} & A_{13} & A_{14} & A_{15} & A_{16} \\ A_{21} & A_{22} & A_{23} & A_{24} & A_{25} & A_{26} \\ A_{31} & A_{32} & A_{33} & A_{34} & A_{35} & A_{36} \\ A_{41} & A_{42} & A_{43} & A_{44} & A_{45} & A_{46} \\ A_{51} & A_{52} & A_{53} & A_{54} & A_{55} & A_{56} \\ A_{61} & A_{62} & A_{63} & A_{64} & A_{65} & A_{66} \end{bmatrix},$$

we need to identify the three additional matrix forms:

$$\widetilde{\boldsymbol{A}} = \begin{bmatrix} A_{11} & A_{12} & A_{13} & A_{14} & A_{15} & A_{16} \\ A_{21} & A_{22} & A_{23} & A_{24} & A_{25} & A_{26} \\ A_{31} & A_{32} & A_{33} & A_{34} & A_{35} & A_{36} \\ 2A_{41} & 2A_{42} & 2A_{43} & 2A_{44} & 2A_{45} & 2A_{46} \\ 2A_{51} & 2A_{52} & 2A_{53} & 2A_{54} & 2A_{55} & 2A_{56} \\ 2A_{61} & 2A_{62} & 2A_{63} & 2A_{64} & 2A_{65} & 2A_{66} \end{bmatrix},$$

$$\overline{\boldsymbol{A}} = \begin{bmatrix} A_{11} & A_{12} & A_{13} & 2A_{14} & 2A_{15} & 2A_{16} \\ A_{21} & A_{22} & A_{23} & 2A_{24} & 2A_{25} & 2A_{26} \\ A_{31} & A_{32} & A_{33} & 2A_{34} & 2A_{35} & 2A_{36} \\ A_{41} & A_{42} & A_{43} & 2A_{44} & 2A_{45} & 2A_{46} \\ A_{51} & A_{52} & A_{53} & 2A_{54} & 2A_{55} & 2A_{56} \\ A_{61} & A_{62} & A_{63} & 2A_{64} & 2A_{65} & 2A_{66} \end{bmatrix},$$

$$\widehat{\boldsymbol{A}} = \begin{bmatrix} A_{11} & A_{12} & A_{13} & 2A_{14} & 2A_{15} & 2A_{16} \\ A_{21} & A_{22} & A_{23} & 2A_{24} & 2A_{25} & 2A_{26} \\ A_{31} & A_{32} & A_{33} & 2A_{34} & 2A_{35} & 2A_{36} \\ 2A_{41} & 2A_{42} & 2A_{43} & 4A_{44} & 4A_{45} & 4A_{46} \\ 2A_{51} & 2A_{52} & 2A_{53} & 4A_{54} & 4A_{55} & 4A_{56} \\ 2A_{61} & 2A_{62} & 2A_{63} & 4A_{64} & 4A_{65} & 4A_{66} \end{bmatrix}.$$

Using these representations, it can easily be shown that the tensorial products can be written using matrix multiplication as

$$\boldsymbol{\sigma} = \boldsymbol{L}\cdot\widetilde{\boldsymbol{\varepsilon}}, \quad \widetilde{\boldsymbol{\varepsilon}} = \widehat{\boldsymbol{M}}\cdot\boldsymbol{\sigma}, \quad \widetilde{\boldsymbol{\varepsilon}}^r = \widetilde{\boldsymbol{T}}\cdot\widetilde{\boldsymbol{\varepsilon}}^0, \quad \boldsymbol{\sigma}^r = \overline{\boldsymbol{H}}\cdot\boldsymbol{\sigma}^0,$$
$$\boldsymbol{\sigma}^T = \widetilde{\boldsymbol{\varepsilon}}^T\cdot\boldsymbol{P}, \quad \widetilde{\boldsymbol{\varepsilon}}^T = \boldsymbol{\sigma}^T\cdot\widehat{\boldsymbol{Q}}, \quad [\widetilde{\boldsymbol{\varepsilon}}^r]^T = [\widetilde{\boldsymbol{\varepsilon}}^0]^T\cdot\overline{\boldsymbol{F}}, \quad [\boldsymbol{\sigma}^r]^T = [\boldsymbol{\sigma}^0]^T\cdot\widetilde{\boldsymbol{G}}.$$

In addition, the dyadic products

$$A_{ijkl} = \sigma_{ij}\varepsilon_{kl}, \quad B_{ijkl} = \varepsilon_{ij}\sigma_{kl}, \quad C_{ijkl} = \varepsilon_{ij}\varepsilon_{kl}, \quad D_{ijkl} = \sigma_{ij}\sigma_{kl},$$

can be written using matrix multiplications as

$$\overline{\boldsymbol{A}} = \boldsymbol{\sigma}\cdot\widetilde{\boldsymbol{\varepsilon}}^T, \quad \widetilde{\boldsymbol{B}} = \widetilde{\boldsymbol{\varepsilon}}\cdot\boldsymbol{\sigma}^T, \quad \widehat{\boldsymbol{C}} = \widetilde{\boldsymbol{\varepsilon}}\cdot\widetilde{\boldsymbol{\varepsilon}}^T, \quad \boldsymbol{D} = \boldsymbol{\sigma}\cdot\boldsymbol{\sigma}^T.$$

An obvious property is that the four matrix representations respect the usual matrix summation and multiplication by a scalar c, i.e.

$$C_{ijkl} = A_{ijkl} + B_{ijkl} \Longrightarrow \overset{\#}{\boldsymbol{C}} = \overset{\#}{\boldsymbol{A}} + \overset{\#}{\boldsymbol{B}},$$
$$B_{ijkl} = cA_{ijkl} \Longrightarrow \overset{\#}{\boldsymbol{B}} = c\overset{\#}{\boldsymbol{A}},$$

where $\overset{\#}{\{\bullet\}} = \{\bullet\}, \widehat{\{\bullet\}}, \widetilde{\{\bullet\}}$ or $\overline{\{\bullet\}}$.

1.1.4.1. *Multiplication identities of fourth-order tensors*

For fourth-order tensors that respect minor symmetries, the multiplication of A_{ijkl} and B_{ijkl} provides the tensor C_{ijkl}, for which it holds that

$$C_{ijkl} = A_{ijmn}B_{mnkl}.$$

In matrix notation, this is written as

$$\boldsymbol{C} = \boldsymbol{A}\cdot\widetilde{\boldsymbol{B}}.$$

Using this relation, we can also show that

$$\boldsymbol{A}\cdot\widetilde{\boldsymbol{B}} = \overline{\boldsymbol{A}}\cdot\boldsymbol{B} = \boldsymbol{C}, \quad \widetilde{\boldsymbol{A}}\cdot\widetilde{\boldsymbol{B}} = \widehat{\boldsymbol{A}}\cdot\boldsymbol{B} = \widetilde{\boldsymbol{C}},$$
$$\overline{\boldsymbol{A}}\cdot\overline{\boldsymbol{B}} = \boldsymbol{A}\cdot\widehat{\boldsymbol{B}} = \overline{\boldsymbol{C}}, \quad \widetilde{\boldsymbol{A}}\cdot\widehat{\boldsymbol{B}} = \widehat{\boldsymbol{A}}\cdot\overline{\boldsymbol{B}} = \widehat{\boldsymbol{C}}.$$

1.1.4.2. *Transpose identities of fourth-order tensors*

From the structure of the matrices, it is easily shown that, if $\boldsymbol{A}$ is the transpose of a 6×6 matrix $\boldsymbol{B}$, then

$$\boldsymbol{B}^T = \boldsymbol{A}, \quad \widehat{\boldsymbol{B}}^T = \widehat{\boldsymbol{A}}, \quad \widetilde{\boldsymbol{B}}^T = \overline{\boldsymbol{A}}, \quad \overline{\boldsymbol{B}}^T = \widetilde{\boldsymbol{A}}.$$

1.1.4.3. *Inversion identities of fourth-order tensors*

For fourth-order tensors that respect minor symmetries (connect second-order symmetric tensors), the inverse of L_{ijkl} is the tensor M_{ijkl}, for which it holds that

$$L_{ijmn}M_{mnkl} = \mathcal{I}_{ijkl} = \frac{1}{2}[\delta_{ik}\delta_{jl} + \delta_{il}\delta_{jk}].$$

In matrix notation, this expression can be written as

$$\boldsymbol{L}\cdot\widetilde{\boldsymbol{M}} = \boldsymbol{\mathcal{I}} = \begin{bmatrix} 1 & 0 & 0 & 0 & 0 & 0 \\ 0 & 1 & 0 & 0 & 0 & 0 \\ 0 & 0 & 1 & 0 & 0 & 0 \\ 0 & 0 & 0 & 0.5 & 0 & 0 \\ 0 & 0 & 0 & 0 & 0.5 & 0 \\ 0 & 0 & 0 & 0 & 0 & 0.5 \end{bmatrix}.$$

The above relation can be written in the following ways:

$$\boldsymbol{L}\cdot\widehat{\boldsymbol{M}} = \overline{\boldsymbol{\mathcal{I}}}, \quad \widehat{\boldsymbol{L}}\cdot\boldsymbol{M} = \widetilde{\boldsymbol{\mathcal{I}}}, \quad \widetilde{\boldsymbol{L}}\cdot\widetilde{\boldsymbol{M}} = \widetilde{\boldsymbol{\mathcal{I}}}, \quad \overline{\boldsymbol{L}}\cdot\overline{\boldsymbol{M}} = \overline{\boldsymbol{\mathcal{I}}},$$

where

$$\widetilde{\boldsymbol{\mathcal{I}}} = \overline{\boldsymbol{\mathcal{I}}} = \begin{bmatrix} 1 & 0 & 0 & 0 & 0 & 0 \\ 0 & 1 & 0 & 0 & 0 & 0 \\ 0 & 0 & 1 & 0 & 0 & 0 \\ 0 & 0 & 0 & 1 & 0 & 0 \\ 0 & 0 & 0 & 0 & 1 & 0 \\ 0 & 0 & 0 & 0 & 0 & 1 \end{bmatrix}.$$

Thus, in Voigt notation, the following properties hold:

$$\boldsymbol{L}^{-1} = \widehat{\boldsymbol{M}}, \quad \widehat{\boldsymbol{L}}^{-1} = \boldsymbol{M}, \quad \widetilde{\boldsymbol{L}}^{-1} = \widetilde{\boldsymbol{M}}, \quad \overline{\boldsymbol{L}}^{-1} = \overline{\boldsymbol{M}}.$$

A summary of the various properties and the connection between indicial and Voigt notation is presented in Table 1.1.

indicial	Voigt	indicial	Voigt
$W = \sigma_{ij}\varepsilon_{ij}$	$W = \boldsymbol{\sigma}^T \cdot \widetilde{\boldsymbol{\varepsilon}}$	$C_{ijkl} = A_{ijkl} + B_{ijkl}$	$\overset{\#}{\boldsymbol{C}} = \overset{\#}{\boldsymbol{A}} + \overset{\#}{\boldsymbol{B}}$
$W = \varepsilon_{ij}\sigma_{ij}$	$W = \widetilde{\boldsymbol{\varepsilon}}^T \cdot \boldsymbol{\sigma}$	$B_{ijkl} = cA_{ijkl}$	$\overset{\#}{\boldsymbol{B}} = c\overset{\#}{\boldsymbol{A}}$
$\sigma_{ij} = L_{ijkl}\varepsilon_{kl}$	$\boldsymbol{\sigma} = \boldsymbol{L} \cdot \widetilde{\boldsymbol{\varepsilon}}$	$C_{ijkl} = A_{ijmn}B_{mnkl}$	$\boldsymbol{C} = \boldsymbol{A} \cdot \widetilde{\boldsymbol{B}} = \overline{\boldsymbol{A}} \cdot \boldsymbol{B}$
$\varepsilon_{ij} = M_{ijkl}\sigma_{kl}$	$\widetilde{\boldsymbol{\varepsilon}} = \widehat{\boldsymbol{M}} \cdot \boldsymbol{\sigma}$	$C_{ijkl} = A_{ijmn}B_{mnkl}$	$\widehat{\boldsymbol{C}} = \widetilde{\boldsymbol{A}} \cdot \widehat{\boldsymbol{B}} = \widehat{\boldsymbol{A}} \cdot \overline{\boldsymbol{B}}$
$\varepsilon^r_{ij} = T_{ijkl}\varepsilon^0_{kl}$	$\widetilde{\boldsymbol{\varepsilon}}^r = \widetilde{\boldsymbol{T}} \cdot \widetilde{\boldsymbol{\varepsilon}}^0$	$C_{ijkl} = A_{ijmn}B_{mnkl}$	$\widetilde{\boldsymbol{C}} = \widetilde{\boldsymbol{A}} \cdot \widetilde{\boldsymbol{B}} = \widehat{\boldsymbol{A}} \cdot \boldsymbol{B}$
$\sigma^r_{ij} = H_{ijkl}\sigma^0_{kl}$	$\boldsymbol{\sigma}^r = \overline{\boldsymbol{H}} \cdot \boldsymbol{\sigma}^0$	$C_{ijkl} = A_{ijmn}B_{mnkl}$	$\overline{\boldsymbol{C}} = \overline{\boldsymbol{A}} \cdot \overline{\boldsymbol{B}} = \boldsymbol{A} \cdot \widehat{\boldsymbol{B}}$
$\sigma_{kl} = \varepsilon_{ij}P_{ijkl}$	$\boldsymbol{\sigma}^T = \widetilde{\boldsymbol{\varepsilon}}^T \cdot \boldsymbol{P}$	$B^T_{ijkl} = A_{ijkl}$	$\boldsymbol{B}^T = \boldsymbol{A}$
$\varepsilon_{kl} = \sigma_{ij}Q_{ijkl}$	$\widetilde{\boldsymbol{\varepsilon}}^T = \boldsymbol{\sigma}^T \cdot \widehat{\boldsymbol{Q}}$	$B^T_{ijkl} = A_{ijkl}$	$\widehat{\boldsymbol{B}}^T = \widehat{\boldsymbol{A}}$
$\sigma^r_{kl} = \sigma^0_{ij}F_{ijkl}$	$[\boldsymbol{\sigma}^r]^T = [\boldsymbol{\sigma}^0]^T \cdot \widetilde{\boldsymbol{F}}$	$B^T_{ijkl} = A_{ijkl}$	$\widetilde{\boldsymbol{B}}^T = \overline{\boldsymbol{A}}$
$\varepsilon^r_{kl} = \varepsilon^0_{ij}G_{ijkl}$	$[\widetilde{\boldsymbol{\varepsilon}}^r]^T = [\widetilde{\boldsymbol{\varepsilon}}^0]^T \cdot \overline{\boldsymbol{G}}$	$B^T_{ijkl} = A_{ijkl}$	$\overline{\boldsymbol{B}}^T = \widetilde{\boldsymbol{A}}$
$A_{ijkl} = \sigma_{ij}\sigma_{kl}$	$\boldsymbol{A} = \boldsymbol{\sigma} \cdot \boldsymbol{\sigma}^T$	$L^{-1}_{ijkl} = M_{ijkl}$	$\boldsymbol{L}^{-1} = \widehat{\boldsymbol{M}}$
$B_{ijkl} = \varepsilon_{ij}\varepsilon_{kl}$	$\widehat{\boldsymbol{B}} = \widetilde{\boldsymbol{\varepsilon}} \cdot \widetilde{\boldsymbol{\varepsilon}}^T$	$L^{-1}_{ijkl} = M_{ijkl}$	$\widehat{\boldsymbol{L}}^{-1} = \boldsymbol{M}$
$C_{ijkl} = \varepsilon_{ij}\sigma_{kl}$	$\widetilde{\boldsymbol{C}} = \widetilde{\boldsymbol{\varepsilon}} \cdot \boldsymbol{\sigma}^T$	$L^{-1}_{ijkl} = M_{ijkl}$	$\widetilde{\boldsymbol{L}}^{-1} = \widetilde{\boldsymbol{M}}$
$D_{ijkl} = \sigma_{ij}\varepsilon_{kl}$	$\overline{\boldsymbol{D}} = \boldsymbol{\sigma} \cdot \widetilde{\boldsymbol{\varepsilon}}^T$	$L^{-1}_{ijkl} = M_{ijkl}$	$\overline{\boldsymbol{L}}^{-1} = \overline{\boldsymbol{M}}$

Table 1.1. *Tensor operations in indicial and Voigt notation. The symbol* $\overset{\#}{\{\bullet\}}$ *denotes one of the four Voigt operators* $\{\bullet\}$, $\widehat{\{\bullet\}}$, $\widetilde{\{\bullet\}}$ *or* $\overline{\{\bullet\}}$

1.1.4.4. *Tensor rotation in Voigt notation*

A second-order tensor $\boldsymbol{A}$ in the global coordinate system $\boldsymbol{x}$ can be rotated in order to obtain the tensor $\boldsymbol{A}'$ in a local coordinate system $\boldsymbol{x}'$ through the relation

$$A'_{ij} = R_{mi}R_{nj}A_{mn}\,.$$

A usual rotation tensor $\boldsymbol{R}$ in classical mechanics is a second-order orthogonal tensor ($\boldsymbol{R}^{-1} = \boldsymbol{R}^T$) with the general form

$$\boldsymbol{R} = \begin{bmatrix} a & b & c \\ d & e & f \\ g & h & i \end{bmatrix}.$$

If a rotation ω is performed around the axis i with i =1, 2 or 3, then the rotation tensor is written [LAI 10]

$$\boldsymbol{R}_1 = \begin{bmatrix} 1 & 0 & 0 \\ 0 & \cos\omega & -\sin\omega \\ 0 & \sin\omega & \cos\omega \end{bmatrix}, \quad \boldsymbol{R}_2 = \begin{bmatrix} \cos\omega & 0 & \sin\omega \\ 0 & 1 & 0 \\ -\sin\omega & 0 & \cos\omega \end{bmatrix},$$

$$\boldsymbol{R}_3 = \begin{bmatrix} \cos\omega & -\sin\omega & 0 \\ \sin\omega & \cos\omega & 0 \\ 0 & 0 & 1 \end{bmatrix},$$

respectively. For a symmetric second-order tensor $\boldsymbol{A}$, a rotation yields another symmetric second order tensor $\boldsymbol{A}'$ and we can identify a fourth-order rotation tensor with minor symmetries that links the two second-order tensors through the relation

$$A'_{ij} = Q_{ijmn} A_{mn}.$$

In the above expression, $Q_{ijmn} = R_{mi} R_{nj}$ denotes the rotation from global to local coordinates and is written in matrix form as

$$\boldsymbol{Q} = \begin{bmatrix} a^2 & d^2 & g^2 & ad & ag & dg \\ b^2 & e^2 & h^2 & be & bh & eh \\ c^2 & f^2 & i^2 & cf & ci & fi \\ ab & de & gh & \frac{1}{2}[bd + ae] & \frac{1}{2}[bg + ah] & \frac{1}{2}[eg + dh] \\ ac & df & gi & \frac{1}{2}[cd + af] & \frac{1}{2}[cg + ai] & \frac{1}{2}[fg + di] \\ bc & ef & hi & \frac{1}{2}[ce + bf] & \frac{1}{2}[ch + bi] & \frac{1}{2}[fh + ei] \end{bmatrix}.$$

Due to the orthogonality of $\boldsymbol{R}$, it can be easily verified that

$$A_{ij} = Q^{-1}_{ijmn} A'_{mn} = Q^T_{ijmn} A'_{mn}.$$

Using the properties of the fourth-order tensors identified previously in this subsection, a rotation of a "stress type" tensor can be expressed in Voigt notation as

$$\boldsymbol{\sigma}' = \overline{\boldsymbol{Q}}\cdot\boldsymbol{\sigma}, \quad \boldsymbol{\sigma} = \overline{\boldsymbol{Q}}^{-1}\cdot\boldsymbol{\sigma}' = \widetilde{\boldsymbol{Q}}^{T}\cdot\boldsymbol{\sigma}'.$$

Similarly, a rotation of a "strain type" tensor can be expressed in Voigt notation as

$$\widetilde{\boldsymbol{\varepsilon}}' = \widetilde{\boldsymbol{Q}}\cdot\widetilde{\boldsymbol{\varepsilon}}, \quad \widetilde{\boldsymbol{\varepsilon}} = \widetilde{\boldsymbol{Q}}^{-1}\cdot\widetilde{\boldsymbol{\varepsilon}}' = \overline{\boldsymbol{Q}}^{T}\cdot\widetilde{\boldsymbol{\varepsilon}}'.$$

With the help of $\boldsymbol{Q}$ and its various Voigt forms, four types of rotations for fourth-order tensors appear:

– For a fourth-order tensor $\boldsymbol{L}$ that connects a "stress type" with a "strain type" tensor, $\boldsymbol{\sigma} = \boldsymbol{L}\cdot\widetilde{\boldsymbol{\varepsilon}}$, it holds that

$$\boldsymbol{\sigma}' = \overline{\boldsymbol{Q}}\cdot\boldsymbol{\sigma} = \overline{\boldsymbol{Q}}\cdot\boldsymbol{L}\cdot\widetilde{\boldsymbol{\varepsilon}} = \overline{\boldsymbol{Q}}\cdot\boldsymbol{L}\cdot\overline{\boldsymbol{Q}}^{T}\cdot\widetilde{\boldsymbol{\varepsilon}}',$$

$$\boldsymbol{\sigma} = \widetilde{\boldsymbol{Q}}^{T}\cdot\boldsymbol{\sigma}' = \widetilde{\boldsymbol{Q}}^{T}\cdot\boldsymbol{L}'\cdot\widetilde{\boldsymbol{\varepsilon}}' = \widetilde{\boldsymbol{Q}}^{T}\cdot\boldsymbol{L}'\cdot\widetilde{\boldsymbol{Q}}\cdot\widetilde{\boldsymbol{\varepsilon}}.$$

These expressions lead to the conclusion that

$$\boldsymbol{L}' = \overline{\boldsymbol{Q}}\cdot\boldsymbol{L}\cdot\overline{\boldsymbol{Q}}^{T} \quad \text{and} \quad \boldsymbol{L} = \widetilde{\boldsymbol{Q}}^{T}\cdot\boldsymbol{L}'\cdot\widetilde{\boldsymbol{Q}}.$$

– For a fourth-order tensor $\boldsymbol{M}$ that connects a "strain type" with a "stress type" tensor, $\widetilde{\boldsymbol{\varepsilon}} = \widehat{\boldsymbol{M}}\cdot\boldsymbol{\sigma}$, it holds that

$$\widetilde{\boldsymbol{\varepsilon}}' = \widetilde{\boldsymbol{Q}}\cdot\widetilde{\boldsymbol{\varepsilon}} = \widetilde{\boldsymbol{Q}}\cdot\widehat{\boldsymbol{M}}\cdot\boldsymbol{\sigma} = \widetilde{\boldsymbol{Q}}\cdot\widehat{\boldsymbol{M}}\cdot\widetilde{\boldsymbol{Q}}^{T}\cdot\boldsymbol{\sigma}',$$

$$\widetilde{\boldsymbol{\varepsilon}} = \overline{\boldsymbol{Q}}^{T}\cdot\widetilde{\boldsymbol{\varepsilon}}' = \overline{\boldsymbol{Q}}^{T}\cdot\widehat{\boldsymbol{M}}'\cdot\boldsymbol{\sigma}' = \overline{\boldsymbol{Q}}^{T}\cdot\widehat{\boldsymbol{M}}'\cdot\overline{\boldsymbol{Q}}\cdot\boldsymbol{\sigma}.$$

These expressions lead to the conclusion that

$$\widehat{\boldsymbol{M}}' = \widetilde{\boldsymbol{Q}}\cdot\widehat{\boldsymbol{M}}\cdot\widetilde{\boldsymbol{Q}}^{T} \quad \text{and} \quad \widehat{\boldsymbol{M}} = \overline{\boldsymbol{Q}}^{T}\cdot\widehat{\boldsymbol{M}}'\cdot\overline{\boldsymbol{Q}}.$$

– For a fourth-order tensor $\boldsymbol{T}$ that connects two "strain type" tensors, $\widetilde{\boldsymbol{\varepsilon}}^{r} = \widetilde{\boldsymbol{T}}\cdot\widetilde{\boldsymbol{\varepsilon}}^{0}$, it holds that

$$\widetilde{\boldsymbol{\varepsilon}}^{r'} = \widetilde{\boldsymbol{Q}}\cdot\widetilde{\boldsymbol{\varepsilon}}^{r} = \widetilde{\boldsymbol{Q}}\cdot\widetilde{\boldsymbol{T}}\cdot\widetilde{\boldsymbol{\varepsilon}}^{0} = \widetilde{\boldsymbol{Q}}\cdot\widetilde{\boldsymbol{T}}\cdot\overline{\boldsymbol{Q}}^{T}\cdot\widetilde{\boldsymbol{\varepsilon}}^{0'},$$

$$\widetilde{\boldsymbol{\varepsilon}}^{r} = \overline{\boldsymbol{Q}}^{T}\cdot\widetilde{\boldsymbol{\varepsilon}}^{r'} = \overline{\boldsymbol{Q}}^{T}\cdot\widetilde{\boldsymbol{T}}'\cdot\widetilde{\boldsymbol{\varepsilon}}^{0'} = \overline{\boldsymbol{Q}}^{T}\cdot\widetilde{\boldsymbol{T}}'\cdot\widetilde{\boldsymbol{Q}}\cdot\widetilde{\boldsymbol{\varepsilon}}^{0}.$$

These expressions lead to the conclusion that

$$\widetilde{\boldsymbol{T}}' = \widetilde{\boldsymbol{Q}} \cdot \widetilde{\boldsymbol{T}} \cdot \overline{\boldsymbol{Q}}^T \quad \text{and} \quad \widetilde{\boldsymbol{T}} = \overline{\boldsymbol{Q}}^T \cdot \widetilde{\boldsymbol{T}}' \cdot \widetilde{\boldsymbol{Q}}.$$

– For a fourth-order tensor $\boldsymbol{H}$ that connects two "stress type" tensors $\boldsymbol{\sigma}^r = \overline{\boldsymbol{H}} \cdot \boldsymbol{\sigma}^0$, it holds that

$$\boldsymbol{\sigma}^{r'} = \overline{\boldsymbol{Q}} \cdot \boldsymbol{\sigma}^r = \overline{\boldsymbol{Q}} \cdot \overline{\boldsymbol{H}} \cdot \boldsymbol{\sigma}^0 = \overline{\boldsymbol{Q}} \cdot \overline{\boldsymbol{H}} \cdot \widetilde{\boldsymbol{Q}}^T \cdot \boldsymbol{\sigma}^{0'},$$

$$\boldsymbol{\sigma}^r = \widetilde{\boldsymbol{Q}}^T \cdot \boldsymbol{\sigma}^{r'} = \widetilde{\boldsymbol{Q}}^T \cdot \overline{\boldsymbol{H}}' \cdot \boldsymbol{\sigma}^{0'} = \widetilde{\boldsymbol{Q}}^T \cdot \overline{\boldsymbol{H}}' \cdot \overline{\boldsymbol{Q}} \cdot \boldsymbol{\sigma}^0.$$

These expressions lead to the conclusion that

$$\overline{\boldsymbol{H}}' = \overline{\boldsymbol{Q}} \cdot \overline{\boldsymbol{H}} \cdot \widetilde{\boldsymbol{Q}}^T \quad \text{and} \quad \overline{\boldsymbol{H}} = \widetilde{\boldsymbol{Q}}^T \cdot \overline{\boldsymbol{H}}' \cdot \overline{\boldsymbol{Q}}.$$

1.1.5. *Special notation for isotropic fourth-order tensors*

An isotropic fourth-order tensor as $\boldsymbol{A}$ can be expressed in the Voigt form as

$$\boldsymbol{A} = \begin{bmatrix} a+\frac{4}{3}b & a-\frac{2}{3}b & a-\frac{2}{3}b & 0 & 0 & 0 \\ a-\frac{2}{3}b & a+\frac{4}{3}b & a-\frac{2}{3}b & 0 & 0 & 0 \\ a-\frac{2}{3}b & a-\frac{2}{3}b & a+\frac{4}{3}b & 0 & 0 & 0 \\ 0 & 0 & 0 & b & 0 & 0 \\ 0 & 0 & 0 & 0 & b & 0 \\ 0 & 0 & 0 & 0 & 0 & b \end{bmatrix},$$

or, in a more compact form as

$$\boldsymbol{A} = 3a\boldsymbol{\mathcal{I}}^h + 2b\boldsymbol{\mathcal{I}}^d, \quad \boldsymbol{\mathcal{I}}^h = \frac{1}{3}\boldsymbol{I} \otimes \boldsymbol{I}, \quad \boldsymbol{\mathcal{I}}^d = \boldsymbol{\mathcal{I}} - \boldsymbol{\mathcal{I}}^h.$$

Clearly, $\boldsymbol{A}$ presents minor and major symmetries and is invertible (see the tensor operations in section 1.1). By the definitions of $\mathcal{I}^h$ and $\mathcal{I}^d$, the useful identities

$$\mathcal{I}^h:\mathcal{I}^h = \mathcal{I}^h, \quad \mathcal{I}^d:\mathcal{I}^d = \mathcal{I}^d, \quad \mathcal{I}^h:\mathcal{I}^d = \mathcal{I}^d:\mathcal{I}^h = \mathbf{0},$$

hold. Hill [HIL 65] has proposed the following abbreviated form for isotropic fourth-order tensors:

$$\boldsymbol{A} = (3a, 2b).$$

The advantage of the abbreviated form is that it permits easy calculations for inverting or multiplying isotropic fourth-order tensors. Considering two tensors $\boldsymbol{A} = (3a, 2b)$ and $\boldsymbol{B} = (3c, 2d)$, the most useful properties are summarized below [QU 06]:

$$\begin{aligned} \mathcal{I} &= (1, 1), \\ \boldsymbol{A} + \boldsymbol{B} &= (3a + 3c, 2b + 2d), \\ \boldsymbol{A}:\boldsymbol{B} &= \boldsymbol{B}:\boldsymbol{A} = (9ac, 4bd), \\ \boldsymbol{A}^{-1} &= (\frac{1}{3a}, \frac{1}{2b}). \end{aligned}$$

These useful forms have been extended to the case of transversely isotropic tensors [WAL 81]. For fully anisotropic tensors, it is more efficient to work with the Voigt notation.

1.2. Tensors in curvilinear coordinates

When dealing with curvilinear coordinates, a distinction between covariant and contravariant tensors is necessary. For illustrative reasons, a subscript denotes a covariant component, while a superscript is used to declare a contravariant component. In a right-handed curvilinear coordinate system $(\vartheta^1, \vartheta^2, \vartheta^3)$, two base vectors exist, the covariant $\boldsymbol{g}_i$, tangential to the coordinate curves passing from a point p, and the contravariant $\boldsymbol{g}^i$, normal to the three coordinate surfaces passing from a point p [KEL 13] (Figure 1.1).

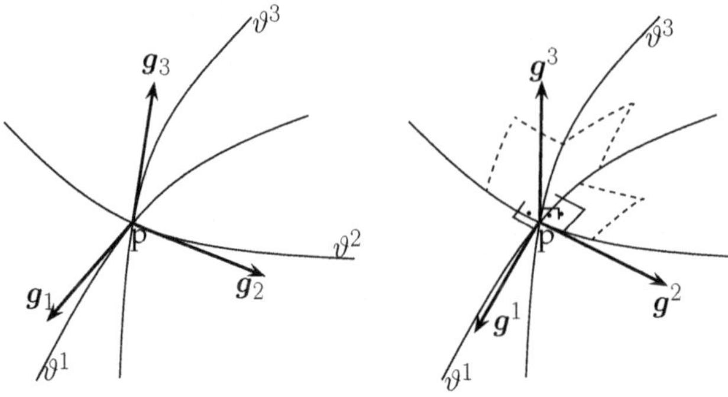

Figure 1.1. *Covariant and contravariant base vectors*

A right-handed Cartesian coordinate system (x^1, x^2, x^3) can be seen as a special case of a curvilinear system with base vectors $\boldsymbol{e}_i = \boldsymbol{e}^i$.

The mixed Kronecker delta is defined as

$$\delta_i^j = \begin{cases} 1, & i = j, \\ 0, & i \neq j. \end{cases}$$

The Cartesian permutation symbol is defined as the third-order tensor $\boldsymbol{\epsilon}$. The covariant (ϵ_{ijk}) and the contravariant (ϵ^{ijk}) components of this tensor are the same and take the values

$$\epsilon_{ijk} = \epsilon^{ijk} = \begin{cases} 1, & \text{if } (i,j,k) \text{ is } (1,2,3),\ (2,3,1) \text{ or } (3,1,2), \\ -1, & \text{if } (i,j,k) \text{ is } (1,3,2),\ (2,1,3) \text{ or } (3,2,1), \\ 0, & \text{if } i = j,\ j = k \text{ or } k = i, \end{cases}$$

with the property $\epsilon^{ijr}\epsilon_{pqr} = \delta_p^i \delta_q^j - \delta_q^i \delta_p^j$.

Table 1.2 summarizes some useful properties for curvilinear coordinate systems. In this table, the terms g_{pq} and g^{pq} express the metric coefficients. Moreover, the symbols ($\otimes$) and ($\cdot$) denote dyadic product and single contraction of vectors, respectively. The determinant of a matrix is denoted by det. The knowledge of the base vectors allows us to perform integral operations in a volume, a surface and a line which are expressed in curvilinear coordinate systems. The infinitesimal volume element is denoted by $\mathrm{d}v$. The infinitesimal surface element $\mathrm{d}s_r$ has a normal vector parallel to $\boldsymbol{g}^r$, while the infinitesimal line element $\mathrm{d}l_r$ has a tangent vector parallel to $\boldsymbol{g}_r$.

$$\boldsymbol{g}_q = \frac{\partial x^p}{\partial \vartheta^q}\boldsymbol{e}_p \qquad \boldsymbol{g}^q = \frac{\partial \vartheta^q}{\partial x^p}\boldsymbol{e}^p$$

$$\boldsymbol{g}_p = g_{pq}\boldsymbol{g}^q \qquad \boldsymbol{g}^p = g^{pq}\boldsymbol{g}_q$$

$$g_{pq} := \boldsymbol{g}_p \cdot \boldsymbol{g}_q = g_{qp} \qquad g^{pq} := \boldsymbol{g}^p \cdot \boldsymbol{g}^q = g^{qp}$$

$$\delta^q_p = \boldsymbol{g}^q \cdot \boldsymbol{g}_p = g^{qs}g_{ps} \qquad \boldsymbol{i} = \delta^q_p \boldsymbol{g}_q \otimes \boldsymbol{g}^p = \boldsymbol{g}_q \otimes \boldsymbol{g}^q$$

$$g = [\boldsymbol{g}_1 \cdot [\boldsymbol{g}_2 \times \boldsymbol{g}_3]]^2 = \det g_{pq} = \frac{1}{\det g^{pq}} > 0$$

$$\epsilon_{ijk}\ g = \epsilon^{pqr} g_{ip}g_{jq}g_{kr} \qquad \epsilon^{ijk}\frac{1}{g} = \epsilon_{pqr}g^{ip}g^{jq}g^{kr}$$

$$\boldsymbol{g}_p \times \boldsymbol{g}_q = \sqrt{g}\epsilon_{pqr}\boldsymbol{g}^r \qquad \boldsymbol{g}^p \times \boldsymbol{g}^q = \frac{1}{\sqrt{g}}\epsilon^{pqr}\boldsymbol{g}_r$$

$$\boldsymbol{\epsilon} = \epsilon_{pqr}\sqrt{g}\ \boldsymbol{g}^p \otimes \boldsymbol{g}^q \otimes \boldsymbol{g}^r = \epsilon^{pqr}\frac{1}{\sqrt{g}}\ \boldsymbol{g}_p \otimes \boldsymbol{g}_q \otimes \boldsymbol{g}_r$$

$$\boldsymbol{\epsilon} = \epsilon^{pqr}\boldsymbol{e}_p \otimes \boldsymbol{e}_q \otimes \boldsymbol{e}_r = \epsilon_{pqr}\boldsymbol{e}^p \otimes \boldsymbol{e}^q \otimes \boldsymbol{e}^r$$

$$\mathrm{d}v = [\boldsymbol{g}_1 \cdot [\boldsymbol{g}_2 \times \boldsymbol{g}_3]]\mathrm{d}\vartheta^1\mathrm{d}\vartheta^2\mathrm{d}\vartheta^3 = \sqrt{g}\mathrm{d}\vartheta^1\mathrm{d}\vartheta^2\mathrm{d}\vartheta^3$$

$$\mathrm{d}s_1 = |\boldsymbol{g}_2 \times \boldsymbol{g}_3|\mathrm{d}\vartheta^2\mathrm{d}\vartheta^3 = \sqrt{g}\sqrt{g^{11}}\mathrm{d}\vartheta^2\mathrm{d}\vartheta^3$$

$$\mathrm{d}s_2 = |\boldsymbol{g}_3 \times \boldsymbol{g}_1|\mathrm{d}\vartheta^3\mathrm{d}\vartheta^1 = \sqrt{g}\sqrt{g^{22}}\mathrm{d}\vartheta^3\mathrm{d}\vartheta^1$$

$$\mathrm{d}s_3 = |\boldsymbol{g}_1 \times \boldsymbol{g}_2|\mathrm{d}\vartheta^1\mathrm{d}\vartheta^2 = \sqrt{g}\sqrt{g^{33}}\mathrm{d}\vartheta^1\mathrm{d}\vartheta^2$$

$$\mathrm{d}l_1 = |\boldsymbol{g}_1|\mathrm{d}\vartheta^1 = \sqrt{g_{11}}\mathrm{d}\vartheta^1$$

$$\mathrm{d}l_2 = |\boldsymbol{g}_2|\mathrm{d}\vartheta^2 = \sqrt{g_{22}}\mathrm{d}\vartheta^2$$

$$\mathrm{d}l_3 = |\boldsymbol{g}_3|\mathrm{d}\vartheta^3 = \sqrt{g_{33}}\mathrm{d}\vartheta^3$$

Table 1.2. *Useful properties for curvilinear coordinate systems*

1.2.1. *Tensor operations*

Every mixed (n, m) order tensor in $\boldsymbol{g}$ base vectors is represented as

$$\boldsymbol{A} = A^{i_1 \dots i_n}_{j_1 \dots j_m}\boldsymbol{g}_{i_1} \otimes \dots \otimes \boldsymbol{g}_{i_n} \otimes \boldsymbol{g}^{j_1} \otimes \dots \otimes \boldsymbol{g}^{j_m}.$$

Consider the curvilinear coordinate systems $(\vartheta^1, \vartheta^2, \vartheta^3)$ and $({}^{\#}\!\vartheta^1, {}^{\#}\!\vartheta^2, {}^{\#}\!\vartheta^3)$. The rotation $\boldsymbol{R}$ and its inverse $\boldsymbol{R}^{-1} = {}^{\#}\!\boldsymbol{R}$, written in the mixed tensor form

$$R_p^q = \frac{\partial \vartheta^q}{\partial\, {}^{\#}\!\vartheta^p}, \quad {}^{\#}\!R_q^p = \frac{\partial\, {}^{\#}\!\vartheta^p}{\partial \vartheta^q},$$

allow us to transform any vector $\boldsymbol{g}_r$ or $\boldsymbol{g}^r$ from one coordinate system to the other through the relations

$${}^{\#}\!\boldsymbol{g}_p = R_p^q \boldsymbol{g}_q, \quad {}^{\#}\!\boldsymbol{g}^p = {}^{\#}\!R_q^p \boldsymbol{g}^q.$$

The transition of the (n, m) order tensor $\boldsymbol{A}$ between the coordinate systems is expressed by the relations

$$\begin{aligned} {}^{\#}\!A_{j_1 \dots j_m}^{i_1 \dots i_n} &= {}^{\#}\!R_{I_1}^{i_1} \dots {}^{\#}\!R_{I_n}^{i_n} R_{j_1}^{J_1} \dots R_{j_m}^{J_m} A_{J_1 \dots J_m}^{I_1 \dots I_n}, \\ A_{j_1 \dots j_m}^{i_1 \dots i_n} &= R_{I_1}^{i_1} \dots R_{I_n}^{i_n}\, {}^{\#}\!R_{j_1}^{J_1} \dots {}^{\#}\!R_{j_m}^{J_m}\, {}^{\#}\!A_{J_1 \dots J_m}^{I_1 \dots I_n}. \end{aligned}$$

The following tensor operations are introduced:

– The (n, m) order tensor $\boldsymbol{A}$, the (n, m) order tensor $\boldsymbol{B}$ and the scalar c produce the (n, m) order tensors

$$\begin{aligned} &\boldsymbol{C} = \boldsymbol{A} + \boldsymbol{B}, \quad \text{with} \quad C_{j_1 \dots j_m}^{i_1 \dots i_n} = A_{j_1 \dots j_m}^{i_1 \dots i_n} + B_{j_1 \dots j_m}^{i_1 \dots i_n}, \\ &\boldsymbol{C} = c\boldsymbol{A}, \quad \text{with} \quad C_{j_1 \dots j_m}^{i_1 \dots i_n} = cA_{j_1 \dots j_m}^{i_1 \dots i_n}. \end{aligned}$$

– The $(n + 1, m)$ order tensor $\boldsymbol{A}$ and the $(p, q + 1)$ order tensor $\boldsymbol{B}$ produce the $(n + p, m + q)$ order tensor

$$\boldsymbol{C} = \boldsymbol{A} \cdot \boldsymbol{B}, \quad \text{with} \quad C_{j_1 \dots j_{m+q}}^{i_1 \dots i_{n+p}} = A_{j_1 \dots j_m}^{i_1 \dots i_n I} B_{I j_{m+1} \dots j_{m+q}}^{i_{n+1} \dots i_{n+p}}.$$

– The $(n, m + 1)$ order tensor $\boldsymbol{A}$ and the $(p + 1, q)$ order tensor $\boldsymbol{B}$ produce the $(n + p, m + q)$ order tensor

$$\boldsymbol{C} = \boldsymbol{A} \cdot \boldsymbol{B}, \quad \text{with} \quad C_{j_1 \dots j_{m+q}}^{i_1 \dots i_{n+p}} = A_{j_1 \dots j_m I}^{i_1 \dots i_n} B_{j_{m+1} \dots j_{m+q}}^{I i_{n+1} \dots i_{n+p}}.$$

– The $(n+2,m)$ order tensor $\boldsymbol{A}$ and the $(p,q+2)$ order tensor $\boldsymbol{B}$ produce the $(n+p,m+q)$ order tensor

$$\boldsymbol{C} = \boldsymbol{A}\!:\!\boldsymbol{B}, \quad \text{with} \quad C^{i_1\ldots i_{n+p}}_{j_1\ldots j_{m+q}} = A^{i_1\ldots i_n IJ}_{j_1\ldots j_m} B^{i_{n+1}\ldots i_{n+p}}_{IJ j_{m+1}\ldots j_{m+q}}.$$

– The $(n,m+2)$ order tensor $\boldsymbol{A}$ and the $(p+2,q)$ order tensor $\boldsymbol{B}$ produce the $(n+p,m+q)$ order tensor

$$\boldsymbol{C} = \boldsymbol{A}\!:\!\boldsymbol{B}, \quad \text{with} \quad C^{i_1\ldots i_{n+p}}_{j_1\ldots j_{m+q}} = A^{i_1\ldots i_n}_{j_1\ldots j_m IJ} B^{IJ i_{n+1}\ldots i_{n+p}}_{j_{m+1}\ldots j_{m+q}}.$$

– The (n,m) order tensor $\boldsymbol{A}$ and the (p,q) order tensor $\boldsymbol{B}$ produce the $(n+p,m+q)$ order tensor

$$\boldsymbol{C} = \boldsymbol{A}\otimes\boldsymbol{B}, \quad \text{with} \quad C^{i_1\ldots i_{n+p}}_{j_1\ldots j_{m+q}} = A^{i_1\ldots i_n}_{j_1\ldots j_m} B^{i_{n+1}\ldots i_{n+p}}_{j_{m+1}\ldots j_{m+q}}.$$

– The $(n+2,m)$ order tensor $\boldsymbol{A}$ and the $(1,0)$ order tensor $\boldsymbol{B}$ produce the $(n+1,m+1)$ order tensors

$$\boldsymbol{C} = \boldsymbol{A}\times\boldsymbol{B}, \quad \text{with} \quad C^{i_1\ldots i_{n+1}}_{j_1\ldots j_{m+1}} = A^{i_1\ldots i_{n+1} I}_{j_1\ldots j_m} B^J \epsilon_{IJ j_{m+1}} \sqrt{g},$$

$$\boldsymbol{C} = \boldsymbol{B}\times\boldsymbol{A}, \quad \text{with} \quad C^{i_1\ldots i_{n+1}}_{j_1\ldots j_{m+1}} = B^I A^{J i_1\ldots i_{n+1}}_{j_2\ldots j_{m+1}} \epsilon_{IJ j_1} \sqrt{g}.$$

– The $(n,m+2)$ order tensor $\boldsymbol{A}$ and the $(0,1)$ order tensor $\boldsymbol{B}$ produce the $(n+1,m+1)$ order tensors

$$\boldsymbol{C} = \boldsymbol{A}\times\boldsymbol{B}, \quad \text{with} \quad C^{i_1\ldots i_{n+1}}_{j_1\ldots j_{m+1}} = A^{i_1\ldots i_n}_{j_1\ldots j_{m+1} I} B_J \frac{\epsilon^{IJ i_{n+1}}}{\sqrt{g}},$$

$$\boldsymbol{C} = \boldsymbol{B}\times\boldsymbol{A}, \quad \text{with} \quad C^{i_1\ldots i_{n+1}}_{j_1\ldots j_{m+1}} = B_I A^{i_2\ldots i_{n+1}}_{J j_1\ldots j_{m+1}} \frac{\epsilon^{IJ i_1}}{\sqrt{g}}.$$

1.2.1.1. *Second-order tensor operations*

For a second-order tensor $\boldsymbol{T}$, we can identify the following transpose forms:

$$\left[T^{ij}\boldsymbol{g}_i\otimes\boldsymbol{g}_j\right]^T = T^{ij}\boldsymbol{g}_j\otimes\boldsymbol{g}_i, \quad \left[T_{ij}\boldsymbol{g}^i\otimes\boldsymbol{g}^j\right]^T = T_{ij}\boldsymbol{g}^j\otimes\boldsymbol{g}^i,$$

$$\left[T_i^j \boldsymbol{g}^i \otimes \boldsymbol{g}_j\right]^T = T_i^j \boldsymbol{g}_j \otimes \boldsymbol{g}^i \,, \quad \left[T_j^i \boldsymbol{g}_i \otimes \boldsymbol{g}^j\right]^T = T_j^i \boldsymbol{g}^j \otimes \boldsymbol{g}_i \,,$$

Moreover, the trace and the determinant of a second-order tensor are defined as

$$\begin{aligned}
\mathrm{tr}\boldsymbol{T} &= \boldsymbol{i}:\boldsymbol{T} = [\boldsymbol{g}_q \otimes \boldsymbol{g}^q]:[T_j^i \boldsymbol{g}_i \otimes \boldsymbol{g}^j] = T_i^i = T^{ij} g_{ij} = T_{ij} g^{ij} \,, \\
\mathrm{det}\boldsymbol{T} &= \frac{[\boldsymbol{T}\cdot\boldsymbol{g}_1]\cdot[[\boldsymbol{T}\cdot\boldsymbol{g}_2]\times[\boldsymbol{T}\cdot\boldsymbol{g}_3]]}{\boldsymbol{g}_1\cdot[\boldsymbol{g}_2\times\boldsymbol{g}_3]} = \frac{g}{6}\epsilon_{ijk}\epsilon_{pqr}T^{ip}T^{jq}T^{kr} \\
&= \frac{1}{6g}\epsilon^{ijk}\epsilon^{pqr}T_{ip}T_{jq}T_{kr} = \frac{1}{6}\epsilon_{ijk}\epsilon^{pqr}T_p^i T_q^j T_r^k \,.
\end{aligned}$$

1.2.2. *Tensor derivatives*

For two second-order tensors $\boldsymbol{A}$ and $\boldsymbol{B}$, the following identities are true:

$$\frac{\partial(\boldsymbol{A}+\boldsymbol{B})}{\partial\vartheta^q} = \frac{\partial\boldsymbol{A}}{\partial\vartheta^q} + \frac{\partial\boldsymbol{B}}{\partial\vartheta^q}, \qquad \frac{\partial(\boldsymbol{A}\Box\boldsymbol{B})}{\partial\vartheta^q} = \frac{\partial\boldsymbol{A}}{\partial\vartheta^q}\Box\boldsymbol{B} + \boldsymbol{A}\Box\frac{\partial\boldsymbol{B}}{\partial\vartheta^q},$$

with $\Box = \cdot\,, :$ or $\otimes$. Considering a scalar ψ, a vector $\boldsymbol{u}$ with covariant components u_i and a second-order tensor $\boldsymbol{T}$ with contravariant components T^{ij}, the following operators are identified:

$$\begin{aligned}
\mathrm{grad}\psi &= \frac{\partial\psi}{\partial\vartheta^j}\boldsymbol{g}^j \,, \\
\mathrm{grad}\boldsymbol{u} &= \frac{\partial\boldsymbol{u}}{\partial\vartheta^j}\otimes\boldsymbol{g}^j = \frac{\partial(u_i\boldsymbol{g}^i)}{\partial\vartheta^j}\otimes\boldsymbol{g}^j = \left[\frac{\partial u_i}{\partial\vartheta^j} - u_k\Gamma_{ij}^k\right]\boldsymbol{g}^i\otimes\boldsymbol{g}^j \,, \\
\mathrm{div}\boldsymbol{u} &= \mathrm{grad}\boldsymbol{u}:\boldsymbol{i} = \left[\frac{\partial u_i}{\partial\vartheta^j} - u_k\Gamma_{ij}^k\right]g^{ij} \,, \\
\mathrm{curl}\boldsymbol{u} &= -\mathrm{grad}\boldsymbol{u}:\boldsymbol{\epsilon} = -\left[\frac{\partial u_i}{\partial\vartheta^j} - u_l\Gamma_{ij}^l\right]\frac{\epsilon^{ijk}}{\sqrt{g}}\boldsymbol{g}_k = -\frac{\partial u_i}{\partial\vartheta^j}\frac{\epsilon^{ijk}}{\sqrt{g}}\boldsymbol{g}_k \,, \\
\mathrm{grad}\boldsymbol{T} &= \frac{\partial\boldsymbol{T}}{\partial\vartheta^j}\otimes\boldsymbol{g}^j = \frac{\partial(T^{ik}\boldsymbol{g}_i\otimes\boldsymbol{g}_k)}{\partial\vartheta^j}\otimes\boldsymbol{g}^j \\
&= \left[\frac{\partial T^{ik}}{\partial\vartheta^j} + T^{lk}\Gamma_{lj}^i + T^{il}\Gamma_{jl}^k\right]\boldsymbol{g}_i\otimes\boldsymbol{g}_k\otimes\boldsymbol{g}^j \,,
\end{aligned}$$

$$\mathrm{div}\boldsymbol{T} = \mathrm{grad}\boldsymbol{T}{:}\boldsymbol{i} = \left[\frac{\partial T^{ij}}{\partial \vartheta^j} + T^{jk}\Gamma^i_{jk} + T^{ik}\Gamma^j_{jk}\right]\boldsymbol{g}_i\,,$$

where $\Gamma^r_{pq} = \dfrac{\partial \boldsymbol{g}_p}{\partial \vartheta^q}\cdot\boldsymbol{g}^r = -\dfrac{\partial \boldsymbol{g}^r}{\partial \vartheta^q}\cdot\boldsymbol{g}_p = \Gamma^r_{qp}$ is the Christoffel symbol of the second kind.

1.2.3. *Transformations between coordinate systems*

In continuum mechanics, it is often necessary to describe the motion of a body with coordinate systems that are not fixed in time. In many occasions, this is not convenient and it is more preferable to represent the body motion through fixed coordinate systems. Differential geometry is a powerful mathematical theory that permits us to identify the correlation between different coordinate systems, either Cartesian or curvilinear.

Consider a body $\mathcal{B}$ and a fixed, Cartesian or curvilinear, coordinate system $(\vartheta^1, \vartheta^2, \vartheta^3)$. Every point on this body is expressed either in the Cartesian coordinate system $\boldsymbol{X}$ with base vectors $\boldsymbol{e}_i = \boldsymbol{e}^i$, or in the Cartesian coordinate system $\boldsymbol{x}$ with base vectors $\boldsymbol{e}'_i = \boldsymbol{e}'^i$ (Figure 1.2). If the three coordinate systems (the fixed one and the two Cartesians) are connected with one to one mappings, then $\boldsymbol{X} := \boldsymbol{X}(\vartheta^1, \vartheta^2, \vartheta^3) = \boldsymbol{Y}(\boldsymbol{x})$ and $\boldsymbol{x} := \boldsymbol{x}(\vartheta^1, \vartheta^2, \vartheta^3) = \boldsymbol{Y}^{-1}(\boldsymbol{X})$, where $\boldsymbol{Y}$ is an invertible, possibly nonlinear, deformation map.

From differential geometry, the properties and definitions of Tables 1.3 and 1.4 hold [STE 08, STE 15, KEL 13]. In these tables, the unit vectors $\overline{\boldsymbol{N}}$, $\widetilde{\boldsymbol{N}}$, $\bar{\boldsymbol{n}}$, $\tilde{\boldsymbol{n}}$, shown in Figure 1.2, define vectors attached to a surface S on the body in the two different coordinate systems $\boldsymbol{X}$ and $\boldsymbol{x}$. In the coordinate system $\boldsymbol{X}$, the infinitesimal volume element is denoted by $\mathrm{d}V$, the infinitesimal surface element is denoted by $\mathrm{d}S$ and the infinitesimal line element is denoted by $\mathrm{d}L$. Accordingly, in the coordinate system $\boldsymbol{x}$, the infinitesimal volume element is denoted by $\mathrm{d}v$, the infinitesimal surface element is denoted by $\mathrm{d}s$ and the infinitesimal line element is denoted by $\mathrm{d}l$. Finally, the operators Γ^r_{pq} denote the Christoffel symbols of the second kind and the upper left indices m and s refer to the coordinate systems $\boldsymbol{X}$ and $\boldsymbol{x}$, respectively.

material configuration
$\boldsymbol{G}_q := \dfrac{\partial X^p}{\partial \vartheta^q}\boldsymbol{e}_p = \dfrac{\partial \boldsymbol{X}}{\partial \vartheta^q} \qquad \boldsymbol{G}^q := \dfrac{\partial \vartheta^q}{\partial X^p}\boldsymbol{e}^p = \dfrac{\partial \vartheta^q}{\partial \boldsymbol{X}}$
$G_{pq} := \boldsymbol{G}_p \cdot \boldsymbol{G}_q \qquad G^{pq} := \boldsymbol{G}^p \cdot \boldsymbol{G}^q$
$\boldsymbol{I} := \delta^q_r \boldsymbol{G}_q \otimes \boldsymbol{G}^r = \boldsymbol{G}_q \otimes \boldsymbol{G}^q = \boldsymbol{G}^q \otimes \boldsymbol{G}_q$
$\delta^q_r = \boldsymbol{G}^q \cdot \boldsymbol{G}_r = G^{qr} G_{pr} \qquad G = \mathrm{Det}(G_{pq})$
$\boldsymbol{G}_p \times \boldsymbol{G}_q = \sqrt{G}\epsilon_{pqr}\boldsymbol{G}^r \qquad \boldsymbol{G}^p \times \boldsymbol{G}^q = \dfrac{1}{\sqrt{G}}\epsilon^{pqr}\boldsymbol{G}_r$
$\boldsymbol{\epsilon} = \epsilon^{pqr}\dfrac{1}{\sqrt{G}}\boldsymbol{G}_p \otimes \boldsymbol{G}_q \otimes \boldsymbol{G}_r = \epsilon_{pqr}\sqrt{G}\boldsymbol{G}^p \otimes \boldsymbol{G}^q \otimes \boldsymbol{G}^r$
$\mathrm{Grad}\{\bullet\} := \dfrac{\partial\{\bullet\}}{\partial \vartheta^q} \otimes \boldsymbol{G}^q = \dfrac{\partial\{\bullet\}}{\partial \boldsymbol{X}}$
$\mathrm{Div}\{\bullet\} := \dfrac{\partial\{\bullet\}}{\partial \vartheta^q} \cdot \boldsymbol{G}^q = \mathrm{Grad}\{\bullet\} : \boldsymbol{I}$
$\mathrm{Curl}\{\bullet\} := -\dfrac{\partial\{\bullet\}}{\partial \vartheta^q} \times \boldsymbol{G}^q = -\mathrm{Grad}\{\bullet\} : \boldsymbol{\epsilon}$
${}^m\Gamma^r_{pq} = \dfrac{\partial \boldsymbol{G}_p}{\partial \vartheta^q} \cdot \boldsymbol{G}^r = -\dfrac{\partial \boldsymbol{G}^r}{\partial \vartheta^q} \cdot \boldsymbol{G}_p = {}^m\Gamma^r_{qp}$
$\dfrac{\partial \boldsymbol{G}_r}{\partial \vartheta^p} = {}^m\Gamma^q_{rp}\boldsymbol{G}_q \qquad \dfrac{\partial \boldsymbol{G}^r}{\partial \vartheta^p} = -{}^m\Gamma^r_{pq}\boldsymbol{G}^q$
$\boldsymbol{F} := \mathrm{Grad}\boldsymbol{x} = \boldsymbol{g}_p \otimes \boldsymbol{G}^p \qquad \boldsymbol{F}^T = \boldsymbol{G}^p \otimes \boldsymbol{g}_p$
$\boldsymbol{F}^{-1} = \boldsymbol{G}_p \otimes \boldsymbol{g}^p \qquad \boldsymbol{F}^{-T} = \boldsymbol{g}^p \otimes \boldsymbol{G}_p$
$J := \mathrm{Det}\boldsymbol{F} = \dfrac{\sqrt{g}}{\sqrt{G}} \qquad \mathrm{Cof}\boldsymbol{F} := \dfrac{\partial J}{\partial \boldsymbol{F}} = J\boldsymbol{F}^{-T}$
$\mathrm{d}v = J\,\mathrm{d}V \qquad \bar{\boldsymbol{n}}\,\mathrm{d}s = \mathrm{Cof}\boldsymbol{F} \cdot \overline{\boldsymbol{N}}\,\mathrm{d}S \qquad \tilde{\boldsymbol{n}}\,\mathrm{d}l = \boldsymbol{F} \cdot \widetilde{\boldsymbol{N}}\,\mathrm{d}L$
$\bar{\boldsymbol{n}} = \dfrac{\mathrm{Cof}\boldsymbol{F} \cdot \overline{\boldsymbol{N}}}{\lvert \mathrm{Cof}\boldsymbol{F} \cdot \overline{\boldsymbol{N}} \rvert} \qquad \tilde{\boldsymbol{n}} = \dfrac{\boldsymbol{F} \cdot \widetilde{\boldsymbol{N}}}{\lvert \boldsymbol{F} \cdot \widetilde{\boldsymbol{N}} \rvert}$
$\overline{\boldsymbol{N}}$ normal ($\perp$) to the surface, $\widetilde{\boldsymbol{N}}$ tangent ($\parallel$) to the curve

Table 1.3. *Tensor properties in the coordinate system* $\boldsymbol{X}(\vartheta^1, \vartheta^2, \vartheta^3)$

spatial configuration
$\boldsymbol{g}_q := \dfrac{\partial x^p}{\partial \vartheta^q}\boldsymbol{e}'_p = \dfrac{\partial \boldsymbol{x}}{\partial \vartheta^q} \qquad \boldsymbol{g}^q := \dfrac{\partial \vartheta^q}{\partial x^p}\boldsymbol{e}'^p = \dfrac{\partial \vartheta^q}{\partial \boldsymbol{x}}$
$g_{pq} := \boldsymbol{g}_p \cdot \boldsymbol{g}_q \qquad g^{pq} := \boldsymbol{g}^p \cdot \boldsymbol{g}^q$
$\boldsymbol{i} := \delta^q_r \boldsymbol{g}_q \otimes \boldsymbol{g}^r = \boldsymbol{g}_q \otimes \boldsymbol{g}^q = \boldsymbol{g}^q \otimes \boldsymbol{g}_q$
$\delta^q_r = \boldsymbol{g}^q \cdot \boldsymbol{g}_r = g^{qr} g_{pr} \qquad g = \det(g_{pq})$
$\boldsymbol{g}_p \times \boldsymbol{g}_q = \sqrt{g}\epsilon_{pqr}\boldsymbol{g}^r \qquad \boldsymbol{g}^p \times \boldsymbol{g}^q = \dfrac{1}{\sqrt{g}}\epsilon^{pqr}\boldsymbol{g}_r$
$\boldsymbol{\epsilon} = \epsilon^{ijk}\dfrac{1}{\sqrt{g}}\boldsymbol{g}_p \otimes \boldsymbol{g}_q \otimes \boldsymbol{g}_r = \epsilon_{pqr}\sqrt{g}\boldsymbol{g}^p \otimes \boldsymbol{g}^q \otimes \boldsymbol{g}^r$
$\mathrm{grad}\,\{\bullet\} := \dfrac{\partial\{\bullet\}}{\partial \vartheta^q} \otimes \boldsymbol{g}^q = \dfrac{\partial\{\bullet\}}{\partial \boldsymbol{x}}$
$\mathrm{div}\,\{\bullet\} := \dfrac{\partial\{\bullet\}}{\partial \vartheta^q} \otimes \boldsymbol{g}^q = \mathrm{grad}\,\{\bullet\} : \boldsymbol{i}$
$\mathrm{curl}\,\{\bullet\} := -\dfrac{\partial\{\bullet\}}{\partial \vartheta^q} \times \boldsymbol{g}^q = -\mathrm{grad}\,\{\bullet\} : \boldsymbol{\epsilon}$
${}^s\Gamma^r_{pq} = \dfrac{\partial \boldsymbol{g}_p}{\partial \vartheta^q} \cdot \boldsymbol{g}^r = -\dfrac{\partial \boldsymbol{g}^r}{\partial \vartheta^q} \cdot \boldsymbol{g}_p = {}^s\Gamma^r_{qp}$
$\dfrac{\partial \boldsymbol{g}_r}{\partial \vartheta^p} = {}^s\Gamma^q_{rp}\boldsymbol{g}_q \qquad \dfrac{\partial \boldsymbol{g}^r}{\partial \vartheta^p} = -{}^s\Gamma^r_{pq}\boldsymbol{g}^q$
$\boldsymbol{f} := \mathrm{grad}\boldsymbol{X} = \boldsymbol{G}_p \otimes \boldsymbol{g}^p \qquad \boldsymbol{f}^T = \boldsymbol{g}^p \otimes \boldsymbol{G}_p$
$\boldsymbol{f}^{-1} = \boldsymbol{g}_p \otimes \boldsymbol{G}^p \qquad \boldsymbol{f}^{-T} = \boldsymbol{G}^p \otimes \boldsymbol{g}_p$
$j := \det\boldsymbol{f} = \dfrac{\sqrt{G}}{\sqrt{g}} \qquad \mathrm{cof}\boldsymbol{f} := \dfrac{\partial j}{\partial \boldsymbol{f}} = j\ \boldsymbol{f}^{-T}$
$\mathrm{d}V = j\,\mathrm{d}v \qquad \overline{\boldsymbol{N}}\,\mathrm{d}S = \mathrm{cof}\boldsymbol{f} \cdot \bar{\boldsymbol{n}}\,\mathrm{d}s \qquad \widetilde{\boldsymbol{N}}\,\mathrm{d}L = \boldsymbol{f} \cdot \tilde{\boldsymbol{n}}\,\mathrm{d}l$
$\overline{\boldsymbol{N}} = \dfrac{\mathrm{cof}\boldsymbol{f} \cdot \bar{\boldsymbol{n}}}{\|\mathrm{cof}\boldsymbol{f} \cdot \bar{\boldsymbol{n}}\|} \qquad \widetilde{\boldsymbol{N}} = \dfrac{\boldsymbol{f} \cdot \tilde{\boldsymbol{n}}}{\|\boldsymbol{f} \cdot \tilde{\boldsymbol{n}}\|}$
$\bar{\boldsymbol{n}}$ normal ($\perp$) to the surface, $\tilde{\boldsymbol{n}}$ tangent ($\parallel$) to the curve

Table 1.4. *Tensor properties in the coordinate system* $\boldsymbol{x}(\vartheta^1, \vartheta^2, \vartheta^3)$

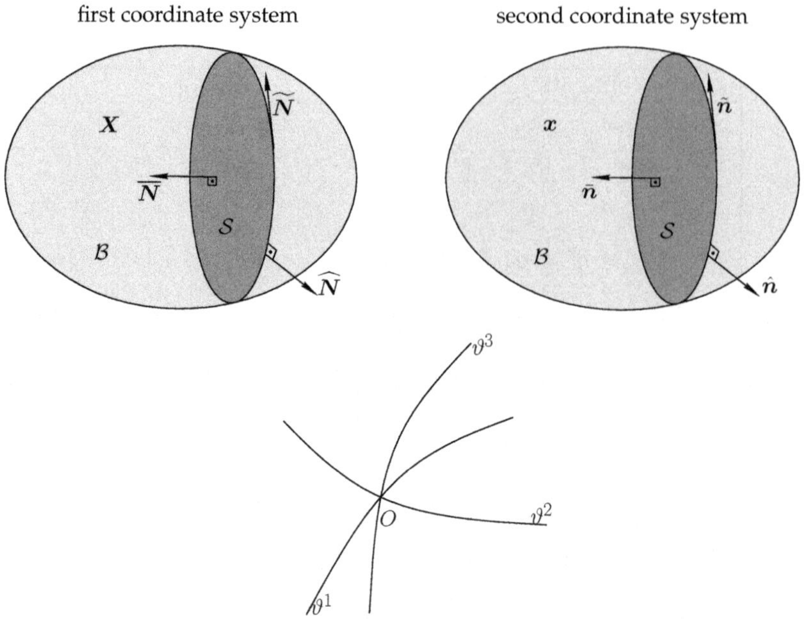

Figure 1.2. *Transformations between two Cartesian coordinate systems $\boldsymbol{X}$ and $\boldsymbol{x}$ with the help of a fixed coordinate system $(\vartheta^1, \vartheta^2, \vartheta^3)$ For a color version of this figure, see www.iste.co.uk/chatzigeorgiou/thermomechanical.zip*

1.2.4. *Several proofs in tensor calculus*

– Considering the two coordinate systems $\boldsymbol{X}$ and $\boldsymbol{x}$ and the properties of Tables 1.3 and 1.4, it is proven that (a) $\text{Div}(\text{Cof}\boldsymbol{F}) = \boldsymbol{0}$, (b) $\dfrac{DJ}{Dt} = J\text{div}\dot{\boldsymbol{x}}$.

PROOF.– Indeed, for the first relation,

$$\begin{aligned}
\frac{\partial \text{Cof}\boldsymbol{F}}{\partial \vartheta^i} &= \frac{\partial J}{\partial \vartheta^i}\boldsymbol{g}^q \otimes \boldsymbol{G}_q + J\frac{\partial \boldsymbol{g}^q}{\partial \vartheta^i} \otimes \boldsymbol{G}_q + J\boldsymbol{g}^q \otimes \frac{\partial \boldsymbol{G}_q}{\partial \vartheta^i} \\
&= \left[\frac{\partial J}{\partial \boldsymbol{F}} : \frac{\partial \boldsymbol{F}}{\partial \vartheta^i}\right] \boldsymbol{g}^q \otimes \boldsymbol{G}_q - J\,{}^s\Gamma^q_{ip}\boldsymbol{g}^p \otimes \boldsymbol{G}_q + J\,{}^m\Gamma^p_{iq}\boldsymbol{g}^q \otimes \boldsymbol{G}_p \\
&= J\left[[\boldsymbol{g}^s \otimes \boldsymbol{G}_s] : \frac{\partial(\boldsymbol{g}_p \otimes \boldsymbol{G}^p)}{\partial \vartheta^i}\right] \boldsymbol{g}^q \otimes \boldsymbol{G}_q - J\,{}^s\Gamma^q_{ip}\boldsymbol{g}^p \otimes \boldsymbol{G}_q \\
&\quad + J\,{}^m\Gamma^p_{iq}\boldsymbol{g}^q \otimes \boldsymbol{G}_p
\end{aligned}$$

$$= J\left[\boldsymbol{G}_p\cdot\frac{\partial \boldsymbol{G}^p}{\partial\vartheta^i} + \boldsymbol{g}^p\cdot\frac{\partial \boldsymbol{g}_p}{\partial\vartheta^i}\right]\boldsymbol{g}^q\otimes\boldsymbol{G}_q - J\,{}^s\Gamma^q_{ip}\boldsymbol{g}^p\otimes\boldsymbol{G}_q$$

$$+J\,{}^m\Gamma^p_{iq}\boldsymbol{g}^q\otimes\boldsymbol{G}_p$$

$$= J\left[-\,{}^m\Gamma^p_{ip} + {}^s\Gamma^p_{ip}\right]\boldsymbol{g}^q\otimes\boldsymbol{G}_q - J\,{}^s\Gamma^q_{ip}\boldsymbol{g}^p\otimes\boldsymbol{G}_q$$

$$+J\,{}^m\Gamma^p_{iq}\boldsymbol{g}^q\otimes\boldsymbol{G}_p\,.$$

Thus,

$$\begin{aligned}\mathrm{Div}(\mathrm{Cof}\boldsymbol{F}) &= \frac{\partial\mathrm{Cof}\boldsymbol{F}}{\partial\vartheta^i}\cdot\boldsymbol{G}^i\\ &= -J\,{}^m\Gamma^p_{ip}\boldsymbol{g}^i + J\,{}^s\Gamma^p_{ip}\boldsymbol{g}^i - J\,{}^s\Gamma^i_{ip}\boldsymbol{g}^p + J\,{}^m\Gamma^i_{iq}\boldsymbol{g}^q = \boldsymbol{0}\,.\end{aligned}$$

For the second relation, we can write

$$\begin{aligned}\frac{DJ}{Dt} &= \frac{\partial J}{\partial\boldsymbol{F}}:\frac{D\boldsymbol{F}}{Dt} = \mathrm{Cof}\boldsymbol{F}:\mathrm{Grad}\dot{\boldsymbol{x}} = [J\boldsymbol{g}^p\otimes\boldsymbol{G}_p]:\left[\frac{\partial\dot{\boldsymbol{x}}}{\partial\vartheta^q}\otimes\boldsymbol{G}^q\right]\\ &= J\left[\frac{\partial\dot{\boldsymbol{x}}}{\partial\vartheta^q}\cdot\boldsymbol{g}^p\right]\delta^q_p = J\frac{\partial\dot{\boldsymbol{x}}}{\partial\vartheta^q}\cdot\boldsymbol{g}^q = J\mathrm{div}\dot{\boldsymbol{x}}.\end{aligned}$$

□

– Considering the two coordinate systems $\boldsymbol{X}$ and $\boldsymbol{x}$ and the properties of Tables 1.3 and 1.4, it is proven that for arbitrary vector $\boldsymbol{a}$ the following identities hold:

1) $\mathrm{Div}(\boldsymbol{a}\cdot\mathrm{Cof}\boldsymbol{F}) = J\mathrm{div}\boldsymbol{a}$,

2) $\mathrm{Curl}(\boldsymbol{a}\cdot\boldsymbol{F}) = \mathrm{curl}\boldsymbol{a}\cdot\mathrm{Cof}\boldsymbol{F}$.

PROOF.– Indeed, for the first identity,

$$\begin{aligned}\mathrm{Div}(\boldsymbol{a}\cdot\mathrm{Cof}\boldsymbol{F}) &= \frac{\partial\boldsymbol{a}}{\partial\vartheta^i}\cdot\mathrm{Cof}\boldsymbol{F}\cdot\boldsymbol{G}^i + \boldsymbol{a}\cdot\mathrm{Div}(\mathrm{Cof}\boldsymbol{F})\\ &= J\frac{\partial\boldsymbol{a}}{\partial\vartheta^i}\cdot\left[\boldsymbol{g}^j\otimes\boldsymbol{G}_j\right]\cdot\boldsymbol{G}^i = J\frac{\partial\boldsymbol{a}}{\partial\vartheta^i}\cdot\boldsymbol{g}^i = J\mathrm{div}\boldsymbol{a}\,.\end{aligned}$$

For the second identity, the left-hand side is written as

$$
\begin{aligned}
\mathrm{Curl}(\boldsymbol{a}\cdot\boldsymbol{F}) &= -\left[\frac{\partial(\boldsymbol{a}\cdot\boldsymbol{F})}{\partial\vartheta^i}\otimes\boldsymbol{G}^i\right]:\left[\epsilon^{pqr}\frac{1}{\sqrt{G}}\,\boldsymbol{G}_p\otimes\boldsymbol{G}_q\otimes\boldsymbol{G}_r\right]\\
&= -\frac{\partial(\boldsymbol{a}\cdot[\boldsymbol{g}_j\otimes\boldsymbol{G}^j])}{\partial\vartheta^i}\cdot\left[\epsilon^{pir}\frac{1}{\sqrt{G}}\,\boldsymbol{G}_p\otimes\boldsymbol{G}_r\right]\\
&= -\frac{\partial(\boldsymbol{a}\cdot\boldsymbol{g}_j)}{\partial\vartheta^i}\epsilon^{jir}\frac{1}{\sqrt{G}}\,\boldsymbol{G}_r+[\boldsymbol{a}\cdot\boldsymbol{g}_j]\epsilon^{pir}\frac{{}^m\Gamma^j_{ip}}{\sqrt{G}}\,\boldsymbol{G}_r\\
&= -J\left[\frac{\partial\boldsymbol{a}}{\partial\vartheta^i}\cdot\boldsymbol{g}_j+\boldsymbol{a}\cdot\frac{\partial\boldsymbol{g}_j}{\partial\vartheta^i}\right]\epsilon^{jir}\frac{1}{\sqrt{g}}\,\boldsymbol{G}_r\\
&= -J\left[\frac{\partial\boldsymbol{a}}{\partial\vartheta^i}\cdot\boldsymbol{g}_j\right]\epsilon^{jir}\frac{1}{\sqrt{g}}\,\boldsymbol{G}_r-J\left[{}^s\Gamma^q_{ij}\boldsymbol{a}\cdot\boldsymbol{g}_q\right]\epsilon^{jir}\frac{1}{\sqrt{g}}\,\boldsymbol{G}_r\\
&= -J\left[\frac{\partial\boldsymbol{a}}{\partial\vartheta^i}\cdot\boldsymbol{g}_j\right]\epsilon^{jir}\frac{1}{\sqrt{g}}\,\boldsymbol{G}_r\,,
\end{aligned}
$$

while the right-hand side is written as

$$
\begin{aligned}
\mathrm{curl}\boldsymbol{a}\cdot\mathrm{Cof}\boldsymbol{F} &= J\left[-\left[\frac{\partial\boldsymbol{a}}{\partial\vartheta^i}\otimes\boldsymbol{g}^i\right]:\left[\frac{\epsilon^{pqr}}{\sqrt{g}}\,\boldsymbol{g}_p\otimes\boldsymbol{g}_q\otimes\boldsymbol{g}_r\right]\right]\cdot[\boldsymbol{g}^s\otimes\boldsymbol{G}_s]\\
&= -J\left[\frac{\partial\boldsymbol{a}}{\partial\vartheta^i}\cdot\boldsymbol{g}_p\right]\epsilon^{pir}\frac{1}{\sqrt{g}}\,\boldsymbol{g}_r\cdot[\boldsymbol{g}^s\otimes\boldsymbol{G}_s]\\
&= -J\left[\frac{\partial\boldsymbol{a}}{\partial\vartheta^i}\cdot\boldsymbol{g}_j\right]\epsilon^{jir}\frac{1}{\sqrt{g}}\,\boldsymbol{G}_r\,.
\end{aligned}
$$

Thus, the two expressions are equivalent. □

– It can be proven that $\boldsymbol{g}^1$ is normal to the surface $\vartheta^1(x^1,x^2,x^3)$=constant.

PROOF.– Consider a point $P=(x_0^1,x_0^2,x_0^3)$ of the surface and a curve C lying on the surface with vector equation $\boldsymbol{r}=(x^1(t)\;\;x^2(t)\;\;x^3(t))^T$ that passes through the point P. The chain rule leads to

$$
\frac{\partial\vartheta^1}{\partial x^1}\frac{\mathrm{d}x^1}{\mathrm{d}t}+\frac{\partial\vartheta^1}{\partial x^2}\frac{\mathrm{d}x^2}{\mathrm{d}t}+\frac{\partial\vartheta^1}{\partial x^3}\frac{\mathrm{d}x^3}{\mathrm{d}t}=0\quad\Rightarrow\quad\boldsymbol{g}^1\cdot\frac{\mathrm{d}\boldsymbol{x}}{\mathrm{d}t}=0\,,
$$

Recalling that $\left.\frac{\mathrm{d}\boldsymbol{x}}{\mathrm{d}t}\right|_{t_0}$ is tangent to the curve at the point P and C is arbitrary, it is concluded that $\boldsymbol{g}^1$ is perpendicular to the tangent vector of any curve C on ϑ^1=constant that passes through P. Thus, $\boldsymbol{g}^1$ is normal to the tangent plane at P. Since P is arbitrary, $\boldsymbol{g}^1$ is eventually normal to the surface $\vartheta^1(x^1, x^2, x^3)$=constant. □

– For a second-order tensor $\boldsymbol{A}$ and using the base vectors $\boldsymbol{g}$, we can show that

$$
\begin{aligned}
\det\boldsymbol{A} &= \frac{g}{6}\epsilon_{ijk}\epsilon_{pqr}A^{ip}A^{jq}A^{kr} = \frac{1}{6g}\epsilon^{ijk}\epsilon^{pqr}A_{ip}A_{jq}A_{kr} \\
&= \frac{1}{6}\epsilon_{ijk}\epsilon^{pqr}A^i_pA^j_qA^k_r = \frac{[\boldsymbol{A}\cdot\boldsymbol{g}_1]\cdot[[\boldsymbol{A}\cdot\boldsymbol{g}_2]\times[\boldsymbol{A}\cdot\boldsymbol{g}_3]]}{\boldsymbol{g}_1\cdot[\boldsymbol{g}_2\times\boldsymbol{g}_3]}\,.
\end{aligned}
$$

PROOF.– The tensor $\boldsymbol{A}$ is generally written as

$$
\begin{aligned}
\boldsymbol{A} &= \widetilde{A}_{ij}\boldsymbol{e}^i\otimes\boldsymbol{e}^j = \widetilde{A}^{ij}\boldsymbol{e}_i\otimes\boldsymbol{e}_j = A_{ij}\boldsymbol{g}^i\otimes\boldsymbol{g}^j \\
&= A^{ij}\boldsymbol{g}_i\otimes\boldsymbol{g}_j = A^i_j\boldsymbol{g}_i\otimes\boldsymbol{g}^j\,.
\end{aligned}
$$

Considering the Cartesian coordinate system, $\boldsymbol{A}$ and its determinant are expressed as

$$
\boldsymbol{A} = \begin{pmatrix} \widetilde{A}_{11} & \widetilde{A}_{12} & \widetilde{A}_{13} \\ \widetilde{A}_{21} & \widetilde{A}_{22} & \widetilde{A}_{23} \\ \widetilde{A}_{31} & \widetilde{A}_{32} & \widetilde{A}_{33} \end{pmatrix},
$$

$$
\begin{aligned}
\det\boldsymbol{A} &= \widetilde{A}_{11}\widetilde{A}_{22}\widetilde{A}_{33} - \widetilde{A}_{11}\widetilde{A}_{23}\widetilde{A}_{32} - \widetilde{A}_{12}\widetilde{A}_{21}\widetilde{A}_{33} + \widetilde{A}_{12}\widetilde{A}_{23}\widetilde{A}_{31} \\
&\quad + \widetilde{A}_{13}\widetilde{A}_{21}\widetilde{A}_{32} - \widetilde{A}_{13}\widetilde{A}_{22}\widetilde{A}_{31}\,.
\end{aligned}
$$

In compact form, the above expression can be written as:

$$
\det\boldsymbol{A} = \epsilon^{pqr}\widetilde{A}_{1p}\widetilde{A}_{2q}\widetilde{A}_{3r}.
$$

Additionally,

$$\epsilon_{123}\det\boldsymbol{A} = \epsilon^{pqr}\widetilde{A}_{1p}\widetilde{A}_{2q}\widetilde{A}_{3r}\,,$$

$$\epsilon_{132}\det\boldsymbol{A} = -\epsilon_{123}\det\boldsymbol{A} = -\epsilon^{prq}\widetilde{A}_{1p}\widetilde{A}_{2r}\widetilde{A}_{3q} = \epsilon^{pqr}\widetilde{A}_{1p}\widetilde{A}_{3q}\widetilde{A}_{2r}\,,$$

$$\epsilon_{122}\det\boldsymbol{A} = 0\,,$$

$$\epsilon^{pqr}\widetilde{A}_{1p}\widetilde{A}_{2q}\widetilde{A}_{2r} = -\epsilon^{prq}\widetilde{A}_{1p}\widetilde{A}_{2q}\widetilde{A}_{2r} = -\epsilon^{pqr}\widetilde{A}_{1p}\widetilde{A}_{2q}\widetilde{A}_{2r} = 0\,,$$

etc., and in the same way it is eventually shown that

$$\epsilon_{ijk}\det\boldsymbol{A} = \epsilon^{pqr}\widetilde{A}_{ip}\widetilde{A}_{jq}\widetilde{A}_{kr}.$$

Thus,

$$\begin{aligned}\frac{\epsilon^{ijk}\epsilon^{pqr}}{6}\widetilde{A}_{ip}\widetilde{A}_{jq}\widetilde{A}_{kr} &= \frac{1}{6}\epsilon^{pqr}[\epsilon^{123}\widetilde{A}_{1p}\widetilde{A}_{2q}\widetilde{A}_{3r} + \epsilon^{231}\widetilde{A}_{2p}\widetilde{A}_{3q}\widetilde{A}_{1r} \\ &\quad +\epsilon^{312}\widetilde{A}_{3p}\widetilde{A}_{1q}\widetilde{A}_{2r} + \epsilon^{132}\widetilde{A}_{1p}\widetilde{A}_{3q}\widetilde{A}_{2r} \\ &\quad +\epsilon^{213}\widetilde{A}_{2p}\widetilde{A}_{1q}\widetilde{A}_{3r} + \epsilon^{321}\widetilde{A}_{3p}\widetilde{A}_{2q}\widetilde{A}_{1r}] \\ &= \frac{1}{6}\epsilon^{pqr}[\widetilde{A}_{1p}\widetilde{A}_{2q}\widetilde{A}_{3r} + \widetilde{A}_{2p}\widetilde{A}_{3q}\widetilde{A}_{1r} + \widetilde{A}_{3p}\widetilde{A}_{1q}\widetilde{A}_{2r} \\ &\quad -\widetilde{A}_{1p}\widetilde{A}_{3q}\widetilde{A}_{2r} - \widetilde{A}_{2p}\widetilde{A}_{1q}\widetilde{A}_{3r} - \widetilde{A}_{3p}\widetilde{A}_{2q}\widetilde{A}_{1r}] \\ &= \frac{1}{6}\epsilon^{pqr}6\widetilde{A}_{1p}\widetilde{A}_{2q}\widetilde{A}_{3r} = \det\boldsymbol{A}\,,\end{aligned}$$

and the determinant of $\boldsymbol{A}$ in Cartesian coordinates is expressed as

$$\det\boldsymbol{A} = \frac{1}{6}\epsilon^{ijk}\epsilon^{pqr}\widetilde{A}_{ip}\widetilde{A}_{jq}\widetilde{A}_{kr} = \frac{1}{6}\epsilon_{ijk}\epsilon_{pqr}\widetilde{A}^{ip}\widetilde{A}^{jq}\widetilde{A}^{kr}\,.$$

Using the relation between coefficients of different coordinate systems, discussed in the tensor operations of this section, we obtain

$$\widetilde{A}^{ij} = \frac{\partial x^i}{\partial \vartheta^m}\frac{\partial x^j}{\partial \vartheta^n}A^{mn} = [\boldsymbol{g}_m \cdot \boldsymbol{e}^i][\boldsymbol{g}_n \cdot \boldsymbol{e}^j]A^{mn}$$

So,

$$\det \boldsymbol{A} = \frac{\epsilon_{ijk}\epsilon_{pqr}}{6}[\boldsymbol{g}_m \cdot \boldsymbol{e}^i][\boldsymbol{g}_n \cdot \boldsymbol{e}^p][\boldsymbol{g}_s \cdot \boldsymbol{e}^j][\boldsymbol{g}_t \cdot \boldsymbol{e}^q][\boldsymbol{g}_u \cdot \boldsymbol{e}^k][\boldsymbol{g}_v \cdot \boldsymbol{e}^r] \\ \times A^{mn} A^{st} A^{uv} .$$

But, using the definition of ϵ [KEL 13, WIK 13],

$$\begin{aligned} \boldsymbol{\epsilon} &= \epsilon_{pqr}\sqrt{g}\, \boldsymbol{g}^p \otimes \boldsymbol{g}^q \otimes \boldsymbol{g}^r = \epsilon^{pqr}\frac{1}{\sqrt{g}}\, \boldsymbol{g}_p \otimes \boldsymbol{g}_q \otimes \boldsymbol{g}_r \\ &= \epsilon^{pqr}\boldsymbol{e}_p \otimes \boldsymbol{e}_q \otimes \boldsymbol{e}_r = \epsilon_{pqr}\boldsymbol{e}^p \otimes \boldsymbol{e}^q \otimes \boldsymbol{e}^r , \end{aligned}$$

it eventually yields that

$$\begin{aligned} \epsilon_{pqr}[\boldsymbol{g}_n \cdot \boldsymbol{e}^p][\boldsymbol{g}_t \cdot \boldsymbol{e}^q][\boldsymbol{g}_v \cdot \boldsymbol{e}^r] &= [[\boldsymbol{\epsilon} \cdot \boldsymbol{g}_v] \cdot \boldsymbol{g}_t] \cdot \boldsymbol{g}_n \\ &= [[[\epsilon_{pqr}\sqrt{g}\, \boldsymbol{g}^p \otimes \boldsymbol{g}^q \otimes \boldsymbol{g}^r] \cdot \boldsymbol{g}_v] \cdot \boldsymbol{g}_t] \cdot \boldsymbol{g}_n \\ &= \epsilon_{ntv}\sqrt{g} , \end{aligned}$$

and similarly

$$\epsilon_{ijk}[\boldsymbol{g}_m \cdot \boldsymbol{e}^i][\boldsymbol{g}_s \cdot \boldsymbol{e}^j][\boldsymbol{g}_u \cdot \boldsymbol{e}^k] = \epsilon_{msu}\sqrt{g} .$$

Thus,

$$\det \boldsymbol{A} = \frac{g}{6}\epsilon_{ijk}\epsilon_{pqr}A^{ip}A^{jq}A^{kr} .$$

Moreover,

$$\begin{aligned} A^i_j \boldsymbol{g}_i \otimes \boldsymbol{g}^j &= g^{jk}A^i_j \boldsymbol{g}_i \otimes \boldsymbol{g}_k \Rightarrow \quad A^{ik} = A^i_j g^{jk} , \\ A_{ij}\boldsymbol{g}^i \otimes \boldsymbol{g}^j &= g^{iq}g^{jp}A_{ij}\boldsymbol{g}_p \otimes \boldsymbol{g}_q \Rightarrow \quad A^{pq} = A_{ij}g^{iq}g^{jp} , \\ \det \boldsymbol{A} &= \frac{g}{6}\epsilon_{ijk}\epsilon_{pqr}A^i_m g^{mp} A^j_n g^{nq} A^k_o g^{or} = \frac{1}{6}\epsilon_{ijk}\epsilon^{mno}A^i_m A^j_n A^k_o . \end{aligned}$$

Similarly,

$$\begin{aligned}\det\boldsymbol{A} &= \frac{g}{6}\epsilon_{ijk}\epsilon_{pqr}A_{mn}g^{mi}g^{np}A_{os}g^{oj}g^{sq}A_{tv}g^{tk}g^{vr}\\ &= \frac{1}{6g}\epsilon^{mot}\epsilon^{nsv}A_{mn}A_{os}A_{tv}\,.\end{aligned}$$

Finally,

$$\begin{aligned}[\boldsymbol{A}\cdot\boldsymbol{g}_1]\cdot[[\boldsymbol{A}\cdot\boldsymbol{g}_2]\times[\boldsymbol{A}\cdot\boldsymbol{g}_3]] &= A_1^iA_2^jA_3^k\boldsymbol{g}_i\cdot[\boldsymbol{g}_j\times\boldsymbol{g}_k]\\ &= \sqrt{g}A_1^iA_2^jA_3^k\epsilon_{jkr}\boldsymbol{g}_i\cdot\boldsymbol{g}^r\\ &= \sqrt{g}\epsilon_{ijk}A_1^iA_2^jA_3^k = \frac{\sqrt{g}}{6}\epsilon_{ijk}\epsilon^{pqr}A_p^iA_q^jA_r^k\,,\end{aligned}$$

and

$$\frac{[\boldsymbol{A}\cdot\boldsymbol{g}_1]\cdot[[\boldsymbol{A}\cdot\boldsymbol{g}_2]\times[\boldsymbol{A}\cdot\boldsymbol{g}_3]]}{\boldsymbol{g}_1\cdot[\boldsymbol{g}_2\times\boldsymbol{g}_3]} = \frac{\sqrt{g}}{6\sqrt{g}}\epsilon_{ijk}\epsilon^{pqr}A_p^iA_q^jA_r^k = \det\boldsymbol{A}\,.$$

□

1.2.5. *Cylindrical and spherical coordinates*

Cylindrical and spherical coordinates appear very frequently in continuum mechanics and micromechanics. Several problems involving cylindrical fibers or spherical particles can be treated analytically expressing the various equations in the corresponding coordinate system.

Cylindrical coordinates (Figure 1.3a) are expressed with the radius r, the angle ω and the vertical axis z. For this coordinate system, the base vectors, the metric coefficients and the Christoffel symbols are shown in Table 1.5.

Using these definitions, the following properties for a vector $\boldsymbol{u}$ and a second-order tensor $\boldsymbol{T}$ hold that

$$\text{grad}\boldsymbol{u} = \frac{\partial u_r}{\partial r}\boldsymbol{e}_r\otimes\boldsymbol{e}_r + \frac{1}{r}\left[\frac{\partial u_r}{\partial\omega} - u_\omega\right]\boldsymbol{e}_r\otimes\boldsymbol{e}_\omega + \frac{\partial u_r}{\partial z}\boldsymbol{e}_r\otimes\boldsymbol{e}_z$$

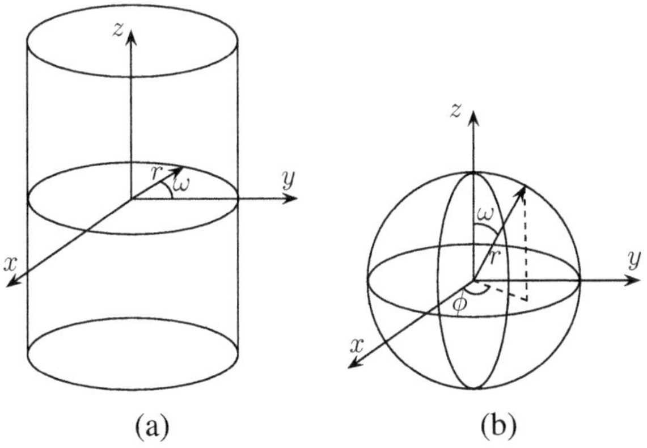

Figure 1.3. *a) Cylindrical and b) spherical coordinate system*

$$\vartheta^1 = r \qquad \vartheta^2 = \omega \qquad \vartheta^3 = z \qquad \boldsymbol{x} = [r\cos\omega \quad r\sin\omega \quad z]^T$$

$$\boldsymbol{g}_1 = \boldsymbol{g}^1 = \cos\omega\boldsymbol{e}_1 + \sin\omega\boldsymbol{e}_2 = \boldsymbol{e}_r$$

$$\boldsymbol{g}_2 = -r\sin\omega\boldsymbol{e}_1 + r\cos\omega\boldsymbol{e}_2 = r\boldsymbol{e}_\omega$$

$$\boldsymbol{g}^2 = \frac{1}{r}\boldsymbol{e}_\omega \qquad \boldsymbol{g}_3 = \boldsymbol{g}^3 = \boldsymbol{e}_3 = \boldsymbol{e}_z$$

$$g_{ij} = \begin{pmatrix} 1 & 0 & 0 \\ 0 & r^2 & 0 \\ 0 & 0 & 1 \end{pmatrix} \qquad g^{ij} = \begin{pmatrix} 1 & 0 & 0 \\ 0 & \frac{1}{r^2} & 0 \\ 0 & 0 & 1 \end{pmatrix} \qquad \sqrt{g} = r$$

$$\Gamma^1_{1p} = \Gamma^1_{p1} = \Gamma^1_{p3} = \Gamma^1_{3p} = 0 \qquad \Gamma^1_{22} = -r$$

$$\Gamma^2_{11} = \Gamma^2_{22} = \Gamma^2_{3p} = \Gamma^2_{p3} = \Gamma^3_{pq} = 0 \qquad \Gamma^2_{12} = \Gamma^2_{21} = \frac{1}{r}$$

Table 1.5. *Cylindrical coordinate system base vectors*

$$+\frac{\partial u_\omega}{\partial r}\boldsymbol{e}_\omega \otimes \boldsymbol{e}_r + \frac{1}{r}\left[\frac{\partial u_\omega}{\partial \omega} + u_r\right]\boldsymbol{e}_\omega \otimes \boldsymbol{e}_\omega + \frac{\partial u_\omega}{\partial z}\boldsymbol{e}_\omega \otimes \boldsymbol{e}_z$$

$$+\frac{\partial u_z}{\partial r}\boldsymbol{e}_z \otimes \boldsymbol{e}_r + \frac{1}{r}\frac{\partial u_z}{\partial \omega}\boldsymbol{e}_z \otimes \boldsymbol{e}_\omega + \frac{\partial u_z}{\partial z}\boldsymbol{e}_z \otimes \boldsymbol{e}_z\,,$$

$$\mathrm{div}\boldsymbol{u} = \frac{\partial u_1}{\partial r} + \frac{1}{r^2}\left[\frac{\partial u_2}{\partial \omega} + ru_1\right] + \frac{\partial u_3}{\partial z}$$

$$= \frac{\partial u_r}{\partial r} + \frac{1}{r}\left[\frac{\partial u_\omega}{\partial \omega} + u_r\right] + \frac{\partial u_z}{\partial z},$$

$$\begin{aligned}\mathrm{curl}\boldsymbol{u} = & \left[\frac{1}{r}\frac{\partial u_z}{\partial \omega} - \frac{\partial u_\omega}{\partial z}\right]\boldsymbol{e}_r + \left[\frac{\partial u_r}{\partial z} - \frac{\partial u_z}{\partial r}\right]\boldsymbol{e}_\omega \\ & -\frac{1}{r}\left[\frac{\partial u_r}{\partial \omega} - u_\omega - r\frac{\partial u_\omega}{\partial r}\right]\boldsymbol{e}_z,\end{aligned}$$

$$\begin{aligned}\mathrm{div}\boldsymbol{T} = & \left[\frac{\partial T^{rr}}{\partial r} + \frac{1}{r}\frac{\partial T^{r\omega}}{\partial \omega} + \frac{T^{rr} - T^{\omega\omega}}{r} + \frac{\partial T^{rz}}{\partial z}\right]\boldsymbol{e}_r \\ & + \left[\frac{\partial T^{\omega r}}{\partial r} + \frac{1}{r}\frac{\partial T^{\omega\omega}}{\partial \omega} + \frac{T^{r\omega} + T^{\omega r}}{r} + \frac{\partial T^{\omega z}}{\partial z}\right]\boldsymbol{e}_\omega \\ & + \left[\frac{\partial T^{zr}}{\partial r} + \frac{1}{r}\frac{\partial T^{z\omega}}{\partial \omega} + \frac{T^{zr}}{r} + \frac{\partial T^{zz}}{\partial z}\right]\boldsymbol{e}_z.\end{aligned}$$

Spherical coordinates (Figure 1.3b) are expressed with the radius r and the two angles ω and ϕ. For this coordinate system, the base vectors, the metric coefficients and the Christoffel symbols are shown in Table 1.6.

Using these definitions, the following properties for a vector $\boldsymbol{u}$ and a second-order tensor $\boldsymbol{T}$ hold in spherical coordinates:

$$\begin{aligned}\mathrm{grad}\boldsymbol{u} = & \frac{\partial u_r}{\partial r}\boldsymbol{e}_r \otimes \boldsymbol{e}_r + \left[\frac{1}{r}\frac{\partial u_r}{\partial \omega} - \frac{u_\omega}{r}\right]\boldsymbol{e}_r \otimes \boldsymbol{e}_\omega + \frac{\partial u_\omega}{\partial r}\boldsymbol{e}_\omega \otimes \boldsymbol{e}_r \\ & + \left[\frac{1}{r\sin\omega}\frac{\partial u_r}{\partial \phi} - \frac{u_\phi}{r}\right]\boldsymbol{e}_r \otimes \boldsymbol{e}_\phi + \frac{\partial u_\phi}{\partial r}\boldsymbol{e}_\phi \otimes \boldsymbol{e}_r \\ & + \left[\frac{1}{r}\frac{\partial u_\omega}{\partial \omega} + \frac{u_r}{r}\right]\boldsymbol{e}_\omega \otimes \boldsymbol{e}_\omega + \frac{1}{r}\frac{\partial u_\phi}{\partial \omega}\boldsymbol{e}_\phi \otimes \boldsymbol{e}_\omega \\ & + \left[\frac{1}{r\sin\omega}\frac{\partial u_\omega}{\partial \phi} - \frac{u_\phi \cos\omega}{r\sin\omega}\right]\boldsymbol{e}_\omega \otimes \boldsymbol{e}_\phi \\ & + \left[\frac{1}{r\sin\omega}\frac{\partial u_\phi}{\partial \phi} + \frac{u_r}{r} + \frac{u_\omega \cos\omega}{r\sin\omega}\right]\boldsymbol{e}_\phi \otimes \boldsymbol{e}_\phi,\end{aligned}$$

$$\vartheta^1 = r \qquad \vartheta^2 = \omega \qquad \vartheta^3 = \phi \qquad \boldsymbol{x} = [r\sin\omega\cos\phi \;\; r\sin\omega\sin\phi \;\; r\cos\omega]^T$$

$$\boldsymbol{g}_1 = \boldsymbol{g}^1 = \sin\omega\cos\phi\boldsymbol{e}_1 + \sin\omega\sin\phi\boldsymbol{e}_2 + \cos\omega\boldsymbol{e}_3 = \boldsymbol{e}_r$$

$$\boldsymbol{g}_2 = r\cos\omega\cos\phi\boldsymbol{e}_1 + r\cos\omega\sin\phi\boldsymbol{e}_2 - r\sin\omega\boldsymbol{e}_3 = r\boldsymbol{e}_\omega$$

$$\boldsymbol{g}_3 = -r\sin\omega\sin\phi\boldsymbol{e}_1 + r\sin\omega\cos\phi\boldsymbol{e}_2 = r\sin\omega\boldsymbol{e}_\phi$$

$$\boldsymbol{g}^2 = \frac{1}{r}\boldsymbol{e}_\omega \qquad \boldsymbol{g}^3 = \frac{1}{r\sin\omega}\boldsymbol{e}_\phi \qquad \sqrt{g} = r^2\sin\omega$$

$$g_{ij} = \begin{pmatrix} 1 & 0 & 0 \\ 0 & r^2 & 0 \\ 0 & 0 & r^2\sin^2\omega \end{pmatrix} \qquad g^{ij} = \begin{pmatrix} 1 & 0 & 0 \\ 0 & \dfrac{1}{r^2} & 0 \\ 0 & 0 & \dfrac{1}{r^2\sin^2\omega} \end{pmatrix}$$

$$\Gamma^1_{1p} = \Gamma^1_{p1} = \Gamma^1_{23} = \Gamma^1_{32} = 0 \qquad \Gamma^1_{22} = -r \qquad \Gamma^1_{33} = -r\sin^2\omega$$

$$\Gamma^2_{11} = \Gamma^2_{13} = \Gamma^2_{31} = \Gamma^2_{23} = \Gamma^2_{32} = \Gamma^2_{22} = 0$$

$$\Gamma^2_{12} = \Gamma^2_{21} = \frac{1}{r} \qquad \Gamma^2_{33} = -\sin\omega\cos\omega \qquad \Gamma^3_{13} = \Gamma^3_{31} = \frac{1}{r}$$

$$\Gamma^3_{11} = \Gamma^3_{12} = \Gamma^3_{21} = \Gamma^3_{22} = \Gamma^3_{33} = 0 \qquad \Gamma^3_{23} = \Gamma^3_{32} = \frac{\cos\omega}{\sin\omega}$$

Table 1.6. *Spherical coordinate system base vectors*

$$\mathrm{div}\boldsymbol{u} = \frac{\partial u_r}{\partial r} + \frac{1}{r}\frac{\partial u_\omega}{\partial \omega} + \frac{1}{r\sin\omega}\frac{\partial u_\phi}{\partial \phi} + \frac{2u_r}{r} + \frac{u_\omega\cos\omega}{r\sin\omega},$$

$$\begin{aligned}\mathrm{curl}\boldsymbol{u} &= \frac{1}{r}\left[\frac{\partial u_\phi}{\partial \omega} - \frac{1}{\sin\omega}\frac{\partial u_\omega}{\partial \phi} + u_\phi\frac{\cos\omega}{\sin\omega}\right]\boldsymbol{e}_r \\ &+ \frac{1}{r}\left[\frac{1}{\sin\omega}\frac{\partial u_r}{\partial \phi} - r\frac{\partial u_\phi}{\partial r} - u_\phi\right]\boldsymbol{e}_\omega \\ &+ \frac{1}{r}\left[r\frac{\partial u_\omega}{\partial r} - \frac{\partial u_r}{\partial \omega} + u_\omega\right]\boldsymbol{e}_\phi,\end{aligned}$$

$$\mathrm{div}\boldsymbol{T} = \mathrm{grad}\boldsymbol{T}\!:\!\boldsymbol{i} = A_r\boldsymbol{e}_r + A_\omega\boldsymbol{e}_\omega + A_\phi\boldsymbol{e}_\phi,$$

where

$$A_r = \frac{\partial T^{rr}}{\partial r} + \frac{1}{r}\frac{\partial T^{r\omega}}{\partial \omega} + \frac{1}{r\sin\omega}\frac{\partial T^{r\phi}}{\partial \phi}$$

$$+\frac{2T^{rr} - T^{\omega\omega} - T^{\phi\phi}}{r} + \frac{T^{r\omega}\cos\omega}{r\sin\omega},$$

$$A_\omega = \frac{\partial T^{\omega r}}{\partial r} + \frac{1}{r}\frac{\partial T^{\omega\omega}}{\partial \omega} + \frac{1}{r\sin\omega}\frac{\partial T^{\omega\phi}}{\partial \phi}$$
$$+\frac{T^{r\omega} + 2T^{\omega r}}{r} + \frac{[T^{\omega\omega} - T^{\phi\phi}]\cos\omega}{r\sin\omega},$$

$$A_\phi = \frac{\partial T^{\phi r}}{\partial r} + \frac{1}{r}\frac{\partial T^{\phi\omega}}{\partial \omega} + \frac{1}{r\sin\omega}\frac{\partial T^{\phi\phi}}{\partial \phi}$$
$$+\frac{T^{r\phi} + 2T^{\phi r}}{r} + \frac{[T^{\omega\phi} + T^{\phi\omega}]\cos\omega}{r\sin\omega}.$$

2

Continuum Mechanics and Constitutive Laws

The study of composites requires an in-depth knowledge of the material response of each individual component that appears in the microstructure. Local conservation laws apply in each material point, and constitutive laws must be formulated in a manner that allows us to determine the local and global response of a heterogeneous medium. This chapter presents a summary of the basic continuum mechanics concepts, i.e. the kinematics, kinetics and conservation laws. Moreover, there is a detailed presentation of the second thermodynamic law and the various energy and dissipation potentials that permit us to identify proper associated constitutive laws. Several examples from classical dissipative materials, i.e. materials in which thermodynamical irreversible processes take place, are presented. Finally, a methodology for identifying material parameters through appropriate experimental protocol is illustrated at the end of the chapter.

Tensor operations and the associated definitions presented in Chapter 1 are frequently utilized. For a detailed exposition of continuum mechanics, the interested reader is referred to Malvern [MAL 69] and Lai *et al.* [LAI 10]. General discussion about the various thermodynamic theories is presented by Maugin [MAU 99].

2.1. Kinematics

A continuum body at time t_0 is considered to be in the undeformed (reference) configuration and occupies the space $\mathcal{D}_0$. At time t, the body has been moved and deformed, occupying the space $\mathcal{D}$ in the deformed (current) configuration. Any point P on this body can be described with the help of a fixed point O in two ways: (1) with a position vector $\boldsymbol{X}$ of the reference configuration and (2) with a position vector $\boldsymbol{x}$ of the current configuration (Figure 2.1).

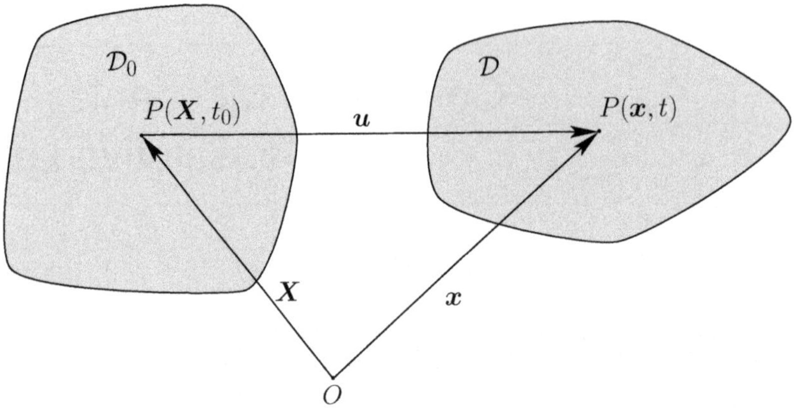

Figure 2.1. *Continuum body at different configurations*

$\boldsymbol{X}$ is usually called the material coordinate because it is attached to a material point and it does not change with time. $\boldsymbol{x}$ is usually called the spatial coordinate because it represents a fixed coordinate system in space. A material point that moves occupies a different position $\boldsymbol{x}$ with time. Both $\boldsymbol{X}$ and $\boldsymbol{x}$ refer to Cartesian coordinate systems and are connected with each other through the nonlinear relation

$$\boldsymbol{x} := \boldsymbol{x}(\boldsymbol{X}, t). \qquad [2.1]$$

The position change of the point P is defined by the displacement vector $\boldsymbol{u}$,

$$\boldsymbol{u} = \boldsymbol{x}(\boldsymbol{X}, t) - \boldsymbol{X}. \qquad [2.2]$$

The velocity of the point is equal to $\dot{\boldsymbol{x}} = \dot{\boldsymbol{u}}$. Moreover, in a continuum body, every material point has density ρ.

Consider $P(\boldsymbol{X}, t_0)$ and a neighbor point $Q(\boldsymbol{X} + \mathrm{d}\boldsymbol{X}, t_0)$. When the body is moved and transformed, these two points will be translated to the positions $P(\boldsymbol{x}, t)$ and $Q(\boldsymbol{x} + \mathrm{d}\boldsymbol{x}, t)$, respectively. The change of the distance vector $\mathrm{d}\boldsymbol{X}$ to $\mathrm{d}\boldsymbol{x}$ is expressed by the relation

$$\mathrm{d}\boldsymbol{x} = \mathrm{Grad}\boldsymbol{x} \cdot \mathrm{d}\boldsymbol{X} = \boldsymbol{F} \cdot \mathrm{d}\boldsymbol{X}, \qquad [2.3]$$

where the $\boldsymbol{F} = \mathrm{Grad}\boldsymbol{x} = \boldsymbol{I} + \mathrm{Grad}\boldsymbol{u}$ is called the deformation gradient and is a second-order tensor. Considering that there is a one-to-one mapping between $\boldsymbol{X}$ and $\boldsymbol{x}$, the deformation gradient is invertible with $\boldsymbol{F}^{-1} = \mathrm{grad}\boldsymbol{X}$. The difference between the lengths of the vectors $\mathrm{d}\boldsymbol{X}$ and $\mathrm{d}\boldsymbol{x}$ is expressed by the relation

$$\begin{aligned} \mathrm{d}\boldsymbol{x} \cdot \mathrm{d}\boldsymbol{x} - \mathrm{d}\boldsymbol{X} \cdot \mathrm{d}\boldsymbol{X} &= [\boldsymbol{F} \cdot \mathrm{d}\boldsymbol{X}] \cdot [\boldsymbol{F} \cdot \mathrm{d}\boldsymbol{X}] - \mathrm{d}\boldsymbol{X} \cdot \mathrm{d}\boldsymbol{X} \\ &= \mathrm{d}\boldsymbol{X} \cdot \left[\boldsymbol{F}^T \cdot \boldsymbol{F} - \boldsymbol{I}\right] \cdot \mathrm{d}\boldsymbol{X} = 2\mathrm{d}\boldsymbol{X} \cdot \boldsymbol{E} \cdot \mathrm{d}\boldsymbol{X}, \end{aligned} \qquad [2.4]$$

where $\boldsymbol{E}$ is called the Green Lagrange strain and is a second-order tensor. This tensor has the following properties:

1) It is zero when rigid body motion occurs. Indeed, a rigid body motion can be characterized by a translation ($\boldsymbol{F} = \boldsymbol{I}$) or a rotation ($\boldsymbol{F}$ orthogonal, i.e. $\boldsymbol{F}^{-1} = \boldsymbol{F}^T$). Both cases lead to $\boldsymbol{E} = \boldsymbol{0}$.

2) It can be determined exclusively by the displacement vector $\boldsymbol{u}$. Indeed, using equations [2.2] and [2.3] yields

$$\begin{aligned} \boldsymbol{E} &= \frac{1}{2}\left[\boldsymbol{F}^T \cdot \boldsymbol{F} - \boldsymbol{I}\right] = \frac{1}{2}\left[[\boldsymbol{I} + \mathrm{Grad}\boldsymbol{u}]^T \cdot [\boldsymbol{I} + \mathrm{Grad}\boldsymbol{u}] - \boldsymbol{I}\right] \\ &= \frac{1}{2}\left[[\mathrm{Grad}\boldsymbol{u}]^T + \mathrm{Grad}\boldsymbol{u} + [\mathrm{Grad}\boldsymbol{u}]^T \cdot \mathrm{Grad}\boldsymbol{u}\right]. \end{aligned} \qquad [2.5]$$

In large deformation processes, there exist several types of strain measures that can be utilized according to the performed analysis and the materials involved [LAI 10]. To establish the conservation laws and express the homogenization framework in large deformations (see Appendices), $\boldsymbol{F}$ and $\boldsymbol{E}$ are sufficient strain measures.

When only small deformations and rotations are accounted for, the quadratic term

$$[\mathrm{Grad}\boldsymbol{u}]^T \cdot \mathrm{Grad}\boldsymbol{u}$$

is very small compared to the first-order terms and can be ignored. In such a case, Grad$\boldsymbol{u}$ obtains small values and it is assumed that there is no distinction between the current and reference configurations, which permits us to utilize only one position vector $\boldsymbol{x}$. Thus, the infinitesimal strain tensor $\boldsymbol{\varepsilon}$ can be defined as

$$\boldsymbol{\varepsilon} = \frac{1}{2}\left[[\mathrm{grad}\boldsymbol{u}]^T + \mathrm{grad}\boldsymbol{u}\right] = \mathrm{grad}_{\mathrm{sym}}\boldsymbol{u}. \qquad [2.6]$$

For the sake of simplicity, the term "small deformations" will in what follows denote both small deformations and rotations.

2.2. Kinetics

A plane S with unit normal vector $\boldsymbol{n}$ crosses the body of Figure 2.2 and passes through an arbitrary internal point P. This plane separates the medium into two portions: the first (designated by II in the figure) lies on the side of the arrow of $\boldsymbol{n}$ and the second (designated by I in the figure) lies on the tail of $\boldsymbol{n}$. Considering the portion I as a free body, there will be on the plane S a resultant force $\Delta\mathcal{F}$ acting on a small area $\Delta\mathcal{A}$ containing P. The traction vector $\boldsymbol{t}$ acting from II to I at the point P on the plane S is defined as the limit of the ratio $\Delta\mathcal{F}/\Delta\mathcal{A}$ as $\Delta\mathcal{A} \to 0$,

$$\boldsymbol{t} = \lim_{\Delta\mathcal{A}\to 0} \frac{\Delta\mathcal{F}}{\Delta\mathcal{A}}. \qquad [2.7]$$

Cauchy's stress principle states the following [LAI 10]:

The traction vector at any given place and time has a common value on all parts of material having a common tangent plane at P and lying on the same side of it. In other words, if $\boldsymbol{n}$ is the unit outward normal (i.e. a vector of unit length pointing outward, away from the material) to the tangent plane, then

$$\boldsymbol{t} := \boldsymbol{t}(\boldsymbol{x}, t, \boldsymbol{n}), \qquad [2.8]$$

where the scalar t denotes time.

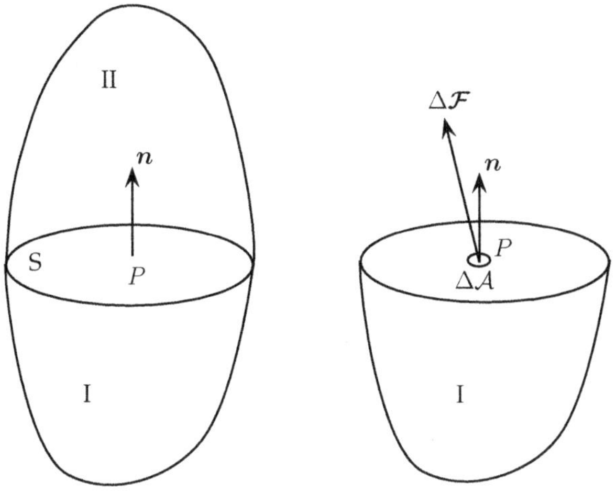

Figure 2.2. *A surface cut in a continuum body under mechanical loading and the force passing through a point P of this surface*

The dependence of the traction vector on the outward normal vector $\boldsymbol{n}$ can be expressed with the help of a linear transformation $\boldsymbol{\sigma}$ as

$$\boldsymbol{t} = \boldsymbol{\sigma}(\boldsymbol{x}, t)\cdot\boldsymbol{n}. \qquad [2.9]$$

The second-order tensor $\boldsymbol{\sigma}$ is called Cauchy's stress tensor.

2.3. Divergence theorem and Reynolds transport theorem

Consider a fixed part of a material like the one shown in Figure 2.3. This part in the reference configuration occupies the space $\mathcal{D}_0$ which is bounded by the surface $\partial\mathcal{D}_0$ with unit vector $\boldsymbol{N}$. Each material point is assigned with a position vector $\boldsymbol{X}$ in $\mathcal{D}_0$. The same part in the current configuration occupies the space $\mathcal{D}$ (moving and deforming, according to the motion of the part) which is bounded by the surface $\partial\mathcal{D}$ with unit vector $\boldsymbol{n}$. Each material point is assigned with a position vector $\boldsymbol{x}$ in $\mathcal{D}$. Both $\boldsymbol{X}$ and $\boldsymbol{x}$ refer to Cartesian coordinate systems.

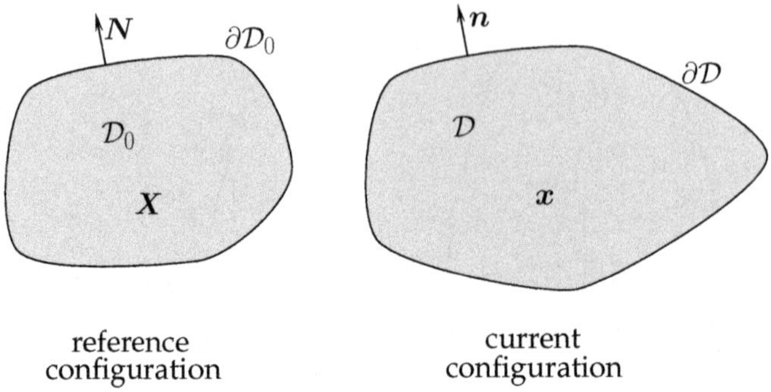

Figure 2.3. *Fixed part of a continuous homogeneous medium: reference and current configuration*

For any tensor $\boldsymbol{A}$ of n_{th} order:

1) The divergence theorem in the current configuration states that [LAI 10]

$$\int_{\mathcal{D}} \mathrm{div}\boldsymbol{A}\,\mathrm{d}v = \int_{\partial\mathcal{D}} \boldsymbol{A}\cdot\boldsymbol{n}\,\mathrm{d}s. \qquad [2.10]$$

In the reference configuration it can be written, analogously,

$$\int_{\mathcal{D}_0} \mathrm{Div}\boldsymbol{A}\,\mathrm{d}V = \int_{\partial\mathcal{D}_0} \boldsymbol{A}\cdot\boldsymbol{N}\,\mathrm{d}S. \qquad [2.11]$$

2) The Reynolds transport theorem in the current configuration states that [LAI 10]

$$\frac{D}{Dt}\int_{\mathcal{D}} \boldsymbol{A}\,\mathrm{d}v = \int_{\mathcal{D}} \left[\frac{D\boldsymbol{A}}{Dt} + \boldsymbol{A}\,\mathrm{div}\dot{\boldsymbol{x}}\right]\mathrm{d}v. \qquad [2.12]$$

In the reference configuration, the time derivative can enter directly inside the integral:

$$\frac{D}{Dt}\int_{\mathcal{D}_0} \boldsymbol{A}\,\mathrm{d}V = \int_{\mathcal{D}_0} \frac{D\boldsymbol{A}}{Dt}\,\mathrm{d}V. \qquad [2.13]$$

2.4. Conservation laws

2.4.1. *Conservation of mass*

The conservation of mass states that [LAI 10]: *the total mass of a fixed part of a material should remain constant at all times.*

Considering the body of Figure 2.3, the conservation of mass is expressed in the current configuration by the relation

$$\frac{D}{Dt}\int_{\mathcal{D}} \rho \, \mathrm{d}v = 0, \qquad [2.14]$$

where ρ denotes the density of the material. Using the Reynolds transport theorem [2.12], the last relation can be written as

$$\int_{\mathcal{D}} \left[\frac{D\rho}{Dt} + \rho \, \mathrm{div}\dot{\boldsymbol{x}} \right] \mathrm{d}v = 0.$$

This expression needs to be valid for every subvolume of the continuum body, therefore the strong form of the conservation of mass is written as

$$\frac{D\rho}{Dt} + \rho \, \mathrm{div}\dot{\boldsymbol{x}} = \frac{D\rho}{Dt} + \rho \, \mathrm{div}\dot{\boldsymbol{u}} = 0. \qquad [2.15]$$

In the reference configuration, the conservation of mass can be derived as follows [KHA 95]. Equation [2.14] states that between current and reference configuration, there is no loss or gain of mass, i.e.

$$\int_{\mathcal{D}_0} \rho_0 \, \mathrm{d}V = \int_{\mathcal{D}} \rho \, \mathrm{d}v.$$

In the above expression ρ_0 denotes the density at the reference configuration, Using the properties of Table 1.3, it is obtained that

$$\int_{\mathcal{D}_0} \rho_0 \, \mathrm{d}V = \int_{\mathcal{D}_0} \rho J \, \mathrm{d}V,$$

where $J = \text{Det}\boldsymbol{F}$. Since the last expression needs to be valid for any subvolume of the body, the conservation of mass in the reference configuration obtains the strong form

$$\rho_0 = \rho J. \qquad [2.16]$$

The last expression leads to the conclusion that ρ_0 is independent of time, since (see section 1.2.4)

$$\frac{D\rho_0}{Dt} = \frac{D\rho}{Dt}J + \rho\frac{DJ}{Dt} = J\left[\frac{D\rho}{Dt} + \rho\,\text{div}\dot{\boldsymbol{x}}\right] = 0. \qquad [2.17]$$

2.4.2. *Conservation of linear momentum*

The conservation of linear momentum states that [LAI 10]: *the total force (surface and body forces) acting on any fixed part of material is equal to the rate of change of linear momentum of the part.*

Considering the body of Figure 2.3, the conservation of linear momentum is expressed in the current configuration by the relation

$$\frac{D}{Dt}\int_{\mathcal{D}} \rho\dot{\boldsymbol{x}}\,\mathrm{d}v = \int_{\mathcal{D}} \rho\boldsymbol{b}\,\mathrm{d}v + \int_{\partial\mathcal{D}} \boldsymbol{t}\,\mathrm{d}s, \qquad [2.18]$$

where $\boldsymbol{b}$ denotes the body forces vector per unit mass. With the help of the definition of the stress tensor [2.9], the divergence theorem [2.10] and the Reynolds transport theorem [2.12], the above expression can be written as

$$\int_{\mathcal{D}}\left[\rho\frac{D\dot{\boldsymbol{x}}}{Dt} + \left[\frac{D\rho}{Dt} + \rho\,\text{div}\dot{\boldsymbol{x}}\right]\dot{\boldsymbol{x}}\right]\mathrm{d}v = \int_{\mathcal{D}} \rho\boldsymbol{b}\,\mathrm{d}v + \int_{\mathcal{D}} \text{div}\boldsymbol{\sigma}\,\mathrm{d}v,$$

or, by using the strong form of the conservation of mass [2.15],

$$\int_{\mathcal{D}}\left[\rho\frac{D\dot{\boldsymbol{x}}}{Dt} - \rho\boldsymbol{b} - \text{div}\boldsymbol{\sigma}\right]\mathrm{d}v = \boldsymbol{0}.$$

The last expression needs to be valid for every subvolume of the continuum body, therefore, the strong form of the conservation of linear momentum is written as

$$\rho \frac{D\dot{\boldsymbol{u}}}{Dt} = \rho \boldsymbol{b} + \operatorname{div}\boldsymbol{\sigma}. \qquad [2.19]$$

In the reference configuration, equation [2.18] can be expressed with the help of Table 1.3 as

$$\frac{D}{Dt}\int_{\mathcal{D}_0} \rho J \dot{\boldsymbol{x}}\, \mathrm{d}V = \int_{\mathcal{D}_0} \rho J \boldsymbol{b}\, \mathrm{d}V + \int_{\partial\mathcal{D}_0} [\boldsymbol{\sigma}\cdot \operatorname{Cof}\boldsymbol{F}]\cdot \boldsymbol{N}\, \mathrm{d}S.$$

Defining the Piola stress tensor $\boldsymbol{\Sigma} := \boldsymbol{\sigma}\cdot\operatorname{Cof}\boldsymbol{F} = J\boldsymbol{\sigma}\cdot\boldsymbol{F}^{-T}$ and the body forces in reference configuration $\boldsymbol{b}_0 := \boldsymbol{b}$, we can use the conservation of mass [2.16] and the divergence theorem [2.11] to write

$$\int_{\mathcal{D}_0} \rho_0 \frac{D\dot{\boldsymbol{x}}}{Dt}\, \mathrm{d}V = \int_{\mathcal{D}_0} \rho_0 \boldsymbol{b}_0\, \mathrm{d}V + \int_{\mathcal{D}_0} \operatorname{Div}\boldsymbol{\Sigma}\, \mathrm{d}V.$$

The strong form of the above relation is expressed as

$$\rho_0 \frac{D\dot{\boldsymbol{u}}}{Dt} = \rho_0 \boldsymbol{b}_0 + \operatorname{Div}\boldsymbol{\Sigma}. \qquad [2.20]$$

2.4.3. *Conservation of angular momentum*

The conservation of angular momentum states that [LAI 10]: *the total moment about a fixed point (surface and body moments) acting on a fixed part of a material is equal to the time rate of change of total angular momentum of the part about the same point.*

Considering the body of Figure 2.3, the conservation of angular momentum is expressed in the current configuration by the relation

$$\frac{D}{Dt}\int_{\mathcal{D}} \boldsymbol{x}\times\rho\dot{\boldsymbol{x}}\, \mathrm{d}v = \int_{\mathcal{D}} \boldsymbol{x}\times\rho\boldsymbol{b}\, \mathrm{d}v + \int_{\partial\mathcal{D}} \boldsymbol{x}\times\boldsymbol{t}\, \mathrm{d}s. \qquad [2.21]$$

With the help of the definition of the stress tensor [2.9], the divergence theorem [2.10] and the Reynolds transport theorem [2.12], the above expression can be written as

$$\int_{\mathcal{D}} \dot{\boldsymbol{x}} \times \rho \dot{\boldsymbol{x}} \, \mathrm{d}v + \int_{\mathcal{D}} \boldsymbol{x} \times \left[\rho \frac{D\dot{\boldsymbol{x}}}{Dt} + \left[\frac{D\rho}{Dt} + \rho \, \mathrm{div}\dot{\boldsymbol{x}} \right] \dot{\boldsymbol{x}} \right] \mathrm{d}v$$
$$= \int_{\mathcal{D}} \boldsymbol{x} \times \rho \boldsymbol{b} \, \mathrm{d}v + \int_{\mathcal{D}} [\boldsymbol{x} \times \mathrm{div}\boldsymbol{\sigma} - \boldsymbol{\epsilon} : \boldsymbol{\sigma}] \, \mathrm{d}v,$$

where $\boldsymbol{\epsilon}$ is the third-order permutation tensor. The first integral of the left-hand side is zero (cross product of a vector with itself vanishes). Using the strong forms of the conservation of mass [2.15] and the conservation of linear momentum [2.19], the conservation of angular momentum is reduced to

$$\int_{\mathcal{D}} \boldsymbol{\epsilon} : \boldsymbol{\sigma} \, \mathrm{d}v = \mathbf{0}.$$

The last expression needs to be valid for every subvolume of the continuum body, therefore the relation $\boldsymbol{\epsilon} : \boldsymbol{\sigma} = \mathbf{0}$ leads to

$$\boldsymbol{\sigma} = \boldsymbol{\sigma}^T. \qquad [2.22]$$

By the definition of the Piola stress tensor, it is provided that $\boldsymbol{\sigma} = J^{-1} \boldsymbol{\Sigma} \cdot \boldsymbol{F}^T$. Thus, in the reference configuration, the conservation of angular momentum leads to

$$\boldsymbol{\Sigma} \cdot \boldsymbol{F}^T = \boldsymbol{F} \cdot \boldsymbol{\Sigma}^T. \qquad [2.23]$$

2.4.4. *Conservation of energy*

The first law of thermodynamics, also known as the conservation of energy, states that [LAG 08]: *the rate of change of the total energy (internal plus kinetic energy) on a fixed part of a material is equal to the rate at which external mechanical work is done to that part by surface tractions and body forces plus the rate at which thermal energy is added by heat flux and heat sources.*

For the body of Figure 2.3, this law is written in the current configuration as [LAI 10]

$$\frac{D}{Dt}\int_{\mathcal{D}} \rho e \,\mathrm{d}v + \frac{D}{Dt}\int_{\mathcal{D}} \frac{1}{2}\rho\dot{\boldsymbol{x}}\cdot\dot{\boldsymbol{x}}\,\mathrm{d}v = \int_{\partial\mathcal{D}} \boldsymbol{t}\cdot\dot{\boldsymbol{x}}\,\mathrm{d}s + \int_{\mathcal{D}} \rho\boldsymbol{b}\cdot\dot{\boldsymbol{x}}\,\mathrm{d}v$$
$$- \int_{\partial\mathcal{D}} \boldsymbol{q}\cdot\boldsymbol{n}\,\mathrm{d}s + \int_{\mathcal{D}} \rho\mathcal{R}\,\mathrm{d}v, \qquad [2.24]$$

where e is the internal energy per unit mass, $\boldsymbol{q}$ is the heat flux vector and $\mathcal{R}$ denotes the heat sources per unit mass. With the help of the definition of the stress tensor [2.9], the divergence theorem [2.10] and the Reynolds transport theorem [2.12], the last expression can be written as

$$\int_{\mathcal{D}} \left[\rho\dot{e} + \rho\frac{D\dot{\boldsymbol{x}}}{Dt}\cdot\dot{\boldsymbol{x}}\right]\mathrm{d}v + \int_{\mathcal{D}} \left[e + \frac{1}{2}\dot{\boldsymbol{x}}\cdot\dot{\boldsymbol{x}}\right]\left[\frac{D\rho}{Dt} + \rho\,\mathrm{div}\dot{\boldsymbol{x}}\right]\mathrm{d}v$$
$$= \int_{\mathcal{D}} \boldsymbol{\sigma}:\mathrm{grad}\dot{\boldsymbol{x}}\,\mathrm{d}v + \int_{\mathcal{D}} [\mathrm{div}\boldsymbol{\sigma} + \rho\boldsymbol{b}]\cdot\dot{\boldsymbol{x}}\,\mathrm{d}v - \int_{\mathcal{D}} \mathrm{div}\boldsymbol{q}\,\mathrm{d}v + \int_{\mathcal{D}} \rho\mathcal{R}\,\mathrm{d}v.$$

Using the strong forms of the conservation of mass [2.15] and the conservation of linear momentum [2.19], the above equation is written as

$$\int_{\mathcal{D}} \rho\dot{e}\,\mathrm{d}v = \int_{\mathcal{D}} [\boldsymbol{\sigma}:\mathrm{grad}\dot{\boldsymbol{x}} - \mathrm{div}\boldsymbol{q} + \rho\mathcal{R}]\,\mathrm{d}v.$$

The final expression needs to be valid for every subvolume of the continuum body, therefore,

$$\rho\dot{e} = \boldsymbol{\sigma}:\mathrm{grad}\dot{\boldsymbol{u}} - \mathrm{div}\boldsymbol{q} + \rho\mathcal{R}. \qquad [2.25]$$

With regard to the reference configuration, equation [2.24], with the help of Table 1.3, can be written in the space $\mathcal{D}_0$ as

$$\frac{D}{Dt}\int_{\mathcal{D}_0} \rho J e\,\mathrm{d}V + \frac{D}{Dt}\int_{\mathcal{D}_0} \frac{1}{2}\rho J\dot{\boldsymbol{x}}\cdot\dot{\boldsymbol{x}}\,\mathrm{d}V = \int_{\partial\mathcal{D}_0} [\boldsymbol{\Sigma}\cdot\boldsymbol{N}]\cdot\dot{\boldsymbol{x}}\,\mathrm{d}S$$

$$+\int_{\mathcal{D}_0} \rho J \boldsymbol{b}_0 \cdot \dot{\boldsymbol{x}} \, \mathrm{d}V - \int_{\partial \mathcal{D}_0} [\boldsymbol{q} \cdot \mathrm{Cof} \boldsymbol{F}] \cdot \boldsymbol{N} \, \mathrm{d}S + \int_{\mathcal{D}_0} \rho J \mathcal{R} \, \mathrm{d}V.$$

Defining (1) the internal energy in the reference configuration $e_0 := e$, (2) the heat flux vector in the reference configuration $\boldsymbol{q}_0 := \boldsymbol{q} \cdot \mathrm{Cof}\boldsymbol{F}$ and (3) the heat sources in the reference configuration $\mathcal{R}_0 := \mathcal{R}$, we can use the conservation of mass [2.16] and the divergence theorem [2.11] to write

$$\int_{\mathcal{D}_0} \rho_0 \dot{e}_0 \, \mathrm{d}V + \int_{\mathcal{D}_0} \rho_0 \frac{D\dot{\boldsymbol{x}}}{Dt} \cdot \dot{\boldsymbol{x}} \, \mathrm{d}V = \int_{\mathcal{D}_0} \boldsymbol{\Sigma} : \mathrm{Grad} \dot{\boldsymbol{x}} \, \mathrm{d}V$$

$$+\int_{\mathcal{D}_0} [\mathrm{Div} \boldsymbol{\Sigma} + \rho_0 \boldsymbol{b}_0] \cdot \dot{\boldsymbol{x}} \, \mathrm{d}V - \int_{\mathcal{D}_0} \mathrm{Div} \boldsymbol{q}_0 \, \mathrm{d}V + \int_{\mathcal{D}_0} \rho_0 \mathcal{R}_0 \, \mathrm{d}V.$$

Using equation [2.20], the final expression leads to the strong form

$$\rho_0 \dot{e}_0 = \boldsymbol{\Sigma} : \dot{\boldsymbol{F}} - \mathrm{Div} \boldsymbol{q}_0 + \rho_0 \mathcal{R}_0. \qquad [2.26]$$

2.4.5. *Entropy inequality*

The second law of thermodynamics, also known as entropy inequality, states that [LAI 10]: *the rate of increase of entropy in a fixed part of material is not less than the influx of entropy across the surface of the part and the entropy supply within the volume.*

For the body of Figure 2.3, this law is written as

$$\frac{D}{Dt} \int_{\mathcal{D}} \rho \varsigma \, \mathrm{d}v \geq - \int_{\partial \mathcal{D}} \frac{\boldsymbol{q}}{\theta} \cdot \boldsymbol{n} \, \mathrm{d}s + \int_{\mathcal{D}} \frac{\rho \mathcal{R}}{\theta} \, \mathrm{d}v, \qquad [2.27]$$

where ς is the specific entropy per unit mass and θ is the absolute temperature (always positive). With the help of the divergence theorem [2.10] and the

Reynolds transport theorem [2.12], the above expression can be written as

$$\int_{\mathcal{D}} \left[\rho \frac{D\varsigma}{Dt} + \varsigma \left[\frac{D\rho}{Dt} + \rho \, \mathrm{div}\dot{\boldsymbol{x}} \right] \right] \mathrm{d}v \geq - \int_{\mathcal{D}} \mathrm{div} \left(\frac{\boldsymbol{q}}{\theta} \right) \mathrm{d}v + \int_{\mathcal{D}} \frac{\rho \mathcal{R}}{\theta} \mathrm{d}v.$$

Using the strong form of the conservation of mass [2.15], the last inequality leads to

$$\int_{\mathcal{D}} \left[\rho \dot{\varsigma} + \mathrm{div} \left(\frac{\boldsymbol{q}}{\theta} \right) - \frac{\rho \mathcal{R}}{\theta} \right] \mathrm{d}v \geq 0.$$

The final expression needs to be valid for every subvolume of the continuum body, therefore,

$$\rho \theta \dot{\varsigma} - \frac{\boldsymbol{q}}{\theta} \cdot \boldsymbol{\nabla\theta} + \mathrm{div}\boldsymbol{q} - \rho \mathcal{R} \geq 0, \quad \boldsymbol{\nabla\theta} = \mathrm{grad}\theta. \qquad [2.28]$$

With regard to the reference configuration, equation [2.27], with the help of Table 1.3, can be written in the space $\mathcal{D}_0$ as

$$\frac{D}{Dt} \int_{\mathcal{D}_0} \rho J \varsigma \, \mathrm{d}V \geq - \int_{\partial \mathcal{D}_0} \frac{\boldsymbol{q}_0}{\theta} \cdot \boldsymbol{N} \, \mathrm{d}S + \int_{\mathcal{D}_0} \frac{\rho_0 \mathcal{R}_0}{\theta} \mathrm{d}V.$$

Defining the specific entropy in the reference configuration $\varsigma_0 := \varsigma$, we can use the conservation of mass [2.16] and the divergence theorem [2.11] to write

$$\int_{\mathcal{D}_0} \rho_0 \dot{\varsigma}_0 \, \mathrm{d}V \geq - \int_{\mathcal{D}_0} \mathrm{Div} \left(\frac{\boldsymbol{q}_0}{\theta} \right) \mathrm{d}V + \int_{\mathcal{D}_0} \frac{\rho_0 \mathcal{R}_0}{\theta} \mathrm{d}V,$$

which leads to the strong form

$$\rho_0 \theta \dot{\varsigma}_0 - \frac{\boldsymbol{q}_0}{\theta} \cdot \boldsymbol{\nabla\theta}_0 + \mathrm{Div}\boldsymbol{q}_0 - \rho_0 \mathcal{R}_0 \geq 0, \quad \boldsymbol{\nabla\theta}_0 = \mathrm{Grad}\theta = \boldsymbol{\nabla\theta} \cdot \boldsymbol{F}. \quad [2.29]$$

2.4.6. *Reduction to small deformations*

Under processes involving only small deformations, the density remains almost constant upon loading [NEM 99b]. Considering $\rho \simeq \rho_0$ and using the symmetric stress tensor $\boldsymbol{\sigma}$ and the symmetric strain tensor $\boldsymbol{\varepsilon}$, the conservation laws in small deformations are summarized in Table 2.1. From this point on and for the remaining part of the book, the discussion is concentrated on small deformations. Some theoretical derivations for large deformation processes are provided in the appendices.

Conservation law	Expression
conservation of mass	$\frac{D\rho}{Dt} \simeq 0$
conservation of linear momentum	$\rho \frac{D\dot{\boldsymbol{u}}}{Dt} = \rho \boldsymbol{b} + \text{div}\boldsymbol{\sigma}$
conservation of angular momentum	$\boldsymbol{\sigma} = \boldsymbol{\sigma}^T$
conservation of energy	$\rho \dot{e} = \boldsymbol{\sigma} : \dot{\boldsymbol{\varepsilon}} - \text{div}\boldsymbol{q} + \rho \mathcal{R}$
entropy inequality	$\theta \rho \dot{\varsigma} - \frac{1}{\theta} \boldsymbol{q} \cdot \boldsymbol{\nabla} \theta + \text{div}\boldsymbol{q} - \rho \mathcal{R} \geq 0$

Table 2.1. *Conservation laws for small deformation processes*

2.5. Constitutive law

Any constitutive law that aims to describe real material response must respect the thermodynamic principles. The various thermodynamic frameworks that exist in the literature, like the thermodynamics of irreversible processes, rational thermodynamics and thermodynamics with internal variables, are discussed extensively by Maugin [MAU 99].

2.5.1. *Thermodynamics considerations*

Under small deformations, the strong form of the first law of thermodynamics can be expressed as

$$\dot{\mathcal{E}} = \boldsymbol{\sigma}:\dot{\varepsilon} - \text{div}\boldsymbol{q} + \rho\mathcal{R}\,, \qquad [2.30]$$

where $\mathcal{E} = \rho e$ is the internal energy per unit volume. Similarly, the second law is written as

$$\theta\dot{\eta} - \frac{\boldsymbol{q}}{\theta}\cdot\boldsymbol{\nabla\theta} + \text{div}\boldsymbol{q} - \rho\mathcal{R} \geq 0, \quad \boldsymbol{\nabla\theta} = \text{grad}\theta\,, \qquad [2.31]$$

where $\eta = \rho\varsigma$ is the specific entropy per unit volume. Combining equations [2.30] and [2.31] yields

$$\gamma = \boldsymbol{\sigma}:\dot{\varepsilon} + \theta\dot{\eta} - \dot{\mathcal{E}} - \frac{\boldsymbol{q}}{\theta}\cdot\boldsymbol{\nabla\theta} \geq 0, \qquad [2.32]$$

where γ is the internal entropy production per unit volume. Thus, the two laws of thermodynamics can be expressed as

$$\begin{aligned} &\mathcal{Q} + r = 0 \quad \text{and} \quad \gamma = \gamma_{\text{loc}} + \gamma_{\text{con}} \geq 0\,, \quad \text{where} \\ &\mathcal{Q} = -\text{div}\boldsymbol{q} + \rho\mathcal{R}, \qquad r = \boldsymbol{\sigma}:\dot{\varepsilon} - \dot{\mathcal{E}}\,, \\ &\gamma_{\text{loc}} = r + \theta\dot{\eta}\,, \qquad \gamma_{\text{con}} = -\frac{\boldsymbol{q}}{\theta}\cdot\boldsymbol{\nabla\theta}. \end{aligned} \qquad [2.33]$$

In the above expressions, r is the difference between the rates of the mechanical work and the internal energy, $\mathcal{Q}$ is the thermal energy rate per unit volume entering or leaving a material point, γ_{loc} is the local entropy production (or intrinsic dissipation) and γ_{con} is the entropy production by heat conduction. Looking more closely at equations $[2.33]_4$ and $[2.33]_5$, it is concluded that

$$\boldsymbol{\sigma}:\dot{\varepsilon} + \theta\dot{\eta} = \gamma_{\text{loc}} + \dot{\mathcal{E}}. \qquad [2.34]$$

The last relation states that: *the sum of the rates of the mechanical and the thermal work equals the sum of the intrinsic dissipation and the rate of change in the internal energy.*

When designing a constitutive law for a material, it is very useful to separate the various mechanisms into categories: elastic or inelastic, reversible or irreversible, dissipative or non-dissipative, etc. Some of these mechanisms are responsible for permanent changes in the material's microstructure. To describe all observable phenomena, it is necessary to express $\mathcal{E}$ in terms of proper variables capable of expressing the material's state under every possible thermomechanical loading path.

2.5.1.1. *Internal energy as a function of the total strain*

In this approach, the internal energy $\mathcal{E}$ is considered a function of the strain tensor $\boldsymbol{\varepsilon}$, the entropy η and the set of internal state variables $\boldsymbol{\xi}$,

$$\mathcal{E} := \mathcal{E}(\boldsymbol{\varepsilon}, \eta, \boldsymbol{\xi})\,. \qquad [2.35]$$

The internal state variables are usually scalars, vectors or second-order tensors and they are very useful for materials whose response depends not only on the level of strain and entropy but also on other parameters such as the history of the loading, microstructural changes, etc. Following the approach of Germain *et al.* [GER 83], the internal energy is assumed to be convex and differentiable with regard to its arguments. The following definitions for the derivatives of the internal energy are postulated:

$$\boldsymbol{\sigma} = \frac{\partial \mathcal{E}}{\partial \boldsymbol{\varepsilon}}, \quad \theta = \frac{\partial \mathcal{E}}{\partial \eta}\,. \qquad [2.36]$$

Instead of working with the internal energy, it is more convenient to introduce the Hemholtz free energy potential Ψ or the Gibbs free energy potential G:

$$\Psi(\boldsymbol{\varepsilon}, \theta, \boldsymbol{\xi}) := \mathcal{E} - \theta\eta\,, \quad \text{or} \quad G(\boldsymbol{\sigma}, \theta, \boldsymbol{\xi}) := \mathcal{E} - \theta\eta - \boldsymbol{\sigma}{:}\boldsymbol{\varepsilon}\,. \qquad [2.37]$$

For the simplicity of the expressions, the following operator is introduced:

$$\frac{\partial\{\bullet\}}{\partial\boldsymbol{\xi}}:\dot{\boldsymbol{\xi}} = \sum_m \frac{\partial\{\bullet\}}{\partial z^m}\dot{z}^m + \sum_n \frac{\partial\{\bullet\}}{\partial \boldsymbol{z}^n}\cdot\dot{\boldsymbol{z}}^n + \sum_q \frac{\partial\{\bullet\}}{\partial \boldsymbol{Z}^q}:\dot{\boldsymbol{Z}}^q, \qquad [2.38]$$

where z^m are scalars, $\boldsymbol{z}^n$ are vectors and $\boldsymbol{Z}^q$ are second-order tensors.

At this point, the Legendre or Legendre–Fenchel transformation is employed. Generally, the Legendre–Fenchel transformation is between convex functions and leads to a maximization problem. Considering the internal energy to be a function of the total strain, specific entropy and internal variables, we can show that

$$\mathrm{d}\mathcal{E} = \frac{\partial\mathcal{E}}{\partial\boldsymbol{\varepsilon}}:\mathrm{d}\boldsymbol{\varepsilon} + \frac{\partial\mathcal{E}}{\partial\eta}\mathrm{d}\eta + \frac{\partial\mathcal{E}}{\partial\boldsymbol{\xi}}:\mathrm{d}\boldsymbol{\xi} \overset{[2.36]}{=} \boldsymbol{\sigma}:\mathrm{d}\boldsymbol{\varepsilon} + \theta\mathrm{d}\eta + \frac{\partial\mathcal{E}}{\partial\boldsymbol{\xi}}:\mathrm{d}\boldsymbol{\xi},$$

$$\begin{aligned}
\mathrm{d}\Psi &= \frac{\partial\Psi}{\partial\boldsymbol{\varepsilon}}:\mathrm{d}\boldsymbol{\varepsilon} + \frac{\partial\Psi}{\partial\theta}\mathrm{d}\theta + \frac{\partial\Psi}{\partial\boldsymbol{\xi}}:\mathrm{d}\boldsymbol{\xi} \overset{[2.37]}{=} \mathrm{d}\mathcal{E} - \eta\mathrm{d}\theta - \theta\mathrm{d}\eta \\
&= \boldsymbol{\sigma}:\mathrm{d}\boldsymbol{\varepsilon} + \theta\mathrm{d}\eta + \frac{\partial\mathcal{E}}{\partial\boldsymbol{\xi}}:\mathrm{d}\boldsymbol{\xi} - \eta\mathrm{d}\theta - \theta\mathrm{d}\eta = \boldsymbol{\sigma}:\mathrm{d}\boldsymbol{\varepsilon} - \eta\mathrm{d}\theta + \frac{\partial\mathcal{E}}{\partial\boldsymbol{\xi}}:\mathrm{d}\boldsymbol{\xi},
\end{aligned}$$

$$\begin{aligned}
\mathrm{d}G &= \frac{\partial G}{\partial\boldsymbol{\sigma}}:\mathrm{d}\boldsymbol{\sigma} + \frac{\partial G}{\partial\theta}\mathrm{d}\theta + \frac{\partial G}{\partial\boldsymbol{\xi}}:\mathrm{d}\boldsymbol{\xi} \\
&\overset{[2.37]}{=} \mathrm{d}\mathcal{E} - \eta\mathrm{d}\theta - \theta\mathrm{d}\eta - \boldsymbol{\sigma}:\mathrm{d}\boldsymbol{\varepsilon} - \boldsymbol{\varepsilon}:\mathrm{d}\boldsymbol{\sigma} \\
&= \boldsymbol{\sigma}:\mathrm{d}\boldsymbol{\varepsilon} + \theta\mathrm{d}\eta + \frac{\partial\mathcal{E}}{\partial\boldsymbol{\xi}}:\mathrm{d}\boldsymbol{\xi} - \eta\mathrm{d}\theta - \theta\mathrm{d}\eta - \boldsymbol{\sigma}:\mathrm{d}\boldsymbol{\varepsilon} - \boldsymbol{\varepsilon}:\mathrm{d}\boldsymbol{\sigma} \\
&= -\boldsymbol{\varepsilon}:\mathrm{d}\boldsymbol{\sigma} - \eta\mathrm{d}\theta + \frac{\partial\mathcal{E}}{\partial\boldsymbol{\xi}}:\mathrm{d}\boldsymbol{\xi}.
\end{aligned}$$

Thus, it can be deduced that[1]

$$\boldsymbol{\sigma} = \frac{\partial\Psi}{\partial\boldsymbol{\varepsilon}}, \quad \eta = -\frac{\partial\Psi}{\partial\theta}, \quad \text{or} \quad \boldsymbol{\varepsilon} = -\frac{\partial G}{\partial\boldsymbol{\sigma}}, \quad \eta = -\frac{\partial G}{\partial\theta}. \qquad [2.39]$$

1 The same result can be obtained by utilizing the methodology of Coleman and Gurtin [COL 67] for thermodynamics with internal state variables, taking into account the issues raised by Lubliner [LUB 72].

The hypothesis of convex $\mathcal{E}$ ensures that the representations [2.39] exist and are unique. Combining equations [2.33] and [2.37] leads to

$$\gamma_{\mathrm{loc}} = \boldsymbol{\sigma}:\dot{\boldsymbol{\varepsilon}} - \eta\dot{\theta} - \dot{\Psi}\,, \quad \text{or} \quad \gamma_{\mathrm{loc}} = -\boldsymbol{\varepsilon}:\dot{\boldsymbol{\sigma}} - \eta\dot{\theta} - \dot{G}\,. \qquad [2.40]$$

Expressing $\dot{\Psi}$ and $\dot{G}$ in terms of their arguments and using [2.39], the last expressions are written as

$$\gamma_{\mathrm{loc}} = -\frac{\partial\Psi}{\partial\boldsymbol{\xi}}:\dot{\boldsymbol{\xi}} = \boldsymbol{\Xi}_{\Psi}:\dot{\boldsymbol{\xi}}\,, \quad \text{or} \quad \gamma_{\mathrm{loc}} = -\frac{\partial G}{\partial\boldsymbol{\xi}}:\dot{\boldsymbol{\xi}} = \boldsymbol{\Xi}_{G}:\dot{\boldsymbol{\xi}}\,. \qquad [2.41]$$

Equation [2.41], in conjunction with the relation $\gamma_{\mathrm{loc}} + \gamma_{\mathrm{con}} \geq 0$, is utilized in order to identify proper evolution equations for the internal state variables. Usually, the mechanical and thermal dissipations are assumed to be decoupled and non-negative, i.e. $\gamma_{\mathrm{loc}} \geq 0$ and $\gamma_{\mathrm{con}} \geq 0$. For this purpose, it is customary in the literature to introduce a dissipation potential [HAL 75, LEM 02]. The properties and characteristics of such a potential are discussed in the following subsection.

Returning back to equations [2.33] and using $[2.39]_2$ (or $[2.39]_4$) to express the rate of entropy as a function of $\dot{\boldsymbol{\varepsilon}}$ (or $\dot{\boldsymbol{\sigma}}$), $\dot{\theta}$ and $\dot{\boldsymbol{\xi}}$, the first law of thermodynamics is written as

$$\begin{aligned} \tilde{c}_v\dot{\theta} - \mathcal{Q} &= -\frac{\partial\Psi}{\partial\boldsymbol{\xi}}:\dot{\boldsymbol{\xi}} + \theta\frac{\partial^2\Psi}{\partial\theta\partial\boldsymbol{\varepsilon}}:\dot{\boldsymbol{\varepsilon}} + \theta\frac{\partial^2\Psi}{\partial\theta\partial\boldsymbol{\xi}}:\dot{\boldsymbol{\xi}}, \\ \tilde{c}_v &= \rho c_v = -\theta\frac{\partial^2\Psi}{\partial\theta^2}, \end{aligned}$$

when Helmholtz free energy potential is considered, or

$$\begin{aligned} \tilde{c}_p\dot{\theta} - \mathcal{Q} &= -\frac{\partial G}{\partial\boldsymbol{\xi}}:\dot{\boldsymbol{\xi}} + \theta\frac{\partial^2 G}{\partial\theta\partial\boldsymbol{\sigma}}:\dot{\boldsymbol{\sigma}} + \theta\frac{\partial^2 G}{\partial\theta\partial\boldsymbol{\xi}}:\dot{\boldsymbol{\xi}}, \\ \tilde{c}_p &= \rho c_p = -\theta\frac{\partial^2 G}{\partial\theta^2}, \end{aligned} \qquad [2.42]$$

when Gibbs free energy potential is considered.

In the last expressions c_v and c_p are specific heat capacities at constant volume and constant pressure respectively.

While the above analysis permits us to design proper material constitutive laws, it does not provide clear answers on energetic considerations. In fact, it is impossible at this stage to identify which part of the internal energy can be recovered and which part is permanently stored due to microstructural changes. To proceed further, certain assumptions are unavoidable. The additional hypothesis here is that the Helmholtz or the Gibbs free energy potential consists of a recoverable and an irrecoverable part,

$$\begin{aligned} \Psi(\boldsymbol{\varepsilon}, \theta, \boldsymbol{\xi}) &= \Psi^{\mathrm{r}}(\boldsymbol{\varepsilon}, \theta, \boldsymbol{\xi}) + \Psi^{\mathrm{ir}}(\theta, \boldsymbol{\xi}), \text{ or} \\ G(\boldsymbol{\sigma}, \theta, \boldsymbol{\xi}) &= G^{\mathrm{r}}(\boldsymbol{\sigma}, \theta, \boldsymbol{\xi}) + G^{\mathrm{ir}}(\theta, \boldsymbol{\xi}) . \end{aligned} \qquad [2.43]$$

The term "recoverable" is utilized in order to indicate that Ψ^{r} (or G^{r}) can return to its initial value during thermomechanical loading paths, even if sometimes the internal variables evolve and cannot be restored[2]. However, the term "irrecoverable" indicates that, when $\boldsymbol{\xi}$ evolves, Ψ^{ir} (or G^{ir}) changes in an irrecoverable manner. At the same temperature level, the final value of Ψ^{ir} (or G^{ir}) is higher than the initial one.

According to equations [2.43], the relation between the stress and strain is defined only from the recoverable part of the free energy potential,

$$\boldsymbol{\sigma} = \frac{\partial \Psi^{\mathrm{r}}}{\partial \boldsymbol{\varepsilon}}, \quad \text{or} \quad \boldsymbol{\varepsilon} = -\frac{\partial G^{\mathrm{r}}}{\partial \boldsymbol{\sigma}} . \qquad [2.44]$$

Equations [2.41] and [2.43] allow us to write

$$\gamma_{\mathrm{loc}} = -\frac{\partial \Psi^{\mathrm{r}}}{\partial \boldsymbol{\xi}} : \dot{\boldsymbol{\xi}} - \frac{\partial \Psi^{\mathrm{ir}}}{\partial \boldsymbol{\xi}} : \dot{\boldsymbol{\xi}}, \quad \text{or} \quad \gamma_{\mathrm{loc}} = -\frac{\partial G^{\mathrm{r}}}{\partial \boldsymbol{\xi}} : \dot{\boldsymbol{\xi}} - \frac{\partial G^{\mathrm{ir}}}{\partial \boldsymbol{\xi}} : \dot{\boldsymbol{\xi}} . \qquad [2.45]$$

At this point, it would be interesting to investigate more deeply the various energetic terms. From [2.43] and $[2.39]_{2,4}$ it becomes tempting to assume that

2 This is, for instance, the case in damage mechanisms: upon stress loading-unloading under constant temperature, the damage parameter evolves and cannot be restored, although the Ψ^{r} recovers its original value (see the continuum damage model case of section 2.5.3).

the specific entropy is the sum of a recoverable and an irrecoverable part:

$$\eta = \eta^{\mathrm{r}} + \eta^{\mathrm{ir}}, \quad \text{where}$$
$$\eta^{\mathrm{r}} = -\frac{\partial \Psi^{\mathrm{r}}}{\partial \theta}, \quad \eta^{\mathrm{ir}} = -\frac{\partial \Psi^{\mathrm{ir}}}{\partial \theta}, \quad \text{or}$$
$$\eta^{\mathrm{r}} = -\frac{\partial G^{\mathrm{r}}}{\partial \theta}, \quad \eta^{\mathrm{ir}} = -\frac{\partial G^{\mathrm{ir}}}{\partial \theta}. \qquad [2.46]$$

The last hypothesis adds another restriction: the entropy produced by the recoverable part of the free energy potential must also return to its original value when Ψ^{r} (or G^{r}) recovers. Under this condition, the internal energy presents a similar decomposition, i.e.

$$\mathcal{E} = \mathcal{E}^{\mathrm{r}} + \mathcal{E}^{\mathrm{ir}}, \quad \mathcal{E}^{\mathrm{r}} = \Psi^{\mathrm{r}} + \theta\eta^{\mathrm{r}}, \quad \mathcal{E}^{\mathrm{ir}} = \Psi^{\mathrm{ir}} + \theta\eta^{\mathrm{ir}}. \qquad [2.47]$$

Moreover, the thermal power can also be decomposed into two terms:

$$\dot{\mathrm{W}}_{\mathrm{t}} = \dot{\mathrm{W}}_{\mathrm{t}}^{\mathrm{r}} + \dot{\mathrm{W}}_{\mathrm{t}}^{\mathrm{ir}}, \quad \dot{\mathrm{W}}_{\mathrm{t}} = \theta\dot{\eta}, \quad \dot{\mathrm{W}}_{\mathrm{t}}^{\mathrm{r}} = \theta\dot{\eta}^{\mathrm{r}}, \quad \dot{\mathrm{W}}_{\mathrm{t}}^{\mathrm{ir}} = \theta\dot{\eta}^{\mathrm{ir}}. \qquad [2.48]$$

However, the total mechanical power can be split into a recoverable part, an irrecoverable part and a dissipative part,

$$\dot{\mathrm{W}}_{\mathrm{m}} = \dot{\mathrm{W}}_{\mathrm{m}}^{\mathrm{r}} + \dot{\mathrm{W}}_{\mathrm{m}}^{\mathrm{ir}} + \dot{\mathrm{W}}_{\mathrm{m}}^{\mathrm{d}},$$
$$\dot{\mathrm{W}}_{\mathrm{m}} = \boldsymbol{\sigma}:\dot{\boldsymbol{\varepsilon}}, \quad \dot{\mathrm{W}}_{\mathrm{m}}^{\mathrm{d}} = \gamma_{\mathrm{loc}},$$
$$\dot{\mathrm{W}}_{\mathrm{m}}^{\mathrm{r}} = \boldsymbol{\sigma}:\dot{\boldsymbol{\varepsilon}} + \frac{\partial \Psi^{\mathrm{r}}}{\partial \boldsymbol{\xi}}:\dot{\boldsymbol{\xi}}, \quad \text{or} \quad \dot{\mathrm{W}}_{\mathrm{m}}^{\mathrm{r}} = \boldsymbol{\sigma}:\dot{\boldsymbol{\varepsilon}} + \frac{\partial G^{\mathrm{r}}}{\partial \boldsymbol{\xi}}:\dot{\boldsymbol{\xi}},$$
$$\dot{\mathrm{W}}_{\mathrm{m}}^{\mathrm{ir}} = \frac{\partial \Psi^{\mathrm{ir}}}{\partial \boldsymbol{\xi}}:\dot{\boldsymbol{\xi}}, \quad \text{or} \quad \dot{\mathrm{W}}_{\mathrm{m}}^{\mathrm{ir}} = \frac{\partial G^{\mathrm{ir}}}{\partial \boldsymbol{\xi}}:\dot{\boldsymbol{\xi}}. \qquad [2.49]$$

Adding $[2.49]_1$ and $[2.48]_1$ yields

$$\dot{\mathrm{W}}_{\mathrm{m}} + \dot{\mathrm{W}}_{\mathrm{t}} = \dot{\mathrm{W}}_{\mathrm{m}}^{\mathrm{d}} + \dot{\mathrm{W}}_{\mathrm{m}}^{\mathrm{r}} + \dot{\mathrm{W}}_{\mathrm{t}}^{\mathrm{r}} + \dot{\mathrm{W}}_{\mathrm{m}}^{\mathrm{ir}} + \dot{\mathrm{W}}_{\mathrm{t}}^{\mathrm{ir}}. \qquad [2.50]$$

Comparing the last expression with [2.34], it becomes evident that

$$\dot{\mathcal{E}} = \dot{\mathcal{E}}^{\mathrm{r}} + \dot{\mathcal{E}}^{\mathrm{ir}}, \quad \dot{\mathcal{E}}^{\mathrm{r}} = \dot{\mathrm{W}}^{\mathrm{r}}_{\mathrm{m}} + \dot{\mathrm{W}}^{\mathrm{r}}_{\mathrm{t}}, \quad \dot{\mathcal{E}}^{\mathrm{ir}} = \dot{\mathrm{W}}^{\mathrm{ir}}_{\mathrm{m}} + \dot{\mathrm{W}}^{\mathrm{ir}}_{\mathrm{t}}. \qquad [2.51]$$

As it is observed by equations [2.45], [2.49] and [2.51], while both the recoverable and irrecoverable parts of the free energy potential may contribute to the production of intrinsic dissipation, only the irrecoverable part can produce stored internal energy. Such a distinction is important for dissipative processes. Damage and viscoelastic mechanisms usually cause dissipation without creating permanent stored energy. However, plasticity and viscoplasticity are accompanied by both dissipation and permanent energy storage in the material.

As a final remark, it is noted that in this approach the internal variables $\boldsymbol{\xi}$ must include the inelastic strain tensors that appear due to nonlinear mechanisms like plasticity, viscoplasticity, etc.

2.5.1.2. *Alternative approaches*

Apart from the above approach, there are additional general thermodynamic formulations in the literature that allow us to distinguish the various energetic mechanisms. The methodology presented by Lemaitre and Chaboche [LEM 02] considers that the total strain is split into an elastic and an inelastic part. Moreover, the stress is thermodynamically connected with the elastic strain. Such formalism is convenient for inelastic mechanisms that produce their own strain (for instance, plastic and viscoplastic materials). A thorough energetic discussion using this type of strain decomposition is presented by Rosakis *et al.* [ROS 00]. In a slightly different version, the entropy is also partitioned into thermoelastic and inelastic parts and the internal energy depends on the thermoelastic strain and thermoelastic entropy [SIM 92, MAU 99].

Chrysochoos and coworkers have proposed another interesting approach [CHR 10, BEN 14]. In their methodology the stress is decomposed into a reversible and an irreversible part. Moreover, the strain is thermodynamically connected with the reversible part of the stress. Such formalism can be convenient for viscoelastic and damage mechanisms, where no partition of the total strain occurs and there is the possibility of identifying "reversible" and "irreversible" parts of the stress. The idea of stress decomposition also permits us to decompose the intrinsic dissipation in a manner that accounts for viscous and non-viscous behavior, as well as fading memory and constant memory effects occuring during fatigue tests [WAC 83].

2.5.2. *Dissipation potentials*

In the thermodynamics of irreversible processes, Onsager [ONS 31] introduced the reciprocity relations which state that the fluxes and forces are linearly dependent on each other. With the introduction of a dissipation function of quadratic form in terms of the fluxes, Onsager formulated the principle of the least dissipation of energy. Ziegler [ZIE 63] extended the idea to a more generalized dissipation function in order to capture nonlinear cases. Moreover, Moreau [MOR 70] introduced the hypothesis of normal dissipativity for the cases of viscosity and plasticity. Following this approach, a general procedure was introduced for the cases of plasticity and viscoplasticity [GER 73, HAL 75, GER 83]. Halphen and Nguyen [HAL 75] have introduced the notion of generalized standard materials, while de Saxcé and Bousshine [SAX 02] have proposed the extended definition of implicit standard materials.

In continuum mechanics, it is customary to proceed with a stronger form of the second law of thermodynamics: it is assumed that the thermal (due to conduction) and mechanical (intrinsic) dissipation need to be both non-negative. The positiveness of γ_{con} is automatically satisfied when the Fourier law between heat fluxes and temperature gradients is considered

$$\boldsymbol{q} = -\boldsymbol{\kappa}\cdot\text{grad}\theta = -\boldsymbol{\kappa}\cdot\boldsymbol{\nabla}\boldsymbol{\theta}\,, \qquad [2.52]$$

with $\boldsymbol{\kappa}$ denoting the thermal conductivity tensor. Thus, the second law of thermodynamics $[2.33]_2$ is reduced to the inequality

$$\gamma_{\text{loc}} \geq 0. \qquad [2.53]$$

According to the results of the previous subsection, the intrinsic dissipation can be written in a compact form as

$$\boldsymbol{\Xi} : \dot{\boldsymbol{\xi}} \geq 0, \qquad [2.54]$$

where the first term denotes thermodynamic forces and the second term denotes thermodynamic fluxes. The dyadic product has the same meaning as that of equation [2.38]. To proceed further, the existence of a pseudo-potential $\phi(\dot{\boldsymbol{\xi}})$ non-negative, convex, closed, such that $\phi(\mathbf{0}) = 0$, is postulated. Then, the dual function $\phi^*(\boldsymbol{\Xi})$ exists and it is given by the Legendre–Fenchel transformation

$$\phi^*(\boldsymbol{\Xi}) = \sup_{\dot{\boldsymbol{\xi}}_a}\{\boldsymbol{\Xi} : \dot{\boldsymbol{\xi}}_a - \phi(\dot{\boldsymbol{\xi}}_a)\}. \qquad [2.55]$$

The dual function is also non-negative, convex, closed and $\phi^*(\mathbf{0}) = 0$ [ROC 70]. The fluxes $\dot{\boldsymbol{\xi}}$ and the forces $\boldsymbol{\Xi}$ are said to be thermodynamically related at one point, if one of the following holds at this point:

1) $\boldsymbol{\Xi} \in \partial\phi(\dot{\boldsymbol{\xi}}), \quad \partial\phi(\dot{\boldsymbol{\xi}}) = \{\boldsymbol{\Xi} \mid \phi(\dot{\boldsymbol{\xi}}^*) \geq \phi(\dot{\boldsymbol{\xi}}) + \boldsymbol{\Xi}(\dot{\boldsymbol{\xi}}^* - \dot{\boldsymbol{\xi}}) \ \forall \dot{\boldsymbol{\xi}}^*\}$;

2) $\dot{\boldsymbol{\xi}} \in \partial\phi^*(\boldsymbol{\Xi}), \quad \partial\phi^*(\boldsymbol{\Xi}) = \{\dot{\boldsymbol{\xi}} \mid \phi^*(\boldsymbol{\Xi}^*) \geq \phi^*(\boldsymbol{\Xi}) + \dot{\boldsymbol{\xi}}(\boldsymbol{\Xi}^* - \boldsymbol{\Xi}) \ \forall \boldsymbol{\Xi}^*\}$;

3) $\phi(\dot{\boldsymbol{\xi}}) + \phi^*(\boldsymbol{\Xi}) = \boldsymbol{\Xi} : \dot{\boldsymbol{\xi}}$,

4) for a given force $\boldsymbol{\Xi}$, the flux $\dot{\boldsymbol{\xi}}$ associated with the force maximizes the quantity $\boldsymbol{\Xi} : \dot{\boldsymbol{\xi}}_a - \phi(\dot{\boldsymbol{\xi}}_a)$;

5) for a given flux $\dot{\boldsymbol{\xi}}$, the force $\boldsymbol{\Xi}$ associated with the flux maximizes the quantity $\boldsymbol{\Xi}_a : \dot{\boldsymbol{\xi}} - \phi^*(\boldsymbol{\Xi}_a)$.

These five statements are equivalent [ROC 70, GER 82]. Note that $\partial\phi$ and $\partial\phi^*$ are called subdifferentials of ϕ and ϕ^*, respectively. In engineering terms, this definition states that, upon all possible choices, the evolution of the dissipative mechanism follows at each instant the path that provides the maximum dissipation. Actually, the maximization of dissipation tends to provide a more stable state for a physical system.

If ϕ and ϕ^* are differentiable[3], then the first two statements change to $\boldsymbol{\Xi} = \partial\phi/\partial\dot{\boldsymbol{\xi}}$ and $\dot{\boldsymbol{\xi}} = \partial\phi^*/\partial\boldsymbol{\Xi}$, respectively. The introduction of this

3 A differentiable convex function $\Phi(\boldsymbol{\Xi})$ on an interval has the property

$$\Phi(\boldsymbol{\Xi}_2) - \Phi(\boldsymbol{\Xi}_1) \geq \frac{\partial\Phi}{\partial\boldsymbol{\Xi}_1} : (\boldsymbol{\Xi}_2 - \boldsymbol{\Xi}_1), \quad \forall \boldsymbol{\Xi}_1, \boldsymbol{\Xi}_2 \text{ in the interval.}$$

pseudo-potential is consistent with the second law of thermodynamics, as

$$\boldsymbol{\Xi} : \dot{\boldsymbol{\xi}} \geq \phi(\dot{\boldsymbol{\xi}}) \geq 0.$$

The following examples from viscoelasticity, viscoplasticity and rate-independent elastoplasticity can be seen as general guidelines for constructing appropriate evolution laws of nonlinear mechanisms in dissipative materials.

2.5.2.1. *Viscoelasticity*

Viscoelastic materials present a reversible response (i.e. they can return to the initial state) which depends on the rate of the applied loading. This mechanism usually appears in polymeric materials. In viscoelasticity, the pseudo-potential $\phi(\dot{\boldsymbol{\xi}}^{\mathrm{v}})$ is a non-negative, closed, differentiable, convex function which vanishes at the origin. In this case,

$$\boldsymbol{\Xi}^{\mathrm{v}} = \frac{\partial \phi}{\partial \dot{\boldsymbol{\xi}}^{\mathrm{v}}}. \qquad [2.56]$$

Of course, we can equivalently formulate an appropriate dual function of the pseudo-potential.

2.5.2.2. *Viscoplasticity*

Viscoplasticity also depends on the rate of the applied loading, but the material response is irreversible and is accompanied by permanent deformation. In metals, this mechanism is linked with the movement of dislocations in the crystalline structure. It has been shown experimentally that plastic strain rate depends on the applied stress [LEM 02], so the use of a dual function $\phi^*(\boldsymbol{\Xi}^{\mathrm{vp}})$ of the pseudo-potential is a more appropriate choice for a viscoplastic material. This pseudo-potential is a non-negative, closed, differentiable, convex function which vanishes at the origin,

$$\dot{\boldsymbol{\xi}}^{\mathrm{vp}} = \frac{\partial \phi^*}{\partial \boldsymbol{\Xi}^{\mathrm{vp}}}. \qquad [2.57]$$

2.5.2.3. *Associated plasticity*

Elastoplasticity is a rate independent phenomenon which, when activated, causes the development of permanent deformation. It is usually considered as the limit case of viscoplasticity, in which the notion of critical stress exists and causes yielding in the material. In metals the development of plastic strains is due to the creation of dislocations in the crystalline structure. The evolution of the permanent deformation may or may not be linked to the criterion under which plasticity occurs, allowing us to categorize the elastoplastic mechanisms into associated and non-associated. In associated plasticity, the dual function of the pseudo-potential is equal to the indicator function I_c:

$$\phi^*(\Xi^{\mathrm{p}}) = I_c(\Xi^{\mathrm{p}}) = \begin{cases} 0, & \Xi^{\mathrm{p}} \in C, \\ +\infty, & \Xi^{\mathrm{p}} \notin C. \end{cases} \qquad [2.58]$$

The domain C in the indicator function is bounded by the surface Φ (proper convex function), which is called the "yield surface",

$$C = \{\Xi^{\mathrm{p}} \mid \Phi(\Xi^{\mathrm{p}}) \leq 0\}. \qquad [2.59]$$

Using this pseudo-potential, we obtain [ROC 70, FRE 02, MAU 92]

$$\partial\phi^*(\Xi^{\mathrm{p}}) = \partial I_c(\Xi^{\mathrm{p}}) = \begin{cases} \{0\}, & \Phi(\Xi^{\mathrm{p}}) < 0, \\ \bigcup\{\lambda\partial\Phi(\Xi^{\mathrm{p}}) \mid \lambda \geq 0\}, & \Phi(\Xi^{\mathrm{p}}) = 0, \\ \emptyset, & \Phi(\Xi^{\mathrm{p}}) > 0. \end{cases}$$

If Φ is differentiable then, according to the second statement,

$$\dot{\xi} = \lambda\frac{\partial\Phi}{\partial\Xi^{\mathrm{p}}}, \qquad \lambda \geq 0, \;\; \Phi(\Xi^{\mathrm{p}}) \leq 0, \;\; \lambda\Phi(\Xi^{\mathrm{p}}) = 0. \qquad [2.60]$$

2.5.2.4. *Associated plasticity and viscoplasticity*

In the case where both (associated) rate-independent plasticity and viscoplasticity are present, the dual of the pseudo-potential (with the properties mentioned above)

$$\phi^*(\Xi) = \Omega(\Xi) + I_c(\Xi), \qquad [2.61]$$

is introduced, where C is given by [2.59], Ω and Φ are convex functions with $dom\Omega = dom\Phi = \Re^n$, Ω is differentiable (moreover, non-negative and vanishes at the origin), Φ is continuous and $\Phi < 0$ has at least one solution. Then [ROC 70],

$$\partial\phi^*(\Xi) = \partial\Omega(\Xi) + \partial I_c(\Xi), \qquad [2.62]$$

and consequently for Φ differentiable [HAL 75]

$$\dot{\boldsymbol{\xi}} = \frac{\partial\Omega}{\partial\Xi} + \lambda\frac{\partial\Phi}{\partial\Xi}, \qquad \lambda \geq 0,\ \ \Phi(\Xi) \leq 0,\ \ \lambda\Phi(\Xi) = 0. \qquad [2.63]$$

2.5.2.5. *Non-associated plasticity*

In the case of non-associated plasticity, we can identify two surfaces [LEM 02]: a potential surface $F^{\mathrm{p}}(\Xi^{\mathrm{p}})$ and a yield surface $\Phi^{\mathrm{p}}(\Xi^{\mathrm{p}})$, for which

$$\Phi^{\mathrm{p}}(\Xi^{\mathrm{p}}) \leq 0, \quad \lambda \geq 0, \quad \Phi^{\mathrm{p}}(\Xi^{\mathrm{p}})\lambda = 0. \qquad [2.64]$$

Consider F^{p} and Φ^{p} as proper convex, differentiable functions and additionally $F^{\mathrm{p}}(\mathbf{0}) = 0$. If $F^{\mathrm{p}} \geq \Phi^{\mathrm{p}}$, then the fluxes $\dot{\boldsymbol{\xi}}^{\mathrm{p}}$ are assumed to be provided by the expression

$$\dot{\boldsymbol{\xi}}^{\mathrm{p}} = \lambda\frac{\partial F^{\mathrm{p}}}{\partial\Xi^{\mathrm{p}}}. \qquad [2.65]$$

This yields

$$\Xi^{\mathrm{p}}:\dot{\boldsymbol{\xi}}^{\mathrm{p}} = \dot{\boldsymbol{\xi}}^{\mathrm{p}}:\Xi^{\mathrm{p}} = \lambda\frac{\partial F^{\mathrm{p}}}{\partial\Xi^{\mathrm{p}}}:\Xi^{\mathrm{p}} \geq \lambda F^{\mathrm{p}}(\Xi^{\mathrm{p}}) \geq \lambda\Phi^{\mathrm{p}}(\Xi^{\mathrm{p}}) = 0. \qquad [2.66]$$

As a remark, we could also utilize appropriate bipotentials based on the theory of implicit standard materials [BOD 01, SAX 02].

2.5.3. *Examples of constitutive laws*

Here, several types of materials are discussed as illustrative examples of how we can identify constitutive laws.

2.5.3.1. *General structure of constitutive laws*

Any type of material behavior law can be properly expressed through the thermodynamic framework by choosing the appropriate free energy potential. The forms of the Helmholtz or the Gibbs energy potentials in equations [2.43] can be considered as general guidelines. Indeed, we can construct a proper constitutive law, bearing in mind the following aspects:

1) In order to investigate the dependence of the Helmholtz free energy potential on the strain, we can utilize the methodology described by Malvern [MAL 69]. In a purely elastic material, the elastic energy Ψ can be written in a Taylor series expanded form

$$\Psi = C_0 + \boldsymbol{C}_1{:}\boldsymbol{\varepsilon} + \frac{1}{2}\boldsymbol{\varepsilon}{:}\boldsymbol{C}{:}\boldsymbol{\varepsilon} + ...$$

where C_0 is a constant, $\boldsymbol{C}_1$ is a second-order symmetric tensor and $\boldsymbol{C}$ is a fourth-order symmetric tensor with major and minor symmetries. Due to the assumption of small deformations, higher order terms under strain can be neglected. The constant C_0 can be arbitrarily selected equal to 0, while $\boldsymbol{C}_1$ can also be 0 as long as there are no prestress conditions inside the material. Thus, the only remaining term is the quadratic. This leads to an expression of the stress as

$$\boldsymbol{\sigma} = \boldsymbol{C}{:}\boldsymbol{\varepsilon}.$$

Considering that a complex material may present N types of mechanism that generate eigenstrains $\boldsymbol{\varepsilon}^{*(q)}$ during thermomechanical loading (for example thermal strains, plastic strains, viscoelastic strains, etc.), the total strain is decomposed additively in an elastic part, $\boldsymbol{\varepsilon}^{\mathrm{e}}$, and the N eigenstrains, $\boldsymbol{\varepsilon}^{*(i)}$,

$$\boldsymbol{\varepsilon} = \boldsymbol{\varepsilon}^{\mathrm{e}} + \sum_{q=1}^{N} \boldsymbol{\varepsilon}^{*(q)}.$$

In such a case, the stress is expressed in terms of the elastic strain, i.e.

$$\boldsymbol{\sigma} = \boldsymbol{C}{:}\left[\boldsymbol{\varepsilon} - \sum_{q=1}^{N} \boldsymbol{\varepsilon}^{*(q)}\right],$$

which in turn leads to a Helmholtz free energy of the form

$$\Psi = \frac{1}{2}\left[\boldsymbol{\varepsilon} - \sum_{q=1}^{N} \boldsymbol{\varepsilon}^{*(q)}\right] :\boldsymbol{C}: \left[\boldsymbol{\varepsilon} - \sum_{q=1}^{N} \boldsymbol{\varepsilon}^{*(q)}\right] + \Psi^{\mathrm{r}}_{(\theta,\boldsymbol{\xi})}(\theta,\boldsymbol{\xi}) + \Psi^{\mathrm{ir}}(\theta,\boldsymbol{\xi}).$$

It becomes evident that $\boldsymbol{\varepsilon}^{*(q)}$ must be subsets of the internal variables $\boldsymbol{\xi}$. Note also that the fourth-order elasticity tensor $\boldsymbol{C}$ may depend on the temperature or on some internal variables. Typical examples of such behavior are the materials undergoing damage mechanisms [LEM 02] and the shape-memory alloys [LAG 08].

2) In thermodynamics, two specific heat capacities are defined according to the applied conditions: the first is under constant pressure (stress) and the second is under constant volume (strain). In the mechanics of solids, the specific heat under constant pressure, c_p, can be obtained through a much simpler experimental procedure than the one under constant volume, c_v. In the constitutive laws described below, the material parameter c_0 is related to the product of the specific heat capacity under constant pressure and the material density. For thermoelastic materials $c_0 = \rho c_p$, while for thermoinelastic materials c_0 and ρc_p may differ due to a term (usually small) arising from the thermoinelastic coupling.

In a purely thermal process, where the mechanical fields and internal variables are ignored, we expect that [FER 37]

$$\frac{\mathrm{d}\eta}{\mathrm{d}\theta} = \frac{c_0}{\theta}.$$

Assuming that at temperature θ_0 the entropy is equal to η_0, the last expression, after integration, yields

$$\eta = c_0 \ln\left(\frac{\theta}{\theta_0}\right) + \eta_0 = -\frac{\mathrm{d}G}{\mathrm{d}\theta},$$

where equation $[2.39]_4$ has been taken into account. In the above purely thermal process, assuming that at the initial state $\theta = \theta_0$, $G(\theta_0)$ is equal to $\mathcal{E}_0 - \eta_0\theta_0$ and the potential G can be expressed in terms of the temperature as

$$G = c_0\left[[\theta - \theta_0] - \theta \ln\left(\frac{\theta}{\theta_0}\right)\right] - \eta_0\theta + \mathcal{E}_0.$$

Since, for solids, c_v and c_p are almost the same, a similar form for the Helmholtz potential is expected. For materials undergoing thermomechanical loading, the last expression can be seen as the purely thermal part of the recoverable free energy potential Ψ^{r} or G^{r}. Note also that the parameters c_0, η_0 and $\mathcal{E}_0$ may depend on the internal variables (for instance in shape-memory alloys, they are assumed to be dependent on the martensite volume fraction [LAG 08]).

Combining the two discussed cases, we can write a proper Helmholtz free energy potential in the general form

$$\begin{aligned}\Psi(\boldsymbol{\varepsilon},\theta,\boldsymbol{\xi}) &= \frac{1}{2}\left[\boldsymbol{\varepsilon}-\sum_{q=1}^{N}\boldsymbol{\varepsilon}^{*(q)}\right]:\boldsymbol{C}(\theta,\boldsymbol{\xi}):\left[\boldsymbol{\varepsilon}-\sum_{q=1}^{N}\boldsymbol{\varepsilon}^{*(q)}\right]\\ &\quad+c_0(\boldsymbol{\xi})\left[[\theta-\theta_0]-\theta\ln\left(\frac{\theta}{\theta_0}\right)\right]-\eta_0(\boldsymbol{\xi})\theta+\mathcal{E}_0(\boldsymbol{\xi})\\ &\quad+\Psi^{\mathrm{r}}_{\boldsymbol{\xi}}(\theta,\boldsymbol{\xi})+\Psi^{\mathrm{ir}}(\theta,\boldsymbol{\xi}).\end{aligned}$$

It is worth noticing that the proposed form of the energy potential is only a guidance for the proper construction of constitutive laws. It is not mandatory that it be followed and perhaps we can design other forms of energy potentials that are thermodynamically consistent. Closing this discussion, the equivalent form of the Gibbs free energy can be expressed as

$$\begin{aligned}G(\boldsymbol{\sigma},\theta,\boldsymbol{\xi}) &= -\frac{1}{2}\boldsymbol{\sigma}:\boldsymbol{C}^{-1}(\theta,\boldsymbol{\xi}):\boldsymbol{\sigma}-\boldsymbol{\sigma}:\sum_{q=1}^{N}\boldsymbol{\varepsilon}^{*(q)}\\ &\quad+c_0(\boldsymbol{\xi})\left[[\theta-\theta_0]-\theta\ln\left(\frac{\theta}{\theta_0}\right)\right]-\eta_0(\boldsymbol{\xi})\theta+\mathcal{E}_0(\boldsymbol{\xi})\\ &\quad+\Psi^{\mathrm{r}}_{\boldsymbol{\xi}}(\theta,\boldsymbol{\xi})+\Psi^{\mathrm{ir}}(\theta,\boldsymbol{\xi}).\end{aligned}$$

2.5.3.2. *Thermoelasticity*

Consider a classical thermoelastic material initially at a state of reference temperature θ_0. The total strain is decomposed into an elastic part, $\boldsymbol{\varepsilon}^{\mathrm{e}}$, and a thermal part, $\boldsymbol{\varepsilon}^{\mathrm{th}}$,

$$\boldsymbol{\varepsilon}=\boldsymbol{\varepsilon}^{\mathrm{e}}+\boldsymbol{\varepsilon}^{\mathrm{th}}, \qquad [2.67]$$

where $\boldsymbol{\varepsilon}^{\text{th}} = \boldsymbol{\alpha}[\theta - \theta_0]$. The Helmholtz free energy potential in this case is written as

$$\Psi(\boldsymbol{\varepsilon}, \theta) = \frac{1}{2}\left[\boldsymbol{\varepsilon} - \boldsymbol{\alpha}[\theta - \theta_0]\right] : \boldsymbol{C} : \left[\boldsymbol{\varepsilon} - \boldsymbol{\alpha}[\theta - \theta_0]\right] \\ + c_0\left[[\theta - \theta_0] - \theta \ln\left(\frac{\theta}{\theta_0}\right)\right] - \eta_0\theta + \mathcal{E}_0, \qquad [2.68]$$

where $\boldsymbol{C}$ is the elasticity tensor, $\boldsymbol{\alpha}$ is the thermal expansion tensor, c_0 is the specific heat capacity at constant pressure multiplied by the material density ($c_0 = \rho c_p$), η_0 is the initial entropy and $\mathcal{E}_0$ is the initial internal energy. Moreover, $\boldsymbol{C}$ is a fourth-order tensor with major and minor symmetries and $\boldsymbol{\alpha}$ is a symmetric second-order tensor. Alternatively, a Gibbs free energy potential of the following form can be utilized

$$G(\boldsymbol{\sigma}, \theta) = -\frac{1}{2}\boldsymbol{\sigma} : \boldsymbol{C}^{-1} : \boldsymbol{\sigma} - \boldsymbol{\sigma} : \boldsymbol{\alpha}[\theta - \theta_0] \\ + c_0\left[[\theta - \theta_0] - \theta \ln\left(\frac{\theta}{\theta_0}\right)\right] - \eta_0\theta + \mathcal{E}_0, \qquad [2.69]$$

which provides exactly the same constitutive material response. Since no mechanical dissipation occurs in thermoelasticity, $\gamma_{\text{loc}} = 0$. According to the discussion in subsection 2.5.1, the use of $[2.39]_{1,2}$, $[2.33]_5$, $[2.42]_1$ and [2.68] yields

$$\boldsymbol{\sigma} = \boldsymbol{C} : \left[\boldsymbol{\varepsilon} - \boldsymbol{\alpha}[\theta - \theta_0]\right], \\ \eta = \boldsymbol{\alpha} : \boldsymbol{\sigma} + c_0 \ln\left(\frac{\theta}{\theta_0}\right) + \eta_0, \\ r = -\theta\dot{\eta} = -c_0\dot{\theta} - \theta\boldsymbol{\alpha} : \dot{\boldsymbol{\sigma}}, \qquad [2.70]$$

while the energy equation [2.33] takes the general form

$$\tilde{c}_v\dot{\theta} - \mathcal{Q} = \theta\frac{\partial^2\Psi}{\partial\theta\partial\boldsymbol{\varepsilon}} : \dot{\boldsymbol{\varepsilon}}, \quad \tilde{c}_v = -\theta\frac{\partial^2\Psi}{\partial\theta^2}.$$

Following the discussion in section 2.5.1, the various work and energy rates are identified as

$$\dot{\mathrm{W}}_{\mathrm{m}} = \dot{\mathrm{W}}^{\mathrm{r}}_{\mathrm{m}} = \boldsymbol{\sigma}:\dot{\boldsymbol{\varepsilon}}, \quad \dot{\mathrm{W}}_{\mathrm{t}} = \dot{\mathrm{W}}^{\mathrm{r}}_{\mathrm{t}} = \theta\dot{\eta} = \theta\boldsymbol{\alpha}:\dot{\boldsymbol{\sigma}} + c_0\dot{\theta},$$
$$\dot{\mathcal{E}} = \dot{\mathcal{E}}^{\mathrm{r}} = \boldsymbol{\sigma}:\dot{\boldsymbol{\varepsilon}} + \theta\boldsymbol{\alpha}:\dot{\boldsymbol{\sigma}} + c_0\dot{\theta}. \qquad [2.71]$$

2.5.3.3. *Viscoelasticity*

In a general viscoelastic material, the total strain $\boldsymbol{\varepsilon}$ is decomposed into an elastic part, $\boldsymbol{\varepsilon}^{\mathrm{e}}$, a viscous part, $\boldsymbol{\varepsilon}^{\mathrm{v}}$, and a thermal part, $\boldsymbol{\varepsilon}^{\mathrm{th}}$,

$$\boldsymbol{\varepsilon} = \boldsymbol{\varepsilon}^{\mathrm{e}} + \boldsymbol{\varepsilon}^{\mathrm{v}} + \boldsymbol{\varepsilon}^{\mathrm{th}}. \qquad [2.72]$$

Consider a viscoelastic material following a Poynting–Thomson type rheological model [MOR 05], initially at a state of reference temperature θ_0 (Figure 2.4). The Helmholtz free energy potential in this case is a function of the total strain tensor $\boldsymbol{\varepsilon}$, the temperature θ and the viscoelastic strain tensor $\boldsymbol{\varepsilon}^{\mathrm{v}}$,

$$\begin{aligned}\Psi &= \Psi^{\mathrm{r}}\left(\boldsymbol{\varepsilon}, \theta, \boldsymbol{\varepsilon}^{\mathrm{v}}\right)\\ &= \frac{1}{2}\left[\boldsymbol{\varepsilon} - \boldsymbol{\varepsilon}^{\mathrm{v}} - \boldsymbol{\alpha}[\theta - \theta_0]\right]:\boldsymbol{C}:\left[\boldsymbol{\varepsilon} - \boldsymbol{\varepsilon}^{\mathrm{v}} - \boldsymbol{\alpha}[\theta - \theta_0]\right]\\ &\quad + \frac{1}{2}\boldsymbol{\varepsilon}^{\mathrm{v}}:\boldsymbol{C}^{\mathrm{v}}:\boldsymbol{\varepsilon}^{\mathrm{v}} + c_0\left[[\theta - \theta_0] - \theta\ln\left(\frac{\theta}{\theta_0}\right)\right] - \eta_0\theta + \mathcal{E}_0, \qquad [2.73]\end{aligned}$$

where $\boldsymbol{C}^{\mathrm{v}}$ is a fourth-order stiffness-type tensor (i.e. having major and minor symmetries) related to viscoelasticity and $\boldsymbol{C}$, $\boldsymbol{\alpha}$, c_0, η_0 and $\mathcal{E}_0$ are material constants similar to those of a thermoelastic material. Moreover, $\boldsymbol{\varepsilon}^{\mathrm{th}} = \boldsymbol{\alpha}[\theta - \theta_0]$. According to the discussion in section 2.5.1, the use of $[2.44]_1$, $[2.46]_{1,2,3}$, $[2.33]_5$, [2.45] and $[2.42]_1$ yields

$$\boldsymbol{\sigma} = \boldsymbol{C}:\left[\boldsymbol{\varepsilon} - \boldsymbol{\varepsilon}^{\mathrm{v}} - \boldsymbol{\alpha}[\theta - \theta_0]\right],$$
$$\eta = \eta^{\mathrm{r}} = \boldsymbol{\alpha}:\boldsymbol{\sigma} + c_0\ln\left(\frac{\theta}{\theta_0}\right) + \eta_0,$$

$$
\begin{aligned}
\gamma_{\mathrm{loc}} &= -\frac{\partial \Psi^{\mathrm{r}}}{\partial \boldsymbol{\varepsilon}^{\mathrm{v}}}:\dot{\boldsymbol{\varepsilon}}^{\mathrm{v}} = \boldsymbol{\sigma}^{\mathrm{v}}:\dot{\boldsymbol{\varepsilon}}^{\mathrm{v}}, \quad \boldsymbol{\sigma}^{\mathrm{v}} = \boldsymbol{\sigma} - \boldsymbol{C}^{\mathrm{v}}:\boldsymbol{\varepsilon}^{\mathrm{v}}, \\
r &= -\theta\dot{\eta} + \gamma_{\mathrm{loc}} = -c_0\dot{\theta} - \theta\boldsymbol{\alpha}:\dot{\boldsymbol{\sigma}} + \gamma_{\mathrm{loc}},
\end{aligned}
\qquad [2.74]
$$

while the energy equation takes the general form

$$
\tilde{c}_v\dot{\theta} - \mathcal{Q} = \gamma_{\mathrm{loc}} + \theta\frac{\partial^2\Psi}{\partial\theta\partial\boldsymbol{\varepsilon}}:\dot{\boldsymbol{\varepsilon}} + \theta\frac{\partial^2\Psi}{\partial\theta\partial\boldsymbol{\varepsilon}^{\mathrm{v}}}:\dot{\boldsymbol{\varepsilon}}^{\mathrm{v}}, \quad \tilde{c}_v = -\theta\frac{\partial^2\Psi}{\partial\theta^2}.
$$

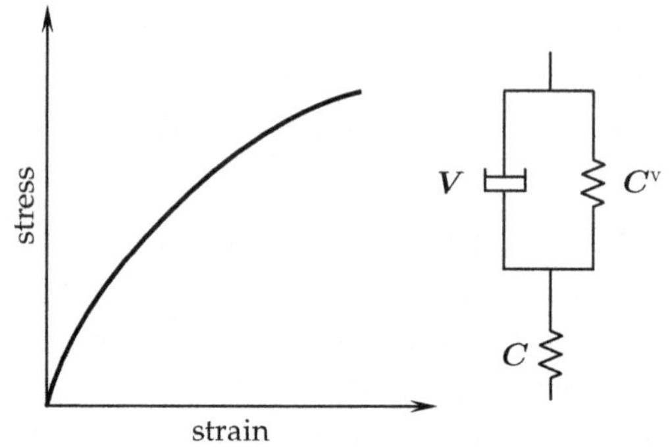

Figure 2.4. *Poynting–Thomson type model: typical mechanical response and rheological scheme consisting of a spring and a Kelvin–Voigt branch connected in series*

In viscoelasticity, it is customary to introduce a dissipation potential that depends on the thermodynamic fluxes. In the examined case, such potential can be written as

$$
\phi(\dot{\boldsymbol{\varepsilon}}^{\mathrm{v}}) = \frac{1}{2}\dot{\boldsymbol{\varepsilon}}^{\mathrm{v}}:\boldsymbol{V}:\dot{\boldsymbol{\varepsilon}}^{\mathrm{v}}, \qquad [2.75]
$$

where $\boldsymbol{V}$ is a fourth-order viscosity tensor with major and minor symmetries. Then, according to the discussion in section 2.5.2, we obtain

$$
\boldsymbol{\sigma}^{\mathrm{v}} = \frac{\partial\phi}{\partial\dot{\boldsymbol{\varepsilon}}^{\mathrm{v}}} = \boldsymbol{V}:\dot{\boldsymbol{\varepsilon}}^{\mathrm{v}}, \qquad [2.76]
$$

which allows us to write the dissipation in the quadratic form $\gamma_{\text{loc}} = \dot{\varepsilon}^{\text{v}}\!:\!\boldsymbol{V}\!:\!\dot{\varepsilon}^{\text{v}}$. Following the discussion in section 2.5.1, the various work and energy rates are identified as

$$\begin{aligned}\dot{\mathrm{W}}^{\mathrm{r}}_{\mathrm{m}} &= \boldsymbol{\sigma}\!:\!\dot{\varepsilon} - \boldsymbol{\sigma}^{\mathrm{v}}\!:\!\dot{\varepsilon}^{\mathrm{v}}, \quad \dot{\mathrm{W}}^{\mathrm{d}}_{\mathrm{m}} = \boldsymbol{\sigma}^{\mathrm{v}}\!:\!\dot{\varepsilon}^{\mathrm{v}}, \quad \dot{\mathrm{W}}^{\mathrm{r}}_{\mathrm{t}} = \theta\dot{\eta}^{\mathrm{r}} = \theta\boldsymbol{\alpha}\!:\!\dot{\boldsymbol{\sigma}} + c_0\dot{\theta}, \\ \dot{\mathcal{E}}^{\mathrm{r}} &= \boldsymbol{\sigma}\!:\!\dot{\varepsilon} - \boldsymbol{\sigma}^{\mathrm{v}}\!:\!\dot{\varepsilon}^{\mathrm{v}} + \theta\boldsymbol{\alpha}\!:\!\dot{\boldsymbol{\sigma}} + c_0\dot{\theta}.\end{aligned} \qquad [2.77]$$

It is worth mentioning that the Maxwell model can be obtained with the above formalism by setting $\boldsymbol{C}^{\mathrm{v}} = \mathbf{0}$. Moreover, the Kelvin–Voigt model can be obtained with the above formalism by considering that $\boldsymbol{C}$ tends to infinity. Indeed, ignoring the thermal expansion contribution, equation $[2.74]_1$ can be written as

$$\varepsilon - \varepsilon^{\mathrm{v}} = \boldsymbol{C}^{-1}\!:\!\boldsymbol{\sigma} \rightarrow \mathbf{0}.$$

This leads to the conclusion that $\varepsilon = \varepsilon^{\mathrm{v}}$. Combining equations $[2.74]_4$ and [2.76] yields

$$\boldsymbol{\sigma} = \boldsymbol{C}^{\mathrm{v}}\!:\!\varepsilon + \boldsymbol{V}\!:\!\dot{\varepsilon}.$$

The last relation expresses the Kelvin–Voigt viscoelastic model.

2.5.3.4. *Elastoplasticity with isotropic hardening*

In a general elastoplastic material (Figure 2.5), the total strain ε is decomposed into an elastic part, ε^{e}, a plastic part, ε^{p}, and a thermal part, $\varepsilon^{\mathrm{th}}$,

$$\varepsilon = \varepsilon^{\mathrm{e}} + \varepsilon^{\mathrm{p}} + \varepsilon^{\mathrm{th}}. \qquad [2.78]$$

The Helmholtz free energy of an elastoplastic material with isotropic hardening can be expressed in terms of the total strain tensor ε, the plastic strain tensor ε^{p}, the temperature θ and an internal variable p (the accumulated plastic strain) related to the plastic mechanism,

$$\begin{aligned}\Psi\left(\varepsilon, \theta, \varepsilon^{\mathrm{p}}, p\right) &= \Psi^{\mathrm{r}}\left(\varepsilon, \theta, \varepsilon^{\mathrm{p}}\right) + \Psi^{\mathrm{ir}}\left(\theta, p\right), \\ \Psi^{\mathrm{r}}\left(\varepsilon, \theta, \varepsilon^{\mathrm{p}}\right) &= \frac{1}{2}\left[\varepsilon - \varepsilon^{\mathrm{p}} - \boldsymbol{\alpha}[\theta - \theta_0]\right]:\boldsymbol{C}\!:\left[\varepsilon - \varepsilon^{\mathrm{p}} - \boldsymbol{\alpha}[\theta - \theta_0]\right]\end{aligned}$$

$$+c_0 \left[[\theta - \theta_0] - \theta \ln\left(\frac{\theta}{\theta_0}\right) \right] - \eta_0 \theta + \mathcal{E}_0,$$

$$\Psi^{\text{ir}}(\theta, p) = F(\theta, p), \qquad [2.79]$$

where $F(\theta, p)$ is a hardening function related to elastoplasticity. Moreover, $\boldsymbol{C}$, $\boldsymbol{\alpha}$, η_0 and $\mathcal{E}_0$ are material constants similar to those of a thermoelastic material and $\boldsymbol{\varepsilon}^{\text{th}} = \boldsymbol{\alpha}[\theta - \theta_0]$. The constant c_0 is exactly equal to ρc_p only if the second derivative of F with respect to the temperature is zero. Equivalently, instead of the Helmholtz potential, we can utilize a Gibbs free energy potential of the form

$$G(\boldsymbol{\sigma}, \theta, \boldsymbol{\varepsilon}^{\text{p}}, p) = G^{\text{r}}(\boldsymbol{\sigma}, \theta, \boldsymbol{\varepsilon}^{\text{p}}) + G^{\text{ir}}(\theta, p),$$

$$G^{\text{r}}(\boldsymbol{\sigma}, \theta, \boldsymbol{\varepsilon}^{\text{p}}) = -\frac{1}{2}\boldsymbol{\sigma}:\boldsymbol{C}^{-1}:\boldsymbol{\sigma} - \boldsymbol{\sigma}: [\boldsymbol{\varepsilon}^{\text{p}} + \boldsymbol{\alpha}[\theta - \theta_0]]$$

$$+c_0 \left[[\theta - \theta_0] - \theta \ln\left(\frac{\theta}{\theta_0}\right) \right] - \eta_0 \theta + \mathcal{E}_0,$$

$$G^{\text{ir}}(\theta, p) = F(\theta, p). \qquad [2.80]$$

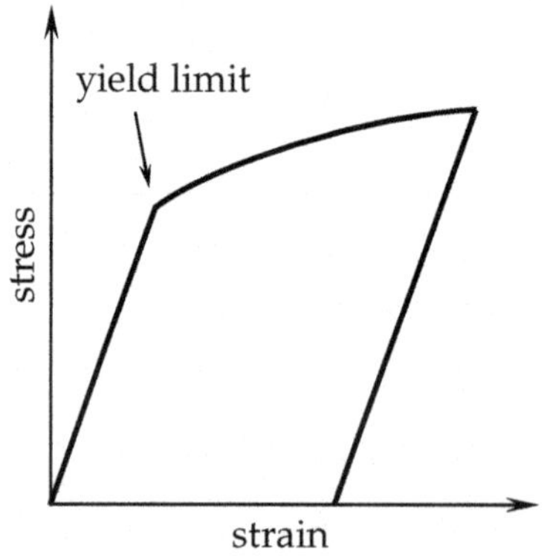

Figure 2.5. *Typical mechanical loading-unloading response of an elastoplastic material*

The two potentials [2.79] and [2.80] provide exactly the same constitutive response for the material. Considering the Hermholtz free energy potential and taking into account the discussion of subsection 2.5.1, the use of $[2.44]_1$, $[2.46]_{1,2,3}$, $[2.33]_5$, [2.45] and $[2.42]_1$ and [2.79] yields

$$\begin{aligned}
\boldsymbol{\sigma} &= \boldsymbol{C}\colon [\boldsymbol{\varepsilon} - \boldsymbol{\varepsilon}^{\mathrm{p}} - \boldsymbol{\alpha}[\theta - \theta_0]], \quad \eta = \eta^{\mathrm{r}} + \eta^{\mathrm{ir}}, \\
\eta^{\mathrm{r}} &= \boldsymbol{\alpha}\colon\boldsymbol{\sigma} + c_0 \ln\left(\frac{\theta}{\theta_0}\right) + \eta_0, \quad \eta^{\mathrm{ir}} = -F_\theta, \\
\gamma_{\mathrm{loc}} &= -\frac{\partial \Psi^{\mathrm{r}}}{\partial \boldsymbol{\varepsilon}^{\mathrm{p}}}\colon\dot{\boldsymbol{\varepsilon}}^{\mathrm{p}} - \frac{\partial \Psi^{\mathrm{ir}}}{\partial p}\dot{p} = \boldsymbol{\sigma}\colon\dot{\boldsymbol{\varepsilon}}^{\mathrm{p}} - F_p \dot{p}, \\
r &= -\theta\dot{\eta} + \gamma_{\mathrm{loc}} = -c_0\dot{\theta} - \theta\boldsymbol{\alpha}\colon\dot{\boldsymbol{\sigma}} + \theta\dot{F}_\theta + \gamma_{\mathrm{loc}},
\end{aligned} \qquad [2.81]$$

while the energy equation takes the general form

$$\tilde{c}_v\dot{\theta} - \mathcal{Q} = \gamma_{\mathrm{loc}} + \theta\frac{\partial^2\Psi}{\partial\theta\partial\boldsymbol{\varepsilon}}\colon\dot{\boldsymbol{\varepsilon}} + \theta\frac{\partial^2\Psi}{\partial\theta\partial\boldsymbol{\varepsilon}^{\mathrm{p}}}\colon\dot{\boldsymbol{\varepsilon}}^{\mathrm{p}} + \theta\frac{\partial^2\Psi}{\partial\theta\partial p}\dot{p}, \quad \tilde{c}_v = -\theta\frac{\partial^2\Psi}{\partial\theta^2}.$$

In the above expressions, $F_p = \partial F/\partial p$ and $F_\theta = \partial F/\partial\theta$. It becomes clear that, for $\boldsymbol{\sigma} = \boldsymbol{0}$ and $\theta = \theta_0$, the recoverable Helmholtz potential and the recoverable entropy obtain the values $\Psi^{\mathrm{r}} = -\eta_0\theta_0 + \mathcal{E}_0$ and $\eta^{\mathrm{r}} = \eta_0$ respectively, independently of the thermomechanical path that the material travels.

In associated von Mises isotropic plasticity, the yield surface takes the form

$$\Phi := \Phi(\boldsymbol{\sigma}, -F_p) = \sigma^{\mathrm{vM}} - F_p - \sigma_Y, \qquad [2.82]$$

and the relations [2.60] provide the evolution equation of the plastic strain

$$\dot{p} = \lambda, \quad \dot{\boldsymbol{\varepsilon}}^{\mathrm{p}} = \lambda\boldsymbol{\Lambda} = \boldsymbol{\Lambda}\dot{p}, \qquad \boldsymbol{\Lambda} = \frac{\partial\Phi}{\partial\boldsymbol{\sigma}} = \frac{3\boldsymbol{\sigma}'}{2\sigma^{\mathrm{vM}}}, \qquad [2.83]$$

along with the Kuhn–Tucker conditions

$$\Phi \leq 0, \quad \dot{p} \geq 0, \quad \Phi\dot{p} = 0. \qquad [2.84]$$

In the above expressions, σ^{vM} is the von Mises stress and $\boldsymbol{\sigma}'$ is the deviatoric stress, given by

$$\sigma^{\mathrm{vM}} = \sqrt{\frac{3}{2}\boldsymbol{\sigma}':\boldsymbol{\sigma}'}, \qquad \boldsymbol{\sigma}' = \boldsymbol{\sigma} - \frac{1}{3}\mathrm{tr}\boldsymbol{\sigma}\boldsymbol{I}. \qquad [2.85]$$

Moreover, σ_Y is the elastic limit (positive constant). Note that the same expressions are obtained using the thermodynamic framework of Lemaitre and Chaboche [LEM 02].

Following the discussion of section 2.5.1, the various work and energy rates are identified as

$$\begin{aligned}
\dot{\mathrm{W}}^{\mathrm{r}}_{\mathrm{m}} &= \boldsymbol{\sigma}:\left[\dot{\boldsymbol{\varepsilon}} - \dot{\boldsymbol{\varepsilon}}^{\mathrm{p}}\right], \quad \dot{\mathrm{W}}^{\mathrm{ir}}_{\mathrm{m}} = F_p\dot{p}, \quad \dot{\mathrm{W}}^{\mathrm{d}}_{\mathrm{m}} = \boldsymbol{\sigma}:\dot{\boldsymbol{\varepsilon}}^{\mathrm{p}} - F_p\dot{p},\\
\dot{\mathrm{W}}^{\mathrm{r}}_{\mathrm{t}} &= \theta\dot{\eta}^{\mathrm{r}} = \theta\boldsymbol{\alpha}:\dot{\boldsymbol{\sigma}} + c_0\dot{\theta}, \quad \dot{\mathrm{W}}^{\mathrm{ir}}_{\mathrm{t}} = \theta\dot{\eta}^{\mathrm{ir}} = -\theta\dot{F}_\theta,\\
\dot{\mathcal{E}}^{\mathrm{r}} &= \boldsymbol{\sigma}:\left[\dot{\boldsymbol{\varepsilon}} - \dot{\boldsymbol{\varepsilon}}^{\mathrm{p}}\right] + \theta\boldsymbol{\alpha}:\dot{\boldsymbol{\sigma}} + c_0\dot{\theta},\\
\dot{\mathcal{E}}^{\mathrm{ir}} &= F_p\dot{p} - \theta\dot{F}_\theta = \frac{\partial\Psi}{\partial p}\dot{p} - \theta\frac{\partial^2\Psi}{\partial p\partial\theta}\dot{p} - \theta\frac{\partial^2\Psi^{\mathrm{ir}}}{\partial\theta^2}\dot{\theta}.
\end{aligned} \qquad [2.86]$$

The term $-\theta\dfrac{\partial^2\Psi^{\mathrm{ir}}}{\partial\theta^2}$ represents the effect of the plastic mechanism on the specific heat capacity. According to experimental observations, such an effect is negligible [ROS 00]. Ignoring this part, the definition of $\dot{\mathcal{E}}^{\mathrm{ir}}$ agrees with the one given by Rosakis *et al.* [ROS 00]. However, Ranc and Chrysochoos [RAN 13] consider as the rate of stored energy only the term $\dfrac{\partial\Psi}{\partial p}\dot{p}$, while the term $\theta\dfrac{\partial^2\Psi}{\partial p\partial\theta}\dot{p}$ is defined as the thermoplastic coupling source.

2.5.3.5. *Elastoplasticity with kinematic hardening*

Kinematic hardening in elastoplastic materials causes tension-compression asymmetry in the mechanical response (Figure 2.6).

The Helmholtz free energy of a J_2 elastoplastic material with linear kinematic hardening can be expressed in terms of the total strain tensor $\boldsymbol{\varepsilon}$, the plastic strain tensor $\boldsymbol{\varepsilon}^{\mathrm{p}}$, the temperature θ and a second order tensor $\boldsymbol{a}$ which is related with the kinematic hardening,

$$\begin{aligned}\Psi &= \Psi^{\mathrm{r}}\left(\boldsymbol{\varepsilon}, \theta, \boldsymbol{\varepsilon}^{\mathrm{p}}, \boldsymbol{a}\right) \\ &= \frac{1}{2}\left[\boldsymbol{\varepsilon} - \boldsymbol{\varepsilon}^{\mathrm{p}} - \boldsymbol{\alpha}[\theta - \theta_0]\right] : \boldsymbol{C} : \left[\boldsymbol{\varepsilon} - \boldsymbol{\varepsilon}^{\mathrm{p}} - \boldsymbol{\alpha}[\theta - \theta_0]\right] \\ &+ c_0\left[[\theta - \theta_0] - \theta \ln\left(\frac{\theta}{\theta_0}\right)\right] - \eta_0\theta + \mathcal{E}_0 + \frac{1}{2}\boldsymbol{a} : \boldsymbol{C}^{\mathrm{p}} : \boldsymbol{a}. \qquad [2.87]\end{aligned}$$

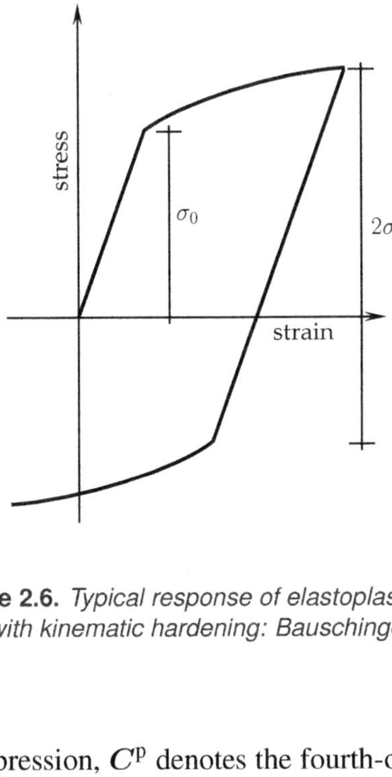

Figure 2.6. *Typical response of elastoplastic materials with kinematic hardening: Bauschinger effect*

In the above expression, $\boldsymbol{C}^{\mathrm{p}}$ denotes the fourth-order plastic tensor related to the kinematic hardening, while $\boldsymbol{C}$, $\boldsymbol{\alpha}$, c_0, η_0 and $\mathcal{E}_0$ are material constants similar to those of a thermoelastic material. Note that while plasticity is an irreversible phenomenon, the kinematic hardening is introduced in the recoverable part of the free energy. This is due to the fact that the back stress tensor $\boldsymbol{X} = \boldsymbol{C}^{\mathrm{p}} : \boldsymbol{a}$ can increase or decrease and, upon a proper mechanical loading path, it can return back to zero. In such cases, all the internal energy introduced due to kinematic hardening is converted into heat (see also the discussion in the work of Lemaitre and Chaboche [LEM 02]). Following the same steps as with the case of elastoplastic materials with isotropic hardening yields

$$\begin{aligned}
\boldsymbol{\sigma} &= \boldsymbol{C}\colon[\boldsymbol{\varepsilon} - \boldsymbol{\varepsilon}^{\mathrm{p}} - \boldsymbol{\alpha}[\theta - \theta_0]]\,,\\
\eta &= \eta^{\mathrm{r}} = \boldsymbol{\alpha}\colon\boldsymbol{\sigma} + c_0 \ln\left(\frac{\theta}{\theta_0}\right) + \eta_0,\\
\gamma_{\mathrm{loc}} &= -\frac{\partial \Psi^{\mathrm{r}}}{\partial \boldsymbol{\varepsilon}^{\mathrm{p}}}\colon\dot{\boldsymbol{\varepsilon}}^{\mathrm{p}} - \frac{\partial \Psi^{\mathrm{r}}}{\partial \boldsymbol{a}}\colon\dot{\boldsymbol{a}} = \boldsymbol{\sigma}\colon\dot{\boldsymbol{\varepsilon}}^{\mathrm{p}} - \boldsymbol{X}\colon\dot{\boldsymbol{a}},\\
r &= -\theta\dot{\eta} + \gamma_{\mathrm{loc}} = -c_0\dot{\theta} - \theta\boldsymbol{\alpha}\colon\dot{\boldsymbol{\sigma}} + \gamma_{\mathrm{loc}},
\end{aligned} \qquad [2.88]$$

while the energy equation takes the general form

$$\begin{aligned}
\tilde{c}_v\dot{\theta} - \mathcal{Q} &= \gamma_{\mathrm{loc}} + \theta\frac{\partial^2 \Psi}{\partial\theta\partial\boldsymbol{\varepsilon}}\colon\dot{\boldsymbol{\varepsilon}} + \theta\frac{\partial^2 \Psi}{\partial\theta\partial\boldsymbol{\varepsilon}^{\mathrm{p}}}\colon\dot{\boldsymbol{\varepsilon}}^{\mathrm{p}} + \theta\frac{\partial^2 \Psi}{\partial\theta\partial\boldsymbol{a}}\colon\dot{\boldsymbol{a}},\\
\tilde{c}_v &= -\theta\frac{\partial^2 \Psi}{\partial\theta^2}.
\end{aligned}$$

Consider a yield surface of the form

$$\Phi := \Phi(\boldsymbol{\sigma}, -\boldsymbol{X}) = J(\boldsymbol{\sigma} - \boldsymbol{X}) - \sigma_Y, \qquad [2.89]$$

where

$$\begin{aligned}
J(\boldsymbol{\sigma} - \boldsymbol{X}) &= \sqrt{\frac{3}{2}[\boldsymbol{\sigma} - \boldsymbol{X}]'\colon[\boldsymbol{\sigma} - \boldsymbol{X}]'},\\
[\boldsymbol{\sigma} - \boldsymbol{X}]' &= \boldsymbol{\sigma} - \boldsymbol{X} - \frac{1}{3}\mathrm{tr}(\boldsymbol{\sigma} - \boldsymbol{X})\boldsymbol{I},
\end{aligned} \qquad [2.90]$$

and σ_Y is the elastic limit (positive constant). Equation [2.60] provides the evolution equations

$$\dot{\boldsymbol{\varepsilon}}^{\mathrm{p}} = \boldsymbol{\Lambda}\dot{p}, \qquad \dot{\boldsymbol{a}} = \dot{\boldsymbol{\varepsilon}}^{\mathrm{p}}, \qquad \boldsymbol{\Lambda} = \frac{3[\boldsymbol{\sigma} - \boldsymbol{X}]'}{2J(\boldsymbol{\sigma} - \boldsymbol{X})}, \qquad [2.91]$$

and the Kuhn–Tucker conditions

$$\Phi \leq 0, \quad \dot{p} \geq 0, \quad \Phi\dot{p} = 0. \qquad [2.92]$$

It is noted that the Lagrange multiplier p in the evolution law represents the accumulated plastic strain.

Following the discussion in section 2.5.1, the various work and energy rates are identified as

$$\begin{aligned} \dot{\mathrm{W}}^{\mathrm{r}}_{\mathrm{m}} &= \boldsymbol{\sigma}\colon[\dot{\boldsymbol{\varepsilon}} - \dot{\boldsymbol{\varepsilon}}^{\mathrm{p}}] + \boldsymbol{X}\colon\dot{\boldsymbol{a}}, \quad \dot{\mathrm{W}}^{\mathrm{d}}_{\mathrm{m}} = \boldsymbol{\sigma}\colon\dot{\boldsymbol{\varepsilon}}^{\mathrm{p}} - \boldsymbol{X}\colon\dot{\boldsymbol{a}}, \\ \dot{\mathrm{W}}^{\mathrm{r}}_{\mathrm{t}} &= \theta\boldsymbol{\alpha}\colon\dot{\boldsymbol{\sigma}} + c_0\dot{\theta}, \quad \dot{\mathcal{E}}^{\mathrm{r}} = \boldsymbol{\sigma}\colon[\dot{\boldsymbol{\varepsilon}} - \dot{\boldsymbol{\varepsilon}}^{\mathrm{p}}] + \boldsymbol{X}\colon\dot{\boldsymbol{a}} + \theta\boldsymbol{\alpha}\colon\dot{\boldsymbol{\sigma}} + c_0\dot{\theta}. \end{aligned} \qquad [2.93]$$

2.5.3.6. *Viscoplasticity*

Compared to the case of elastoplasticity, viscoplasticity presents the same characteristics except that the yield surface is substituted by a dissipation potential. A J_2 isotropic viscoplastic material with isotropic hardening is rather characterized by a family of equipotential surfaces, where the von Mises stress σ^{vM} is connected not only with the accumulated plastic strain p but also with its rate $\dot{p}$. In such cases the constitutive law arises as an extension of the corresponding elastoplastic case and has an equipotential surface of the form [LEM 02]

$$\dot{p} = \begin{cases} 0, & \sigma^{\mathrm{vM}} - F_p - \sigma_Y < 0, \\ \left[\dfrac{\sigma^{\mathrm{vM}} - F_p - \sigma_Y}{K_a}\right]^{N_a}, & \sigma^{\mathrm{vM}} - F_p - \sigma_Y \geq 0, \end{cases} \qquad [2.94]$$

where K_a and N_a are material constants. The above expression permits us to write a pseudo-yield surface of the form

$$\begin{aligned} &\Phi := \Phi(\boldsymbol{\sigma}, -F_p) = \sigma^{\mathrm{vM}} - F_p - \sigma_Y - K_a\dot{p}^{1/N_a} \leq 0, \\ &\quad \dot{p} \geq 0, \quad \Phi\dot{p} = 0. \end{aligned} \qquad [2.95]$$

The rest of the equations that describe the constitutive law, as well as the energy rates, are similar to those of a J_2 isotropic elastoplastic material with isotropic hardening.

2.5.3.7. *Isotropic continuum damage model*

The damage mechanism in materials causes a degradation in the mechanical properties and is observed in uniaxial tests as a reduction of the Young's modulus upon unloading (Figure 2.7).

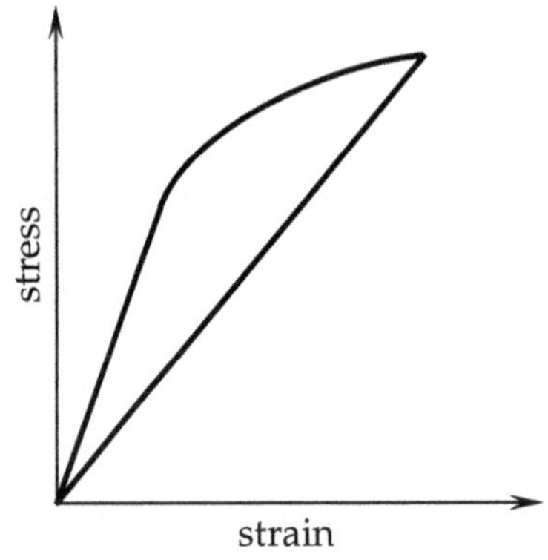

Figure 2.7. *Typical mechanical loading-unloading response of a damaged material*

A simple thermodynamic potential that accounts for isotropic damage can be expressed in the form

$$\begin{aligned}\Psi &= \Psi^{\mathrm{r}}(\boldsymbol{\varepsilon},\theta,d)\\ &= \frac{1}{2}\left[\boldsymbol{\varepsilon}-\boldsymbol{\alpha}[\theta-\theta_0]\right]:[1-d]\boldsymbol{C}:\left[\boldsymbol{\varepsilon}-\boldsymbol{\alpha}[\theta-\theta_0]\right]\\ &\quad +c_0\left[[\theta-\theta_0]-\theta\ln\left(\frac{\theta}{\theta_0}\right)\right]-\eta_0\theta+\mathcal{E}_0,\end{aligned} \qquad [2.96]$$

where d is the scalar damage parameter and $\boldsymbol{C}$, $\boldsymbol{\alpha}$, c_0, η_0, $\mathcal{E}_0$ are material constants similar to those of a thermoelastic material. According to the discussion of section 2.5.1, the use of $[2.44]_1$, $[2.46]_{1,2,3}$, $[2.33]_5$, [2.45] and $[2.42]_1$ yields

$$\boldsymbol{\sigma}=[1-d]\boldsymbol{C}:\left[\boldsymbol{\varepsilon}-\boldsymbol{\alpha}[\theta-\theta_0]\right],\quad \eta=\eta^{\mathrm{r}}=\boldsymbol{\alpha}:\boldsymbol{\sigma}+c_0\ln\left(\frac{\theta}{\theta_0}\right)+\eta_0,$$

$$\begin{aligned} \gamma_{\text{loc}} &= -\frac{\partial \Psi^{\text{r}}}{\partial d}\dot{d} = Y\dot{d}, \\ r &= -\theta\dot{\eta} + \gamma_{\text{loc}} = -c_0\dot{\theta} - \theta\boldsymbol{\alpha}{:}\dot{\boldsymbol{\sigma}} + \gamma_{\text{loc}}, \\ Y &= \frac{1}{2}\left[\boldsymbol{\varepsilon} - \boldsymbol{\alpha}[\theta - \theta_0]\right]{:}\boldsymbol{C}{:}\left[\boldsymbol{\varepsilon} - \boldsymbol{\alpha}[\theta - \theta_0]\right], \end{aligned} \quad [2.97]$$

while the energy equation takes the general form

$$\tilde{c}_v\dot{\theta} - \mathcal{Q} = \gamma_{\text{loc}} + \theta\frac{\partial^2\Psi}{\partial\theta\partial\boldsymbol{\varepsilon}}{:}\dot{\boldsymbol{\varepsilon}} + \theta\frac{\partial^2\Psi}{\partial\theta\partial d}\dot{d}, \quad \tilde{c}_v = -\theta\frac{\partial^2\Psi}{\partial\theta^2}.$$

From the above expressions, it is clear that, for $\boldsymbol{\sigma} = \mathbf{0}$ and $\theta = \theta_0$ the Helmholtz potential and the entropy obtain the values $\Psi = -\eta_0\theta_0 + \mathcal{E}_0$ and $\eta = \eta_0$ respectively, independently of the thermomechanical path that the material travels.

Like the case of elastoplasticity, a damage criterion of the form

$$\Phi := \Phi(Y) = Y - k(d), \quad [2.98]$$

can be postulated, where k is a known function of d. According to the discussion in section 2.5.2, the evolution damage law can be provided by the maximum dissipation principle as $\dot{d} = \lambda$, accompanied by the Kuhn–Tucker conditions

$$\Phi \leq 0, \quad \dot{d} \geq 0, \quad \Phi\dot{d} = 0. \quad [2.99]$$

Moreover, following the discussion of section 2.5.1, the various energy rate terms are identified as

$$\begin{aligned} \dot{\text{W}}^{\text{r}}_{\text{m}} &= \boldsymbol{\sigma}{:}\dot{\boldsymbol{\varepsilon}} - Y\dot{d}, \quad \dot{\text{W}}^{\text{d}}_{\text{m}} = Y\dot{d}, \quad \dot{\text{W}}^{\text{r}}_{\text{t}} = \theta\boldsymbol{\alpha}{:}\dot{\boldsymbol{\sigma}} + c_0\dot{\theta}, \\ \dot{\mathcal{E}}^{\text{r}} &= \boldsymbol{\sigma}{:}\dot{\boldsymbol{\varepsilon}} - Y\dot{d} + \theta\boldsymbol{\alpha}{:}\dot{\boldsymbol{\sigma}} + c_0\dot{\theta}. \end{aligned} \quad [2.100]$$

It is important to mention that the material state undergoes permanent changes once damage evolves. In this formalism, d can only increase and cannot return back to zero. Still, energetically speaking, the internal energy can be restored to its original value.

2.5.3.8. *Elastoplasticity with damage*

Combining the cases of the J_2 elastoplastic law with isotropic hardening and the isotropic continuum damage law, the Helmholtz free energy obtains the form

$$\begin{aligned}
\Psi\left(\boldsymbol{\varepsilon}, \theta, \boldsymbol{\varepsilon}^{\mathrm{p}}, d, p\right) &= \Psi^{\mathrm{r}}\left(\boldsymbol{\varepsilon}, \theta, \boldsymbol{\varepsilon}^{\mathrm{p}}, d\right) + \Psi^{\mathrm{ir}}\left(\theta, p\right), \\
\Psi^{\mathrm{r}}\left(\boldsymbol{\varepsilon}, \theta, \boldsymbol{\varepsilon}^{\mathrm{p}}, d\right) &= \frac{1-d}{2}\left[\boldsymbol{\varepsilon} - \boldsymbol{\varepsilon}^{\mathrm{p}} - \boldsymbol{\alpha}[\theta - \theta_0]\right] : \boldsymbol{C} : \left[\boldsymbol{\varepsilon} - \boldsymbol{\varepsilon}^{\mathrm{p}} - \boldsymbol{\alpha}[\theta - \theta_0]\right] \\
&\quad + c_0\left[[\theta - \theta_0] - \theta \ln\left(\frac{\theta}{\theta_0}\right)\right] - \eta_0\theta + \mathcal{E}_0, \\
\Psi^{\mathrm{ir}}\left(\theta, p\right) &= F(\theta, p).
\end{aligned}$$

Similar analysis with the previous examples leads to the following results:

$$\begin{aligned}
\boldsymbol{\sigma} &= [1-d]\boldsymbol{C} : \left[\boldsymbol{\varepsilon} - \boldsymbol{\varepsilon}^{\mathrm{p}} - \boldsymbol{\alpha}[\theta - \theta_0]\right], \\
\eta &= \eta^{\mathrm{r}} + \eta^{\mathrm{ir}} = \boldsymbol{\alpha} : \boldsymbol{\sigma} + c_0 \ln\left(\frac{\theta}{\theta_0}\right) + \eta_0 - F_\theta, \\
\gamma_{\mathrm{loc}} &= -\frac{\partial \Psi^{\mathrm{r}}}{\partial \boldsymbol{\varepsilon}^{\mathrm{p}}} : \dot{\boldsymbol{\varepsilon}}^{\mathrm{p}} - \frac{\partial \Psi^{\mathrm{r}}}{\partial d}\dot{d} - \frac{\partial \Psi^{\mathrm{ir}}}{\partial p}\dot{p} = \boldsymbol{\sigma} : \dot{\boldsymbol{\varepsilon}}^{\mathrm{p}} + Y\dot{d} - F_p\dot{p}, \\
Y &= \frac{1}{2}\left[\boldsymbol{\varepsilon} - \boldsymbol{\varepsilon}^{\mathrm{p}} - \boldsymbol{\alpha}[\theta - \theta_0]\right] : \boldsymbol{C} : \left[\boldsymbol{\varepsilon} - \boldsymbol{\varepsilon}^{\mathrm{p}} - \boldsymbol{\alpha}[\theta - \theta_0]\right], \\
r &= -\theta\dot{\eta} + \gamma_{\mathrm{loc}} = -c_0\dot{\theta} - \theta\boldsymbol{\alpha} : \dot{\boldsymbol{\sigma}} + \theta\dot{F}_\theta + \gamma_{\mathrm{loc}},
\end{aligned}$$

while the energy equation takes the general form

$$\begin{aligned}
\tilde{c}_v\dot{\theta} - \mathcal{Q} &= \gamma_{\mathrm{loc}} + \theta\frac{\partial^2\Psi}{\partial\theta\partial\boldsymbol{\varepsilon}} : \dot{\boldsymbol{\varepsilon}} + \theta\frac{\partial^2\Psi}{\partial\theta\partial\boldsymbol{\varepsilon}^{\mathrm{p}}} : \dot{\boldsymbol{\varepsilon}}^{\mathrm{p}} + \theta\frac{\partial^2\Psi}{\partial\theta\partial d}\dot{d} + \theta\frac{\partial^2\Psi}{\partial\theta\partial p}\dot{p}, \\
\tilde{c}_v &= -\theta\frac{\partial^2\Psi}{\partial\theta^2}.
\end{aligned}$$

In terms of internal variables, propagation criteria and evolution laws, we have to know if the two mechanisms (plasticity and damage) are developed independently or not. In the case of ductile damage, the plastic multiplier p is related to the evolution of d [LEM 02].

The energy and work rates for this material are written as

$$\dot{\mathrm{W}}^{\mathrm{r}}_{\mathrm{m}} = \boldsymbol{\sigma}\colon [\dot{\varepsilon} - \dot{\varepsilon}^{\mathrm{p}}] - Y\dot{d}, \quad \dot{\mathrm{W}}^{\mathrm{ir}}_{\mathrm{m}} = F_p\dot{p}, \quad \dot{\mathrm{W}}^{\mathrm{d}}_{\mathrm{m}} = \boldsymbol{\sigma}\colon\dot{\varepsilon}^{\mathrm{p}} + Y\dot{d} - F_p\dot{p},$$
$$\dot{\mathrm{W}}^{\mathrm{r}}_{\mathrm{t}} = \theta\boldsymbol{\alpha}\colon\dot{\boldsymbol{\sigma}} + c_0\dot{\theta}, \quad \dot{\mathrm{W}}^{\mathrm{ir}}_{\mathrm{t}} = -\theta\dot{F}_\theta,$$
$$\dot{\mathcal{E}}^{\mathrm{r}} = \boldsymbol{\sigma}\colon [\dot{\varepsilon} - \dot{\varepsilon}^{\mathrm{p}}] - Y\dot{d} + \theta\boldsymbol{\alpha}\colon\dot{\boldsymbol{\sigma}} + c_0\dot{\theta}, \quad \dot{\mathcal{E}}^{\mathrm{ir}} = F_p\dot{p} - \theta\dot{F}_\theta.$$

2.6. Parameter identification for an elastoplastic material

Identifying material parameters for constitutive laws of dissipative materials is usually a challenging problem that requires (1) appropriate experimental protocol that provides sufficient data for the calibration of the designed model and (2) advanced numerical tools for the parameter identification. With complicated constitutive laws it is sometimes impossible to identify a unique set of parameters, since an intercorrelation between them can occur, mainly due to the interplay between the various physical mechanisms that the proposed model attempts to simulate.

The scope of this section is to provide a typical identification strategy that we can follow in order to obtain material parameters for an isotropic elastoplastic material under fully coupled thermomechanical process. The described procedure can be seen as a useful strategy for parameter identification in constitutive laws with strong nonlinearities and thermomechanical couplings.

While the real behavior of the material is three-dimensional, isotropy permits us to simplify significantly the identification process, since, during uniaxial loading conditions, only one stress and one strain component are important. Thus, a simplified version of the Helmholtz free energy potential can be defined, which depends on the (scalar's) total strain ε, the temperature θ, the plastic strain ε^{p} and the accumulated plastic strain p,

$$\Psi\left(\varepsilon, \theta, \varepsilon^{\mathrm{p}}, p\right) = \frac{E}{2}\left[\varepsilon - \varepsilon^{\mathrm{p}} - \alpha[\theta - \theta_0]\right]^2 + c_0\left[[\theta - \theta_0] - \theta\ln\left(\frac{\theta}{\theta_0}\right)\right]$$
$$+ \frac{k_Y}{1/n_Y + 1} p^{1/n_Y + 1}. \qquad [2.101]$$

In the above expression, E is the Young's modulus, α is the thermal expansion coefficient and $c_0 = \rho c_p$ is the specific heat capacity at constant pressure multiplied by the material density. Moreover, a typical power-law type for the irrecoverable part of the Helmholtz energy has been chosen [LEM 02], in which k_Y and n_Y are plastic hardening-related material parameters. Following the usual procedure, the stress σ, the intrinsic dissipation $\gamma_{\rm loc}$ and the energetic term r can be identified as

$$\begin{aligned} \sigma &= E\left[\varepsilon - \varepsilon^{\rm p} - \alpha[\theta - \theta_0]\right], \\ \gamma_{\rm loc} &= \sigma\dot{\varepsilon}^{\rm p} - k_Y p^{1/n_Y}\dot{p}, \\ r &= -c_0\dot{\theta} - \theta\alpha\dot{\sigma} + \gamma_{\rm loc}. \end{aligned} \qquad [2.102]$$

The yield surface for this material is expressed as

$$\Phi = |\sigma| - \sigma_Y - k_Y p^{1/n_Y} \leq 0, \qquad [2.103]$$

leading to the evolution equation

$$\dot{\varepsilon}^{\rm p} = \frac{\sigma}{|\sigma|}\dot{p}, \qquad [2.104]$$

with $\dot{p} \geq 0$ and $\Phi\dot{p} = 0$. Combining all the above expressions, the fully coupled thermomechanical problem of a specimen under uniaxial tensile test is written as

$$\begin{aligned} \sigma &= E\left[\varepsilon - \varepsilon^{\rm p} - \alpha[\theta - \theta_0]\right] = \text{constant}, \quad \dot{\varepsilon}^{\rm p} \geq 0, \\ \sigma &= \sigma_Y + k_Y[\varepsilon^{\rm p}]^{1/n_Y} \quad \text{when } \dot{\varepsilon}^{\rm p} > 0 \text{ (nonlinear response)}, \\ c_0\dot{\theta} &= \kappa\, \text{div}\,(\text{grad}\theta) - \theta\alpha\dot{\sigma} + \sigma_Y\dot{\varepsilon}^{\rm p}, \end{aligned} \qquad [2.105]$$

where κ is the (isotropic) thermal conductivity. In the above equations, body forces and heat sources are ignored.

2.6.1. *Identification of mechanical parameters*

The mechanical properties of an isotropic elastoplastic material can be identified through uniaxial strain controlled tests in dogbone specimens under very slow rates, which (i) eliminate inertia effects and (ii) allow the material to stay at almost constant temperature, since there is sufficient time for the specimen to cool down when dissipation occurs. The actual material's displacement can be measured with an extensometer or with a digital image correlation system, while the force is recorded by the traction machine (Figure 2.8(a)). Isothermal conditions can be further ensured by installing the mechanical test machine into a climate chamber that controls the temperature during the experiments (Figure 2.8(b)).

a) b)

Figure 2.8. *a) Traction machine for uniaxial strain or stress controlled experiments (INSTRON Structural Testing Systems). b) Climate chamber. The actual specimen's displacement is monitored by a digital image correlation (DIC) system. For a color version of this figure, see www.iste.co.uk/chatzigeorgiou/thermomechanical.zip*

The Young's modulus E is identified, as usual, by the slope of the stress–strain curve during unloading (Figure 2.9). The plasticity-related properties σ_Y, k_Y and n_Y can be obtained through an optimization process like the nonlinear least squares method or the more advanced Levenberg–Marquardt method. Both techniques are based on the minimization of a cost function. A cost function appropriate for the specific constitutive law can be written as

$$Z = \sum_{r=1}^{N} \frac{\left[\varepsilon_{\text{exp}}^{\text{p}(r)} - \varepsilon_{\text{num}}^{\text{p}(r)}(\sigma)\right]^2}{\left[\varepsilon_{\text{exp}}^{\text{p}(r)}\right]^2}. \qquad [2.106]$$

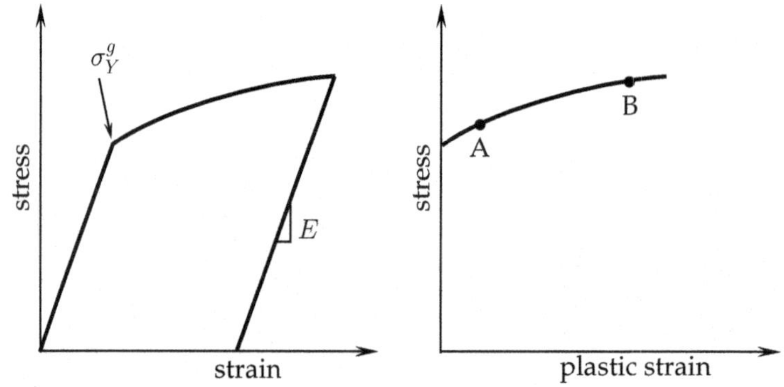

Figure 2.9. *Experimentally obtained stress–strain curve in uniaxial test and stress-plastic strain curve after removing the elastic strains. The points A and B allow the indentification of the guess values for k_Y^g and n_Y^g*

In the above expression, N is the number of available experimental points from a stress–strain curve in which plastic strain occurs (i.e. in the nonlinear region of the curve), $\varepsilon_{\text{exp}}^{\text{p}(r)} = \varepsilon^{(r)} - \sigma^{(r)}/E$ is the plastic strain obtained directly from the experimental results and

$$\varepsilon_{\text{num}}^{\text{p}(r)}(\sigma) = \left[\frac{\sigma^{(r)} - \sigma_Y}{k_Y}\right]^{n_Y} \qquad [2.107]$$

is the numerically obtained plastic strain, which is a function of the stress and plastic-related parameters. Due to the nonlinear nature of the optimization problem, the Newton–Raphson method can be employed. The latter requires the Jacobian of the cost function, as well as a set of initial guess values (σ_Y^g, k_Y^g, n_Y^g) for the plastic parameters. Especially, the initial guess values are essential for obtaining good parameters estimation at the end of the iterative process. A good guess value for the elastic limit, σ_Y^g, can be chosen directly from the stress–strain curve by observing approximately the stress level at which the response becomes nonlinear and the plastic strains are non-zero (Figure 2.9). After this choice, two points at the nonlinear regime of the curve are sufficient for obtaining a good estimation for k_Y^g and n_Y^g. Indeed, taking the natural logarithm of [2.107] yields

$$\frac{1}{n_Y}\ln\varepsilon^{\mathrm{p}} + \ln k_Y = \ln\left(\sigma - \sigma_Y\right). \qquad [2.108]$$

The last expression is linear with respect to $1/n_Y$ and $\ln k_Y$.

2.6.2. *Identification of thermal parameters*

The specific heat capacity c_p can be measured using a differential scanning calorimeter (DSC, Figure 2.10a). The DSC, with the help of a computer, records the evolution of the heat flow Q/t with respect to temperature (or time) in a material sample (Figure 2.10b). It is noted that

$$Q = \int [\mathcal{Q} \times \text{volume}]\mathrm{d}t. \qquad [2.109]$$

With proper techniques [HÖH 03], the differential scanning calorimeter also provides the evolution of the specific heat capacity with respect to temperature. In many materials, c_p varies slowly with respect to the temperature, but on certain occasions it presents important changes at narrow temperature ranges. The latter is the case, for instance, for polymers at the glass transition temperature or for shape-memory alloys during the phase transition between austenite and martensite.

With regard to thermal conductivity, there are several experimental devices that permit its proper evaluation through heat flow measurements. In a typical direct heat flow meter, a material sample of cross-section A is placed between a hot plate (heater) and a cold plate (coolant), as depicted in Figure 2.11. When steady-state conditions are achieved, the heat flow Q/t (Q is given by [2.109]) through the cross-section A is measured with the help of heat flux transducers between two points with distance L. Knowing the temperature drop $\Delta\theta =$

$\theta_1 - \theta_2$ between these two points, the thermal conductivity is computed by the relation

$$\kappa = \frac{Q/[tA]}{\Delta\theta/L}. \qquad [2.110]$$

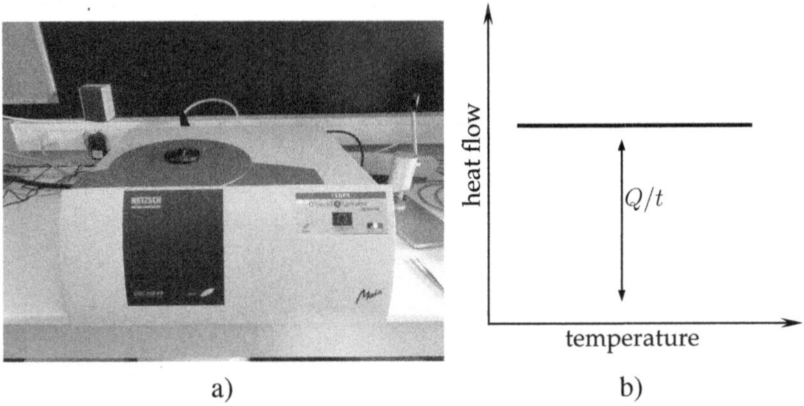

Figure 2.10. *a) Differential scanning calorimeter (NETZSCH DSC 200 F3 Maia) and b) heat flow versus temperature. For a color version of this figure, see www.iste.co.uk/chatzigeorgiou/thermomechanical.zip*

Usually, κ varies slowly with respect to temperature and it is frequently considered constant in solids.

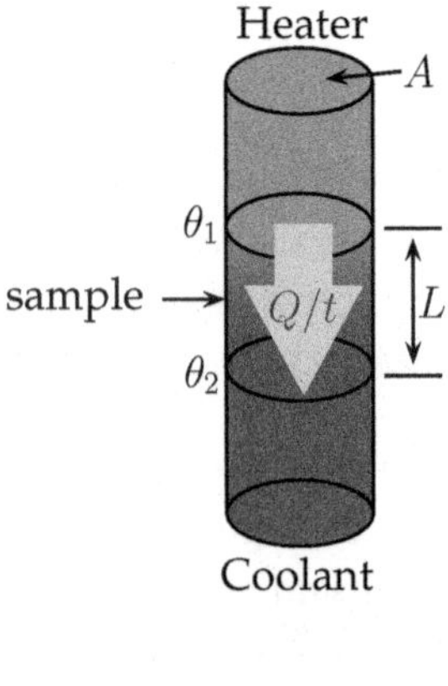

Figure 2.11. *Experimental measurement procedure for obtaining thermal conductivity. For a color version of this figure, see www.iste.co.uk/chatzigeorgiou/thermomechanical.zip*

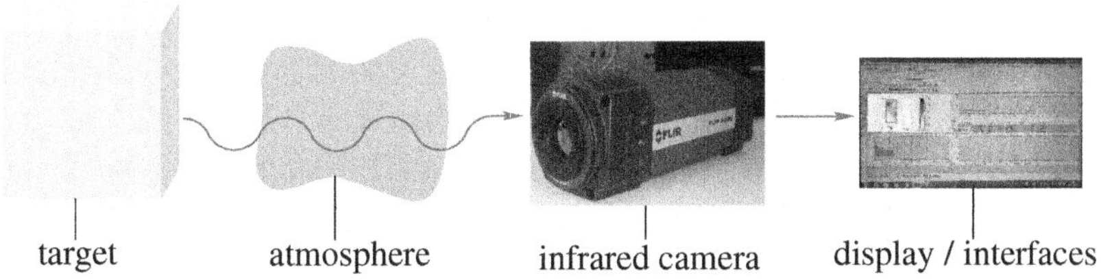

Figure 2.12. *Infrared measurement system (infrared camera: FLIR A325). For a color version of this figure, see www.iste.co.uk/chatzigeorgiou/thermomechanical.zip*

During fully coupled thermomechanical conditions, a specimen is placed into a MTS machine that monitors the mechanical loading and the surface temperature of the specimen is recorded with an infrared camera (Figure 2.12). To avoid the influence of the extensometer on the temperature field, a digital image correlation (DIC) system can be used to monitor the displacement field at the other side of the specimen (see Figure 2.8b).

While a typical traction test guarantees the uniaxiality of the mechanical conditions, the energy equation $[2.105]_4$ retains its 3-D form, causing many difficulties in identifying the real three-dimensional thermal conditions. For a dogbone specimen and under certain reasonable hypotheses, $[2.105]_4$ can be simplified and written in the 0-D form [CHR 89, BER 07]:

$$c_0 \left[\frac{\partial \bar{\theta}}{\partial t} + \frac{\bar{\theta}}{\tau_{\text{th}}} \right] = - \left[\bar{\theta} + \theta_0 \right] \alpha \dot{\sigma} + \sigma_Y \dot{\varepsilon}^{\text{p}}, \qquad [2.111]$$

where θ_0 is a reference temperature, τ_{th} is a characteristic time constant of thermal losses and $\bar{\theta} = \theta - \theta_0$ is the difference between actual and reference temperature. The 0-D expression considers as θ the average temperature at the specimen's surface (Figure 2.13a). The constant τ_{th} can be estimated in the following way: a stress-free specimen is initially heated rapidly up to a certain level. Then, the decrease in temperature is recorded with the infrared camera and the temperature difference versus time curve is obtained (Figure 2.13b). Under these conditions, [2.111] is reduced to

$$\frac{\partial \bar{\theta}}{\partial t} = -\frac{\bar{\theta}}{\tau_{\text{th}}} \quad \text{or} \quad \bar{\theta} = \bar{\theta}_{\text{i}} e^{-t/\tau_{\text{th}}}, \qquad [2.112]$$

where $\bar{\theta}_{\rm i}$ is the initial temperature rise. From the second expression of [2.112], it becomes evident that for $t = \tau_{\rm th}$, $\bar{\theta} = \bar{\theta}_i/e \approx 0.37\bar{\theta}_i$, or inversely, a line parallel to $0.37\bar{\theta}_i$ crosses the temperature difference versus time curve at a point which corresponds to t equal to $\tau_{\rm th}$ (Figure 2.13b). It can also be observed from the first expression of [2.112] that, at time $t = 0$, the slope $\dfrac{\partial\bar{\theta}}{\partial t}$ is equal to $-\tan\phi$, where ϕ is the angle indicated in Figure 2.13b. Thus, the tangent of the $\bar{\theta}$-t curve at $t = 0$ crosses the time axis at $t = \tau_{\rm th}$.

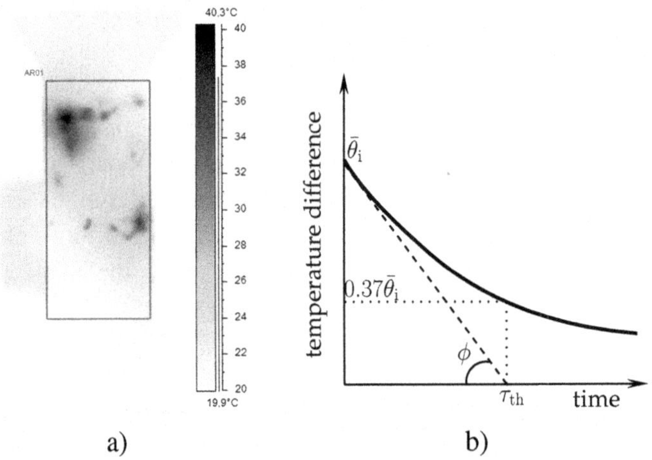

Figure 2.13. *a) Illustrative example of temperature profile at the surface of a specimen and b) temperature difference versus time in experiment for evaluation of τ_{th}. For a color version of this figure, see www.iste.co.uk/chatzigeorgiou/thermomechanical.zip*

2.6.3. *Identification of thermomechanical parameters*

The thermal expansion coefficient α can be measured with the help of a dilatometer (Figure 2.14a). In this apparatus, a sample of the examined material is heated under zero stress and the change in the sample's displacement as a function of the temperature is recorded (Figure 2.14b). Then, the thermal expansion coefficient is calculated by the formula

$$\alpha = \frac{\text{slope of displacement-temperature curve}}{\text{initial sample's size}}. \qquad [2.113]$$

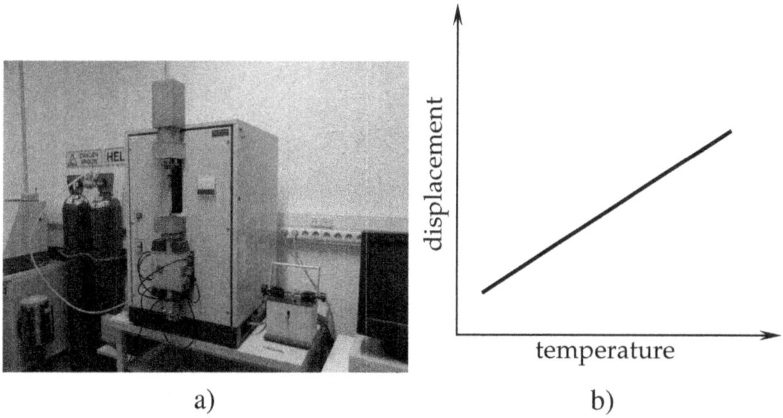

Figure 2.14. *a) Dilatometer (Adamel Lhomargy DT1000) and b) displacement versus temperature curve. For a color version of this figure, see www.iste.co.uk/chatzigeorgiou/thermomechanical.zip*

In many materials, α does not vary significantly at usual temperature ranges.

3

Computational Methods

While the laws of continuum mechanics, thermodynamic principles and deduced constitutive laws are sufficient to describe a material's behavior, most of the times they lead to systems of equations that are strongly nonlinear. In these cases, it is unavoidable to develop and utilize advanced computational tools to simulate the materials response. To this end, this chapter is devoted to the description of various numerical techniques that are robust and efficient. Most of the discussion focuses on the so-called "return mapping algorithms" which are frequently used in the studies of nonlinear dissipative materials. Extensive discussions about these algorithms can be found in the literature [SIM 98, QID 00].

3.1. Thermomechanical problem in weak form

A continuum body occupies the space $\mathcal{B}$ with volume V and bounded by a surface $\partial\mathcal{B}$, and at each point of this surface a normal unit vector $\boldsymbol{n}$ is defined. Each material point in $\mathcal{B}$ is assigned with a position vector $\boldsymbol{x}$. The body is subjected to mechanical and thermal loading conditions (Figure 3.1). Introducing $L^2(\mathcal{B})$ as the Lebesgue space of all functions $f(\boldsymbol{x})$ with $\left[\int_{\mathcal{B}} |f(\boldsymbol{x})|^2 \mathrm{d}\boldsymbol{x}\right]^{\frac{1}{2}} < \infty$, two spaces can be defined in this body:

– The space of displacement test functions $\boldsymbol{\omega}_{\varepsilon}$

$$\mathcal{V}_{\varepsilon} = \left\{\boldsymbol{\omega}_{\varepsilon} \mid \boldsymbol{\omega}_{\varepsilon}\, , \mathrm{grad}\boldsymbol{\omega}_{\varepsilon} \in L^2(\mathcal{B})\, ,\;\; \boldsymbol{\omega}_{\varepsilon} = \mathbf{0} \text{ on } \partial\mathcal{B}^{\mathrm{EB}}\right\},$$

with $\partial\mathcal{B}^{\mathrm{EB}} \subset \partial\mathcal{B}$ denoting the part of the boundary surface where the displacements vector $\boldsymbol{u}$ is prescribed.

– The space of temperature test functions ω_θ

$$\mathcal{V}_\theta = \left\{\omega_\theta \mid \omega_\theta \,, \mathrm{grad}\omega_\theta \in L^2(\mathcal{B}) \,,\ \ \omega_\theta = 0 \text{ on } \partial\mathcal{B}^{\mathrm{TB}}\right\},$$

with $\partial\mathcal{B}^{\mathrm{TB}} \subset \partial\mathcal{B}$ denoting the part of the boundary surface where the temperature θ is prescribed. The definitions of these two spaces are equivalent to the assertion that the displacement and temperature are continuous.

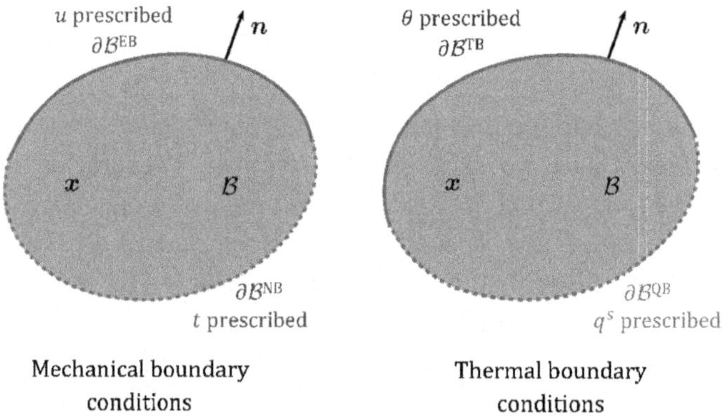

Figure 3.1. *Continuum body: mechanical and thermal boundary conditions. For a color version of this figure, see www.iste.co.uk/chatzigeorgiou/thermomechanical.zip*

The conservation of linear momentum [2.19] when considering small deformations is written as

$$\mathrm{div}\boldsymbol{\sigma} + \rho\boldsymbol{b} = \rho\frac{D\dot{\boldsymbol{u}}}{Dt}.$$

Multiplying the last expression with the displacement test functions and integrating over the volume of the body leads to

$$\int_{\mathcal{B}} \boldsymbol{\omega}_\varepsilon \cdot \mathrm{div}\boldsymbol{\sigma}\,\mathrm{d}v + \int_{\mathcal{B}} \boldsymbol{\omega}_\varepsilon \cdot \rho\boldsymbol{b}\,\mathrm{d}v = \int_{\mathcal{B}} \boldsymbol{\omega}_\varepsilon \cdot \rho\frac{D\dot{\boldsymbol{u}}}{Dt}\,\mathrm{d}v.$$

Using the chain rule yields

$$\int_{\mathcal{B}} \operatorname{div}(\boldsymbol{\omega}_\varepsilon \cdot \boldsymbol{\sigma})\,\mathrm{d}v - \int_{\mathcal{B}} \operatorname{grad}\boldsymbol{\omega}_\varepsilon : \boldsymbol{\sigma}\,\mathrm{d}v + \int_{\mathcal{B}} \boldsymbol{\omega}_\varepsilon \cdot \rho \boldsymbol{b}\,\mathrm{d}v$$
$$= \int_{\mathcal{B}} \boldsymbol{\omega}_\varepsilon \cdot \rho \frac{D\dot{\boldsymbol{u}}}{Dt}\,\mathrm{d}v.$$

Utilizing the divergence theorem [2.10], the weak form of the linear momentum equation is obtained as

$$\int_{\mathcal{B}} \boldsymbol{\omega}_\varepsilon \cdot \rho \frac{D\dot{\boldsymbol{u}}}{Dt}\,\mathrm{d}v + \int_{\mathcal{B}} \operatorname{grad}\boldsymbol{\omega}_\varepsilon : \boldsymbol{\sigma}\,\mathrm{d}v = \int_{\partial\mathcal{B}^{\mathrm{NB}}} \boldsymbol{\omega}_\varepsilon \cdot \boldsymbol{t}\,\mathrm{d}s + \int_{\mathcal{B}} \boldsymbol{\omega}_\varepsilon \cdot \rho \boldsymbol{b}\,\mathrm{d}v\,,$$
$$\forall \boldsymbol{\omega}_\varepsilon \in \mathcal{V}_\varepsilon\,. \qquad [3.1]$$

In the last expression, $\boldsymbol{t} = \boldsymbol{\sigma} \cdot \boldsymbol{n}$ is the external traction vector, acting on the surface $\partial\mathcal{B}^{\mathrm{NB}} \subset \partial\mathcal{B}$ that the overall body is subjected to.

The multiplication of the energy equation [2.30] with the temperature test function and the integration over the volume of the body leads to

$$\int_{\mathcal{B}} \omega_\theta \operatorname{div}\boldsymbol{q}\,\mathrm{d}v - \int_{\mathcal{B}} \omega_\theta \rho \mathcal{R}\,\mathrm{d}v = \int_{\mathcal{B}} \omega_\theta r\,\mathrm{d}v\,,$$

where it is recalled that $r = \boldsymbol{\sigma} : \dot{\boldsymbol{\varepsilon}} - \dot{E}$ is the difference between the rates of the mechanical work and the internal energy. Application of the chain rule in the last expression yields

$$\int_{\mathcal{B}} \operatorname{div}(\omega_\theta \boldsymbol{q})\,\mathrm{d}v - \int_{\mathcal{B}} \operatorname{grad}\omega_\theta \cdot \boldsymbol{q}\,\mathrm{d}v - \int_{\mathcal{B}} \omega_\theta \rho \mathcal{R}\,\mathrm{d}v = \int_{\mathcal{B}} \omega_\theta r\,\mathrm{d}v.$$

Finally, by utilizing the divergence theorem [2.10], the weak form of the energy equation is written as

$$-\int_{\mathcal{B}} \operatorname{grad}\omega_\theta \cdot \boldsymbol{q}\,\mathrm{d}v = \int_{\partial\mathcal{B}^{\mathrm{QB}}} \omega_\theta q^s\,\mathrm{d}s + \int_{\mathcal{B}} \omega_\theta \rho \mathcal{R}\,\mathrm{d}v + \int_{\mathcal{B}} \omega_\theta r\,\mathrm{d}v\,,$$
$$\forall \omega_\theta \in \mathcal{V}_\theta\,, \qquad [3.2]$$

where $\partial\mathcal{B}^{\mathrm{QB}} \subset \partial\mathcal{B}$ denotes the part of the boundary surface where heat fluxes $q^s = -\boldsymbol{q}\cdot\boldsymbol{n}$ are prescribed.

Both equations [3.1] and [3.2] are accompanied by the constitutive laws of the material. For the numerical implementation of the problem, appropriate thermomechanical tangent moduli are required. The following quantities need to be identified:

$$\boldsymbol{D}^{\varepsilon} = \frac{\partial \boldsymbol{\sigma}}{\partial \boldsymbol{\varepsilon}}, \quad \boldsymbol{D}^{\theta} = \frac{\partial \boldsymbol{\sigma}}{\partial \theta}, \quad \boldsymbol{R}^{\varepsilon} = \frac{\partial r}{\partial \boldsymbol{\varepsilon}}, \quad R^{\theta} = \frac{\partial r}{\partial \theta}.$$

With regard to the heat fluxes vector, it is customary to use the Fourier law

$$\boldsymbol{q} = -\boldsymbol{k}\cdot\mathrm{grad}\theta = -\boldsymbol{k}\cdot\boldsymbol{\nabla\theta}\,, \qquad [3.3]$$

where $\boldsymbol{k}$ denotes the thermal conductivity tensor, which in certain occasions is temperature dependent.

3.2. Computational procedure

The procedure described below is iterative and consists of three main steps (Figure 3.2):

1) In the first step, equations [3.1] and [3.2] are solved. The solution of this system provides the total strains $\boldsymbol{\varepsilon}$ and the temperature θ at all points of the structure. These computations are taken care of by the *FE algorithm*.

2) In the second step, at each material point, a return mapping algorithm is utilized [ORT 86, SIM 98]. This procedure is split into two parts:

– In the first part, the internal variables $\boldsymbol{\xi}$ of the material do not evolve and only generation of thermoelastic strains is considered (*thermoelastic prediction*). Thus, $\boldsymbol{\xi}$ is kept fixed, while the initial guess for the stress is computed directly by the $\boldsymbol{\varepsilon}$ and θ provided by the finite element analysis.

– In the second part, the error in the stress is corrected by identifying the actual change in the internal variables (*inelastic correction*). Thus, $\boldsymbol{\varepsilon}$ and θ are kept fixed, while $\boldsymbol{\xi}$ evolves.

With regard to the heat flux: since the thermal conductivity is, at most, a function of temperature, we can compute $\boldsymbol{q}$ by applying directly the Fourier law. These calculations are taken care of by the *constitutive law algorithm.*

3) In order to return and perform calculations in step 1, appropriate tangent moduli are required at each material point. These are computed by applying small, arbitrary perturbations in $\boldsymbol{\varepsilon}$, θ and $\boldsymbol{\nabla\theta}$, using the instantaneous response obtained from the second step. This part is taken care of by the *tangent moduli algorithm.*

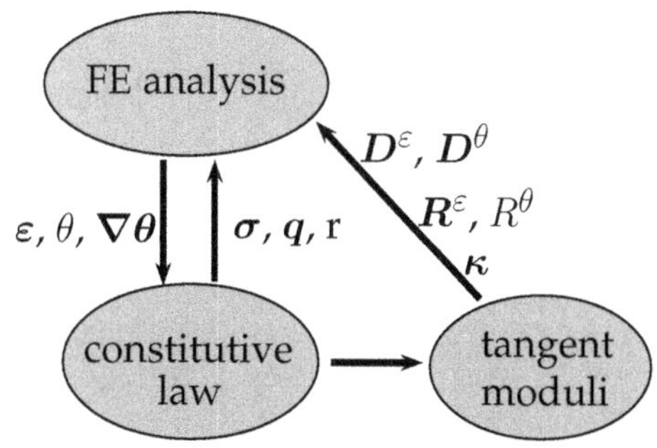

Figure 3.2. *Computational scheme*

3.2.1. *Preliminaries*

The proposed numerical scheme includes three types of increments of a quantity x: (i) increment in time, denoted by Δx, (ii) increment during the Newton–Raphson scheme in the FE calculations, denoted by $\eth x$ and (iii) increment during the Newton–Raphson scheme in the material constitutive law and during the identification of the tangent moduli, denoted by the symbol δx.

In a backward Euler fully implicit numerical scheme, the value of a given quantity x is updated from the previous time step n to the current $n+1$ per

$$x^{(n+1)} = x^{(n)} + \Delta x^{(n+1)}.$$

The time derivative of a quantity x is computed by the relation

$$\dot{x}^{(n+1)}\Delta t = \Delta x^{(n+1)}, \quad \text{with} \quad \Delta t = t^{(n+1)} - t^{(n)}.$$

Such an implicit relation is usually solved iteratively during the FE calculations, and the current value is updated from iteration m to iteration $m+1$ per

$$\begin{aligned} x^{(n+1)(m+1)} &= x^{(n+1)(m)} + \eth x^{(n+1)(m)} \quad \text{or} \\ \Delta x^{(n+1)(m+1)} &= \Delta x^{(n+1)(m)} + \eth x^{(n+1)(m)}. \end{aligned}$$

Additionally, the time derivative of a quantity x is numerically discretized according to the relation

$$\dot{x}^{(n+1)(m)}\Delta t = x^{(n+1)(m)} - x^{(n)} = \Delta x^{(n+1)(m)}.$$

The increment of a function $f(x, \dot{x})$ can be written as

$$\begin{aligned} \eth f^{(n+1)(m)} &= \left[\frac{\partial f}{\partial x}\right]^{(n+1)(m)} \eth x^{(n+1)(m)} \\ &+ \frac{1}{\Delta t}\left[\frac{\partial f}{\partial \dot{x}}\right]^{(n+1)(m)} \eth x^{(n+1)(m)}. \end{aligned}$$

The convergence in the FE analysis is reached once the increments $\eth$ are close to zero. When passing from the FE calculations to the material constitutive law, another iterative scheme is required. In this second scheme, the current value is updated from iteration k to iteration $k+1$ per

$$\begin{aligned} x^{(n+1)(m+1)(k+1)} &= x^{(n+1)(m+1)(k)} + \delta x^{(n+1)(m+1)(k)} \quad \text{or} \\ \Delta x^{(n+1)(m+1)(k+1)} &= \Delta x^{(n+1)(m+1)(k)} + \delta x^{(n+1)(m+1)(k)}. \end{aligned}$$

Additionally, the time derivative of a quantity x is numerically discretized according to the relation

$$\dot{x}^{(n+1)(m+1)(k)}\Delta t = x^{(n+1)(m+1)(k)} - x^{(n)} = \Delta x^{(n+1)(m+1)(k)}.$$

The increment of a function $f(x, \dot{x})$ can be written as

$$\delta f^{(n+1)(m+1)(k)} = \left[\frac{\partial f}{\partial x}\right]^{(n+1)(m+1)(k)} \delta x^{(n+1)(m+1)(k)} + \frac{1}{\Delta t}\left[\frac{\partial f}{\partial \dot{x}}\right]^{(n+1)(m+1)(k)} \delta x^{(n+1)(m+1)(k)}.$$

The convergence in the constitutive law is reached once the increments δ are close to zero.

3.2.2. *Finite element solution*

Table 3.1 describes the solution strategy required to solve the equilibrium (i.e. the conservation of linear momentum neglecting inertia forces) and the energy equation when a nonlinear material response is considered. This step is taken care of by the finite element solver.

3.2.3. *Constitutive law algorithm*

As it is depicted in Table 3.1, the FE analysis provides all the quantities at time step n and also the increments of strain, temperature and temperature gradient at the step $(n + 1)(m + 1)$. As it is already discussed, when passing to the constitutive law, the return mapping algorithm is split into two parts:

1) Initially, it is assumed that no evolution of the internal variables occurs, thus that the material behaves linearly. This allows us to consider a thermoelastic prediction of all the fields. In such prediction, the stress tensor, the quantity r and the heat flux vector are computed for

$$\begin{aligned}
\boldsymbol{\varepsilon}^{(n+1)(m+1)} &= \boldsymbol{\varepsilon}^{(n)} + \Delta\boldsymbol{\varepsilon}^{(n+1)(m+1)},\\
\theta^{(n+1)(m+1)} &= \theta^{(n)} + \Delta\theta^{(n+1)(m+1)},\\
\boldsymbol{\nabla\theta}^{(n+1)(m+1)} &= \boldsymbol{\nabla\theta}^{(n)} + \Delta\boldsymbol{\nabla\theta}^{(n+1)(m+1)},\\
\boldsymbol{\xi}^{(n+1)(m+1)} &= \boldsymbol{\xi}^{(n)}.
\end{aligned}$$

1. At time step n everything is known. At time step $n+1$ and iteration step $m=0$ set

$$\boldsymbol{u}^{(n+1)(0)}=\boldsymbol{u}^{(n)},\quad \boldsymbol{\varepsilon}^{(n+1)(0)}=\boldsymbol{\varepsilon}^{(n)},\quad \theta^{(n+1)(0)}=\theta^{(n)},\quad \boldsymbol{\sigma}^{(n+1)(0)}=\boldsymbol{\sigma}^{(n)},$$
$$\boldsymbol{q}^{(n+1)(0)}=\boldsymbol{q}^{(n)},\quad r^{(n+1)(0)}=r^{(n)},\quad \boldsymbol{\kappa}^{(n+1)(0)}=\boldsymbol{\kappa}^{(n)},\quad \boldsymbol{D}^{\varepsilon(n+1)(0)}=\boldsymbol{D}^{\varepsilon(n)},$$
$$\boldsymbol{D}^{\theta(n+1)(0)}=\boldsymbol{D}^{\theta(n)},\quad \boldsymbol{R}^{\varepsilon(n+1)(0)}=\boldsymbol{R}^{\varepsilon(n)},\quad R^{\theta(n+1)(0)}=R^{\theta(n)}.$$

2. Compute the virtual increment of displacements, $\eth\boldsymbol{u}^{(n+1)(m)}$, and the virtual increment of temperature, $\eth\theta^{(n+1)(m)}$, from the system of equations

$$\begin{aligned}
\int_{\mathcal{B}} \mathrm{grad}\boldsymbol{\omega}_\varepsilon : \tilde{\boldsymbol{\sigma}}^{(n+1)(m)}\,\mathrm{d}v &= \int_{\partial\mathcal{B}^{\mathrm{NB}}} \boldsymbol{\omega}_\varepsilon\cdot\boldsymbol{t}^{(n+1)}\,\mathrm{d}s + \int_{\mathcal{B}} \boldsymbol{\omega}_\varepsilon\cdot\rho\boldsymbol{b}^{(n+1)}\,\mathrm{d}v\,,\\
\int_{\mathcal{B}} \mathrm{grad}\omega_\theta\cdot\tilde{\boldsymbol{Q}}^{(n+1)(m)}\,\mathrm{d}v &= \int_{\partial\mathcal{B}^{\mathrm{QB}}} \omega_\theta q^{s(n+1)}\,\mathrm{d}s\\
&\quad + \int_{\mathcal{B}} \omega_\theta\left[\tilde{r}^{(n+1)(m)}+\rho\mathcal{R}^{(n+1)}\right]\mathrm{d}v\,,\\
\tilde{\boldsymbol{\sigma}}^{(n+1)(m)} &= \boldsymbol{D}^{\varepsilon(n+1)(m)}:\mathrm{grad}\eth\boldsymbol{u}^{(n+1)(m)}\\
&\quad +\boldsymbol{D}^{\theta(n+1)(m)}\eth\theta^{(n+1)(m)}+\boldsymbol{\sigma}^{(n+1)(m)},\\
\tilde{r}^{(n+1)(m)} &= \boldsymbol{R}^{\varepsilon(n+1)(m)}:\mathrm{grad}\eth\boldsymbol{u}^{(n+1)(m)}\\
&\quad +R^{\theta(n+1)(m)}\eth\theta^{(n+1)(m)}+r^{(n+1)(m)}.\\
\tilde{\boldsymbol{Q}}^{(n+1)(m)} &= \boldsymbol{k}^{(n+1)(m)}\cdot\mathrm{grad}\eth\theta^{(n+1)(m)}-\boldsymbol{q}^{(n+1)(m)},
\end{aligned}$$

3. Update the quantities

$$\boldsymbol{u}^{(n+1)(m+1)}=\boldsymbol{u}^{(n+1)(m)}+\eth\boldsymbol{u}^{(n+1)(m)},\quad \eth\boldsymbol{\varepsilon}^{(n+1)(m)}=\mathrm{grad}_{\mathrm{sym}}\eth\boldsymbol{u}^{(n+1)(m)},$$
$$\eth\boldsymbol{\nabla\theta}^{(n+1)(m)}=\mathrm{grad}\eth\theta^{(n+1)(m)},\quad \boldsymbol{\nabla\theta}^{(n+1)(m+1)}=\boldsymbol{\nabla\theta}^{(n+1)(m)}+\eth\boldsymbol{\nabla\theta}^{(n+1)(m)},$$
$$\boldsymbol{\varepsilon}^{(n+1)(m+1)}=\boldsymbol{\varepsilon}^{(n+1)(m)}+\eth\boldsymbol{\varepsilon}^{(n+1)(m)},\quad \theta^{(n+1)(m+1)}=\theta^{(n+1)(m)}+\eth\theta^{(n+1)(m)},$$

and the corresponding Δ increments.

4. At each material point:
   a) provide $\theta^{(n)}$, $\boldsymbol{\varepsilon}^{(n)}$, $\boldsymbol{\sigma}^{(n)}$, $\boldsymbol{\xi}^{(n)}$, $\Delta\theta^{(n+1)(m+1)}$, $\Delta\boldsymbol{\varepsilon}^{(n+1)(m+1)}$, $\Delta\boldsymbol{\nabla\theta}^{(n+1)(m+1)}$ in the **constitutive law algorithm** and compute $\boldsymbol{\sigma}^{(n+1)(m+1)}$, $\boldsymbol{\xi}^{(n+1)(m+1)}$, $r^{(n+1)(m+1)}$ and $\boldsymbol{q}^{(n+1)(m+1)}$.
   b) proceed to the **tangent moduli algorithm** and obtain $\boldsymbol{D}^{\varepsilon(n+1)(m+1)}$, $\boldsymbol{D}^{\theta(n+1)(m+1)}$, $\boldsymbol{R}^{\varepsilon(n+1)(m+1)}$, $R^{\theta(n+1)(m+1)}$ and $\boldsymbol{\kappa}^{(n+1)(m+1)}$.

5. If the convergence criterion is satisfied
   then update the values of all variables at time step $n+1$,
   set $n=n+1$ and return to step 1 for the next time increment,
   else set $m=m+1$ and return to step 2.

Table 3.1. *FE algorithm*

Then, the various criteria of the constitutive law are checked in order to clarify if the material remains thermoelastic or if a nonlinear mechanism is activated.

2) If the thermoelastic prediction indicates that the stress and the other fields do not respect the constitutive law restrictions (for instance in an elastoplastic material, the yield criterion is violated), then an inelastic correction is required. At this stage, the internal variables $\boldsymbol{\xi}$ evolve, while the strain, temperature and temperature gradient are fixed, i.e.

$$\delta\boldsymbol{\varepsilon}^{(n+1)(m+1)(k)} = 0, \quad \delta\theta^{(n+1)(m+1)(k)} = 0, \quad \delta\boldsymbol{\nabla\theta}^{(n+1)(m+1)(k)} = 0,$$

where k denotes the increment number during the inelastic correction step. In this context, the stresses and the internal variables can be expressed in the incremental forms

$$\begin{aligned}\boldsymbol{\sigma}^{(n+1)(m+1)(k+1)} &= \boldsymbol{\sigma}^{(n+1)(m+1)(k)} + \delta\boldsymbol{\sigma}^{(n+1)(m+1)(k)},\\ \boldsymbol{\xi}^{(n+1)(m+1)(k+1)} &= \boldsymbol{\xi}^{(n+1)(m+1)(k)} + \delta\boldsymbol{\xi}^{(n+1)(m+1)(k)}.\end{aligned}$$

At this point, every constitutive law restriction obeying the general relation of the form

$$\boldsymbol{\Phi}(\boldsymbol{\sigma}, \dot{\boldsymbol{\sigma}}, \boldsymbol{\varepsilon}, \dot{\boldsymbol{\varepsilon}}, \theta, \dot{\theta}, \boldsymbol{\xi}, \dot{\boldsymbol{\xi}}) = \mathbf{0},$$

is written in a linearized, incremental manner. Specifically:

a) The stress–strain relation and the internal variables evolution laws are expressed as

$$\delta\boldsymbol{\Phi}^{(n+1)(m+1)(k)} = \mathbf{0}. \qquad [3.4]$$

b) The inelastic criteria, like for instance the yield criterion in elastoplasticity, are written in the Newton–Raphson iterative form

$$\boldsymbol{\Phi}^{(n+1)(m+1)(k+1)} \simeq \boldsymbol{\Phi}^{(n+1)(m+1)(k)} + \delta\boldsymbol{\Phi}^{(n+1)(m+1)(k)} = \mathbf{0}. \qquad [3.5]$$

In both cases,

$$
\begin{aligned}
\delta\boldsymbol{\Phi}^{(n+1)(m+1)(k)} = & \left[\frac{\partial\boldsymbol{\Phi}}{\partial\boldsymbol{\sigma}}\right]^{(n+1)(m+1)(k)} : \delta\boldsymbol{\sigma}^{(n+1)(m+1)(k)} \\
& + \frac{1}{\Delta t}\left[\frac{\partial\boldsymbol{\Phi}}{\partial\dot{\boldsymbol{\sigma}}}\right]^{(n+1)(m+1)(k)} : \delta\boldsymbol{\sigma}^{(n+1)(m+1)(k)} \\
& + \left[\frac{\partial\boldsymbol{\Phi}}{\partial\boldsymbol{\xi}}\right]^{(n+1)(m+1)(k)} : \delta\boldsymbol{\xi}^{(n+1)(m+1)(k)} \\
& + \frac{1}{\Delta t}\left[\frac{\partial\boldsymbol{\Phi}}{\partial\dot{\boldsymbol{\xi}}}\right]^{(n+1)(m+1)(k)} : \delta\boldsymbol{\xi}^{(n+1)(m+1)(k)}. \qquad [3.6]
\end{aligned}
$$

During the inelastic correction step, the term $\boldsymbol{\Phi}^{(n+1)(m+1)(k)}$ of the inelastic criteria is not zero. Actually, the satisfaction of all inelastic criteria, within an acceptable tolerance, indicates the convergence of the numerical algorithm.

Finally, it is noted that the heat flux vector is computed directly, without iterative procedure, through the Fourier law

$$
\boldsymbol{q}^{(n+1)(m+1)} = -\boldsymbol{\kappa}^{(n+1)(m+1)} \cdot \boldsymbol{\nabla}\boldsymbol{\theta}^{(n+1)(m+1)}.
$$

The general steps for the constitutive law implementation are described in Table 3.2.

NOTE.– In several occasions, the evolution equations of the internal variables are complicated, like for instance the plastic strain rate evolution law in classical plasticity, where the deviatoric part of stress and its norm appear in the equation. Direct implementation of [3.4] and [3.6] in such evolution equations causes significant complexity on the linearized system, and sometimes requires the introduction of appropriate residuals (like in the closest point projection scheme [SIM 98]). A strategy to avoid this complexity is the partial implementation of [3.4] and [3.6] in certain equations (like in the convex cutting plane scheme [SIM 98]), which does not guarantee unconditional stability of the algorithm, but can provide very efficient scheme with low computational cost. An extensive discussion about this issue is provided later in the chapter.

1. At time step n everything is known. At time step $n+1$, $\Delta\boldsymbol{\varepsilon}^{(n+1)(m+1)}$, $\Delta\theta^{(n+1)(m+1)}$ and $\Delta\boldsymbol{\nabla}\boldsymbol{\theta}^{(n+1)(m+1)}$ are known.
2. At iteration step $k=0$ set
$\boldsymbol{\varepsilon}^{(n+1)(m+1)} = \boldsymbol{\varepsilon}^{(n)} + \Delta\boldsymbol{\varepsilon}^{(n+1)(m+1)}, \qquad \theta^{(n+1)(m+1)} = \theta^{(n)} + \Delta\theta^{(n+1)(m+1)},$
$\boldsymbol{\xi}^{(n+1)(m+1)(0)} = \boldsymbol{\xi}^{(n)}, \qquad \boldsymbol{\nabla}\boldsymbol{\theta}^{(n+1)(m+1)} = \boldsymbol{\nabla}\boldsymbol{\theta}^{(n)} + \Delta\boldsymbol{\nabla}\boldsymbol{\theta}^{(n+1)(m+1)},$
and compute $\boldsymbol{\sigma}^{(n+1)(m+1)(0)}$.
3. If the material is in thermoelastic state then go to step 6, else continue to step 4.
4. If the material is in inelastic state, then proceed iteratively using the Newton-Raphson scheme in order to calculate the increment of $\delta\boldsymbol{\xi}^{(n+1)(m+1)(k)}$. Update the internal variables $\boldsymbol{\xi}^{(n+1)(m+1)(k+1)}$ and calculate the corrected stress $\boldsymbol{\sigma}^{(n+1)(m+1)(k+1)}$.
5. If the inelasticity criteria are less than a tolerance then go to step 6, else set $k=k+1$ and return to step 4.
6. Compute the $r^{(n+1)(m+1)}$.
7. Compute the thermal conductivity $\boldsymbol{\kappa}^{(n+1)(m+1)}$ at the temperature $\theta^{(n+1)(m+1)}$ and update the heat flux using the Fourier law $\boldsymbol{q}^{(n+1)(m+1)} = -\boldsymbol{\kappa}^{(n+1)(m+1)} \cdot \boldsymbol{\nabla}\boldsymbol{\theta}^{(n+1)(m+1)}$.

Table 3.2. *Constitutive law algorithm*

3.2.4. *Tangent moduli algorithm*

As was depicted in Table 3.1, the finite element analysis requires the computation of appropriate tangent moduli in order to proceed to the next FE iteration step. The computations for the evaluation of the tangent moduli are performed at a specific time step. In this regard, any increment that appears in the calculations is in the same spirit with the virtual $\eth\{\bullet\}$ increments. The main differences with the *constitutive law algorithm* procedure are that now (i) the virtual increments of total strain and temperature are considered non-zero, in the sense of arbitrary perturbations and (ii) there is no iterative process, since the actual values of every variable are given at time step $n+1$ and increment $m+1$.

To proceed with the numerical calculations, any relation

$$\boldsymbol{\Phi}(\boldsymbol{\sigma}, \dot{\boldsymbol{\sigma}}, \boldsymbol{\varepsilon}, \dot{\boldsymbol{\varepsilon}}, \theta, \dot{\theta}, \boldsymbol{\xi}, \dot{\boldsymbol{\xi}}) = \mathbf{0},$$

can be linearized and re-expressed incrementally as

$$\eth\, \mathbf{\Phi}^{(n+1)(m+1)} = \mathbf{0},$$

or, in expanded form,

$$\begin{aligned}
&\left[\frac{\partial \mathbf{\Phi}}{\partial \boldsymbol{\sigma}}\right]^{(n+1)(m+1)} : \eth\boldsymbol{\sigma}^{(n+1)(m+1)} + \left[\frac{\partial \mathbf{\Phi}}{\partial \dot{\boldsymbol{\sigma}}}\right]^{(n+1)(m+1)} : \frac{\eth\boldsymbol{\sigma}^{(n+1)(m+1)}}{\Delta t} \\
&+ \left[\frac{\partial \mathbf{\Phi}}{\partial \boldsymbol{\varepsilon}}\right]^{(n+1)(m+1)} : \eth\boldsymbol{\varepsilon}^{(n+1)(m+1)} + \left[\frac{\partial \mathbf{\Phi}}{\partial \dot{\boldsymbol{\varepsilon}}}\right]^{(n+1)(m+1)} : \frac{\eth\boldsymbol{\varepsilon}^{(n+1)(m+1)}}{\Delta t} \\
&+ \left[\frac{\partial \mathbf{\Phi}}{\partial \theta}\right]^{(n+1)(m+1)} \eth\theta^{(n+1)(m+1)} + \left[\frac{\partial \mathbf{\Phi}}{\partial \dot{\theta}}\right]^{(n+1)(m+1)} \frac{\eth\theta^{(n+1)(m+1)}}{\Delta t} \\
&+ \left[\frac{\partial \mathbf{\Phi}}{\partial \boldsymbol{\xi}}\right]^{(n+1)(m+1)} : \eth\boldsymbol{\xi}^{(n+1)(m+1)} + \left[\frac{\partial \mathbf{\Phi}}{\partial \dot{\boldsymbol{\xi}}}\right]^{(n+1)(m+1)} : \frac{\eth\boldsymbol{\xi}^{(n+1)(m+1)}}{\Delta t} \\
&= \mathbf{0}.
\end{aligned}$$

This linearization holds for the inelasticity criteria too. The final goal of this process is to identify the relations

$$\begin{aligned}
\eth\boldsymbol{\sigma}^{(n+1)(m+1)} &= \boldsymbol{D}^{\varepsilon(n+1)(m+1)} : \eth\boldsymbol{\varepsilon}^{(n+1)(m+1)} \\
&\quad + \boldsymbol{D}^{\theta(n+1)(m+1)} \eth\theta^{(n+1)(m+1)}, \\
\eth\boldsymbol{r}^{(n+1)(m+1)} &= \boldsymbol{R}^{\varepsilon(n+1)(m+1)} : \eth\boldsymbol{\varepsilon}^{(n+1)(m+1)} \\
&\quad + \boldsymbol{R}^{\theta(n+1)(m+1)} \eth\theta^{(n+1)(m+1)}.
\end{aligned}$$

Usually, the thermal conductivity varies slowly with temperature, thus the $m+1$ increment of the heat flux is approximately given by

$$\eth\boldsymbol{q}^{(n+1)(m+1)} \simeq -\boldsymbol{\kappa}^{(n+1)(m+1)} \cdot \eth\boldsymbol{\nabla}\boldsymbol{\theta}^{(n+1)(m+1)}.$$

The numerical algorithm for the computation of the thermomechanical tangent moduli is described in Table 3.3.

1. At time step $n+1$ and iteration step $m+1$, all the necessary variables and their Δ increments are known from the **constitutive law algorithm**.
2. Compute the thermomechanical tangent moduli $\boldsymbol{D}^{\varepsilon(n+1)(m+1)}$, $\boldsymbol{D}^{\theta(n+1)(m+1)}$, $\boldsymbol{R}^{\varepsilon(n+1)(m+1)}$, $R^{\theta(n+1)(m+1)}$ and $\boldsymbol{\kappa}^{(n+1)(m+1)}$.

Table 3.3. *Tangent moduli algorithm*

In order to avoid lengthy expressions, in the following, the iteration numbers will be omitted. Any quantity x denotes the $x^{(n+1)(m+1)(k)}$, the increment Δx denotes the $\Delta x^{(n+1)(m+1)(k)}$, the increment δx denotes the $\delta x^{(n+1)(m+1)(k)}$ and the increment $\eth x$ denotes the $\eth x^{(n+1)(m+1)}$.

3.3. General algorithm in thermoelasticity

For a thermoelastic material, the set of equations that express the material constitutive law can be written in the general, discretized form (see section 2.5.3)

$$\boldsymbol{\sigma} = \boldsymbol{C}\colon [\boldsymbol{\varepsilon} - \boldsymbol{\alpha}[\theta - \theta_0]], \quad \eta = \boldsymbol{\alpha}\colon\boldsymbol{\sigma} + c_0 \ln\left(\frac{\theta}{\theta_0}\right) + \eta_0,$$
$$r = -\theta\frac{\Delta\eta}{\Delta t}. \qquad [3.7]$$

According to the constitutive law algorithm methodology, since both strain and temperature are known from the FE analysis, equation $[3.7]_1$ is sufficient to provide the new value of stress without the necessity of an iteration step. The thermomechanical tangent moduli $\boldsymbol{D}^{\varepsilon}$ and $\boldsymbol{D}^{\theta}$ are obtained from the stress–strain relation, written in incremental form as

$$\eth\boldsymbol{\sigma} = \boldsymbol{C}\colon [\eth\boldsymbol{\varepsilon} - \boldsymbol{\alpha}\eth\theta] \quad \text{(tangent moduli algorithm)}, \qquad [3.8]$$

which allows us to write

$$\eth\boldsymbol{\sigma} = \boldsymbol{D}^{\varepsilon}\colon\eth\boldsymbol{\varepsilon} + \boldsymbol{D}^{\theta}\eth\theta, \quad \boldsymbol{D}^{\varepsilon} = \boldsymbol{C}, \quad \boldsymbol{D}^{\theta} = -\boldsymbol{D}^{\varepsilon}\colon\boldsymbol{\alpha}. \qquad [3.9]$$

However, the incremental form of the energetic term r is written as

$$\eth r = -\eth\theta\frac{\Delta\eta}{\Delta t} - \frac{\theta}{\Delta t}\eth\eta \quad \text{(tangent moduli algorithm).} \qquad [3.10]$$

Using equations [3.7] and [3.9], we obtain

$$\begin{aligned} \eth r &= \boldsymbol{R}^{\varepsilon}\!:\!\eth\boldsymbol{\varepsilon} + R^{\theta}\eth\theta, \\ \boldsymbol{R}^{\varepsilon} &= -\frac{\theta}{\Delta t}\boldsymbol{\alpha}\!:\!\boldsymbol{D}^{\varepsilon}, \\ R^{\theta} &= -\frac{1}{\Delta t}\left[\Delta\eta + c_0 + \theta\boldsymbol{\alpha}\!:\!\boldsymbol{D}^{\theta}\right]. \end{aligned} \qquad [3.11]$$

3.4. General algorithms in elastoplasticity

For an elastoplastic material with isotropic hardening, the set of equations that express the material constitutive law can be written in the general, discretized form (see section 2.5.3)

$$\begin{aligned} \boldsymbol{\sigma} &= \boldsymbol{C}\!:\left[\boldsymbol{\varepsilon} - \boldsymbol{\varepsilon}^{\mathrm{p}} - \boldsymbol{\alpha}[\theta - \theta_0]\right], \\ \eta &= \boldsymbol{\alpha}\!:\!\boldsymbol{\sigma} + c_0 \ln\left(\frac{\theta}{\theta_0}\right) + \eta_0 - F_{\theta}, \\ &\Delta\boldsymbol{\varepsilon}^{\mathrm{p}} = \boldsymbol{\Lambda}(\boldsymbol{\sigma})\Delta p, \\ &\Phi(\boldsymbol{\sigma}, p, \theta) \leq 0, \quad \Delta p \geq 0, \quad \Phi\Delta p = 0, \\ \gamma_{\mathrm{loc}} &= \boldsymbol{\sigma}\!:\frac{\Delta\boldsymbol{\varepsilon}^{\mathrm{p}}}{\Delta t} - F_p\frac{\Delta p}{\Delta t}, \\ r &= -\theta\frac{\Delta\eta}{\Delta t} + \gamma_{\mathrm{loc}}, \end{aligned} \qquad [3.12]$$

where

$$F_p = \frac{\partial F}{\partial p}, \quad F_{\theta} = \frac{\partial F}{\partial \theta}, \quad F_{pp} = \frac{\partial^2 F}{\partial p^2}, \quad F_{p\theta} = \frac{\partial^2 F}{\partial p \partial \theta}, \quad F_{\theta\theta} = \frac{\partial^2 F}{\partial \theta^2}. \qquad [3.13]$$

In the above expressions, the backward Euler time integration scheme has been utilized. Using the framework presented in the previous section, the stress–strain relation is written in incremental form as

$$\begin{aligned} \delta\boldsymbol{\sigma} &= -\boldsymbol{C}:\delta\boldsymbol{\varepsilon}^{\mathrm{p}} \quad \text{(constitutive law algorithm)}, \\ \eth\boldsymbol{\sigma} &= \boldsymbol{C}:\left[\eth\boldsymbol{\varepsilon} - \eth\boldsymbol{\varepsilon}^{\mathrm{p}} - \alpha\eth\theta\right] \quad \text{(tangent moduli algorithm)}. \end{aligned} \qquad [3.14]$$

For the yield criterion, the linearization gives

$$\begin{aligned} \delta\Phi &= \frac{\partial\Phi}{\partial\boldsymbol{\sigma}}:\delta\boldsymbol{\sigma} + \frac{\partial\Phi}{\partial p}\delta p \quad \text{(constitutive law algorithm)}, \\ \eth\Phi &= \frac{\partial\Phi}{\partial\boldsymbol{\sigma}}:\eth\boldsymbol{\sigma} + \frac{\partial\Phi}{\partial p}\eth p + \frac{\partial\Phi}{\partial\theta}\eth\theta \quad \text{(tangent moduli algorithm)}. \end{aligned} \qquad [3.15]$$

The incremental forms of the local dissipation and the energetic term r are also written as

$$\begin{aligned} \eth\gamma_{\mathrm{loc}} &= \eth\boldsymbol{\sigma}:\frac{\Delta\boldsymbol{\varepsilon}^{\mathrm{p}}}{\Delta t} + \boldsymbol{\sigma}:\frac{\eth\boldsymbol{\varepsilon}^{\mathrm{p}}}{\Delta t} - F_{pp}\frac{\Delta p}{\Delta t}\eth p \\ &\quad - F_{p\theta}\frac{\Delta p}{\Delta t}\eth\theta - F_{p}\frac{\eth p}{\Delta t} \quad \text{(tangent moduli algorithm)}, \\ \eth r &= -\eth\theta\frac{\Delta\eta}{\Delta t} - \frac{\theta}{\Delta t}\eth\eta + \eth\gamma_{\mathrm{loc}} \quad \text{(tangent moduli algorithm)}. \end{aligned} \qquad [3.16]$$

If no plastic strain occurs during the increment under consideration, then the material behaves elastically. In this case $\Delta\boldsymbol{\varepsilon}^{\mathrm{p}} = \mathbf{0}$, $\Delta p = 0$ and all the necessary quantities are calculated in a similar manner with the one described for the thermoelastic materials. Attention is required to not neglect certain terms (if they are nonzero) arising from the derivatives of F.

If plastic strain does occur during the increment under consideration, then an iterative scheme is necessary in order to obtain $\Delta\boldsymbol{\varepsilon}^{\mathrm{p}}$ and Δp. In this chapter two ways of linearization are considered [SIM 98] and are discussed in the sections that follow.

3.4.1. *Closest point projection*

In the closest point projection algorithm, the discrete evolution equation is written as

$$\boldsymbol{Rs} = -\Delta\boldsymbol{\varepsilon}^{\mathrm{p}} + \boldsymbol{\Lambda}\Delta p, \qquad [3.17]$$

where $\boldsymbol{Rs}$ denotes the residual from the numerical mismatch during the computations. In the constitutive law algorithm, the incremental form of the above expression is written as

$$\boldsymbol{Rs} + \delta\boldsymbol{Rs} = \boldsymbol{0}, \qquad \delta\boldsymbol{Rs} = -\delta\boldsymbol{\varepsilon}^{\mathrm{p}} + \boldsymbol{\Lambda}\delta p + \Delta p\frac{\partial\boldsymbol{\Lambda}}{\partial\boldsymbol{\sigma}}:\delta\boldsymbol{\sigma}. \qquad [3.18]$$

As it is already mentioned in the previous section, during the inelastic correction step for obtaining the stress, the total strain and the temperature at the current time step are considered constant such that $\delta\boldsymbol{\varepsilon} = \boldsymbol{0}$ and $\delta\theta = 0$, respectively. Thus, the increment of stress is expressed as

$$\delta\boldsymbol{\sigma} = -\boldsymbol{C}:\delta\boldsymbol{\varepsilon}^{\mathrm{p}}, \quad \text{or} \quad \boldsymbol{C}^{-1}:\delta\boldsymbol{\sigma} = -\boldsymbol{Rs} - \boldsymbol{\Lambda}\delta p - \Delta p\frac{\partial\boldsymbol{\Lambda}}{\partial\boldsymbol{\sigma}}:\delta\boldsymbol{\sigma}. \qquad [3.19]$$

However, the yield criterion leads to the equation

$$\Phi + \delta\Phi = \Phi + \frac{\partial\Phi}{\partial\boldsymbol{\sigma}}:\delta\boldsymbol{\sigma} + \frac{\partial\Phi}{\partial p}\delta p = 0. \qquad [3.20]$$

Combining equations [3.19] and [3.20] leads to

$$\delta p = \frac{1}{A^{\mathrm{p}}}\left[\Phi - \frac{\partial\boldsymbol{\Lambda}}{\partial\boldsymbol{\sigma}}:\boldsymbol{\mathcal{C}}:\boldsymbol{Rs}\right], \qquad \boldsymbol{\mathcal{C}} = \left[\boldsymbol{C}^{-1} + \Delta p\frac{\partial\boldsymbol{\Lambda}}{\partial\boldsymbol{\sigma}}\right]^{-1},$$

$$A^{\mathrm{p}} = \frac{\partial\Phi}{\partial\boldsymbol{\sigma}}:\boldsymbol{\mathcal{C}}:\boldsymbol{\Lambda} - \frac{\partial\Phi}{\partial p}. \qquad [3.21]$$

In addition, equations [3.18] and $[3.19]_1$ provide the plastic strains increment as a function of the increment of p per

$$\delta\boldsymbol{\varepsilon}^{\mathrm{p}} = \boldsymbol{C}^{-1}:\boldsymbol{\mathcal{C}}:\left[\boldsymbol{Rs} + \boldsymbol{\Lambda}\delta p\right]. \qquad [3.22]$$

In the iterative scheme, the scalar p and the plastic strains are updated using their increments, i.e. $p^{\text{new}} = p + \delta p$ and $\boldsymbol{\varepsilon}^{\text{p new}} = \boldsymbol{\varepsilon}^{\text{p}} + \delta\boldsymbol{\varepsilon}^{\text{p}}$, while the new stress is computed from the analytical expression $[3.12]_1$. The convergence of the Newton–Raphson and the actual solution of the nonlinear problem are achieved when the absolute value of Φ and the norm of $\boldsymbol{Rs}$ are less than a tolerance.

With regard to the thermomechanical tangent moduli, the closest point projection approach can provide the consistent forms of them. When plastic strain occurs, the linearized form of the evolution equation $[3.18]_2$, the stress–strain relation [3.14] and the expanded yield criterion [3.15] provide the following linear system of equations (at the FE iteration increment)

$$\begin{aligned} \eth\boldsymbol{\sigma} &= \boldsymbol{C}\colon\left[\eth\boldsymbol{\varepsilon} - \eth\boldsymbol{\varepsilon}^{\text{p}} - \boldsymbol{\alpha}\eth\theta\right], \\ \eth\boldsymbol{Rs} = \boldsymbol{0} \quad &\Rightarrow \quad \eth\boldsymbol{\varepsilon}^{\text{p}} = \boldsymbol{\Lambda}\eth p + \Delta p\frac{\partial\boldsymbol{\Lambda}}{\partial\boldsymbol{\sigma}}\colon\eth\boldsymbol{\sigma}, \\ \eth\Phi = 0 \quad &\Rightarrow \quad \frac{\partial\Phi}{\partial\boldsymbol{\sigma}}\colon\eth\boldsymbol{\sigma} + \frac{\partial\Phi}{\partial p}\eth p + \frac{\partial\Phi}{\partial\theta}\eth\theta = 0. \end{aligned} \qquad [3.23]$$

This system leads to

$$\begin{aligned} &\eth p = \boldsymbol{\mathcal{P}}^{\varepsilon}\colon\eth\boldsymbol{\varepsilon} + \mathcal{P}^{\theta}\eth\theta, \qquad \eth\boldsymbol{\sigma} = \boldsymbol{D}^{\varepsilon}\colon\eth\boldsymbol{\varepsilon} + \boldsymbol{D}^{\theta}\eth\theta, \\ &\boldsymbol{\mathcal{P}}^{\varepsilon} = \frac{1}{A^{\text{p}}}\frac{\partial\Phi}{\partial\boldsymbol{\sigma}}\colon\boldsymbol{\mathcal{C}}, \quad \mathcal{P}^{\theta} = -\boldsymbol{\mathcal{P}}^{\varepsilon}\colon\boldsymbol{\alpha} + \frac{1}{A^{\text{p}}}\frac{\partial\Phi}{\partial\theta}, \\ &\boldsymbol{D}^{\varepsilon} = \boldsymbol{\mathcal{C}} - [\boldsymbol{\mathcal{C}}\colon\boldsymbol{\Lambda}]\otimes\boldsymbol{\mathcal{P}}^{\varepsilon}, \quad \boldsymbol{D}^{\theta} = -\boldsymbol{\mathcal{C}}\colon\boldsymbol{\alpha} - \boldsymbol{\mathcal{C}}\colon\boldsymbol{\Lambda}\mathcal{P}^{\theta}, \\ &\boldsymbol{\mathcal{C}} = \left[\boldsymbol{C}^{-1} + \Delta p\frac{\partial\boldsymbol{\Lambda}}{\partial\boldsymbol{\sigma}}\right]^{-1}, \quad A^{\text{p}} = \frac{\partial\Phi}{\partial\boldsymbol{\sigma}}\colon\boldsymbol{\mathcal{C}}\colon\boldsymbol{\Lambda} - \frac{\partial\Phi}{\partial p}. \end{aligned} \qquad [3.24]$$

For the thermal tangent moduli, equation $[3.16]_1$ can be written as

$$\eth\gamma_{\text{loc}} = \boldsymbol{\Gamma}^{\varepsilon}\colon\eth\boldsymbol{\varepsilon} + \Gamma^{\theta}\eth\theta, \qquad [3.25]$$

with

$$\boldsymbol{\Gamma}^{\varepsilon} = \frac{\widetilde{\Delta\boldsymbol{\varepsilon}^{\text{p}}}}{\Delta t}\colon\boldsymbol{D}^{\varepsilon} + \frac{1}{\Delta t}\left[\boldsymbol{\sigma}\colon\boldsymbol{\Lambda} - F_{pp}\Delta p - F_p\right]\boldsymbol{\mathcal{P}}^{\varepsilon},$$

$$\Gamma^{\theta} = \frac{\widetilde{\Delta\varepsilon^{\mathrm{p}}}}{\Delta t}:\boldsymbol{D}^{\theta} + \frac{1}{\Delta t}\left[\boldsymbol{\sigma}:\boldsymbol{\Lambda} - F_{pp}\Delta p - F_p\right]\mathcal{P}^{\theta} - \frac{\Delta p}{\Delta t}F_{p\theta},$$

$$\frac{\widetilde{\Delta\varepsilon^{\mathrm{p}}}}{\Delta t} = \frac{\Delta\varepsilon^{\mathrm{p}}}{\Delta t} + \frac{\Delta p}{\Delta t}\boldsymbol{\sigma}:\frac{\partial\boldsymbol{\Lambda}}{\partial\boldsymbol{\sigma}}, \qquad [3.26]$$

which allows equation $[3.16]_2$ to be expressed as

$$\eth r = \boldsymbol{R}^{\varepsilon}:\eth\boldsymbol{\varepsilon} + R^{\theta}\eth\theta, \qquad [3.27]$$

with

$$\boldsymbol{R}^{\varepsilon} = -\frac{\theta}{\Delta t}\left[\boldsymbol{\alpha}:\boldsymbol{D}^{\varepsilon} - F_{p\theta}\boldsymbol{\mathcal{P}}^{\varepsilon}\right] + \boldsymbol{\Gamma}^{\varepsilon},$$

$$R^{\theta} = -\frac{1}{\Delta t}\left[\Delta\eta + c_0 + \theta\boldsymbol{\alpha}:\boldsymbol{D}^{\theta} - \theta F_{p\theta}\mathcal{P}^{\theta} - \theta F_{\theta\theta}\right] + \Gamma^{\theta}. \qquad [3.28]$$

Note that while the closest point projection accounts for the change in the plastic strain direction during an increment, the appearance of the fourth-order tensor $\frac{\partial\boldsymbol{\Lambda}}{\partial\boldsymbol{\sigma}}$ requires from the algorithm a significant amount of computational time per increment.

3.4.2. *Convex cutting plane*

In the convex cutting plane algorithm, the incremental form of the evolution equation is written as

$$\delta\boldsymbol{\varepsilon}^{\mathrm{p}} = \boldsymbol{\Lambda}\delta p. \qquad [3.29]$$

As was already mentioned in the previous section, during the inelastic correction step for obtaining the stress, the total strain and the temperature at the current time step are considered constant such that $\delta\boldsymbol{\varepsilon} = \mathbf{0}$ and $\delta\theta = 0$, respectively. Thus, the increment of stress is expressed as

$$\delta\boldsymbol{\sigma} = -\boldsymbol{C}:\delta\boldsymbol{\varepsilon}^{\mathrm{p}} = -\boldsymbol{C}:\boldsymbol{\Lambda}\delta p. \qquad [3.30]$$

However, the yield criterion leads to the equation

$$\Phi + \delta\Phi = \Phi + \frac{\partial\Phi}{\partial\boldsymbol{\sigma}}:\delta\boldsymbol{\sigma} + \frac{\partial\Phi}{\partial p}\delta p = 0. \qquad [3.31]$$

Combining equations [3.30] and [3.31] leads to

$$\delta p = \frac{\Phi}{A^{\mathrm{p}}}, \qquad A^{\mathrm{p}} = \frac{\partial\Phi}{\partial\boldsymbol{\sigma}}:\boldsymbol{C}:\boldsymbol{\Lambda} - \frac{\partial\Phi}{\partial p}. \qquad [3.32]$$

The last expression and equation [3.29] permit us to compute the corrected value of the scalar p and the plastic strains through their increments, i.e. $p^{\mathrm{new}} = p + \delta p$ and $\boldsymbol{\varepsilon}^{\mathrm{p\ new}} = \boldsymbol{\varepsilon}^{\mathrm{p}} + \boldsymbol{\Lambda}\delta p$. The corrected stress is computed from the analytical expression $[3.12]_1$. The convergence of the Newton–Raphson and the actual solution of the nonlinear problem are achieved when the absolute value of Φ is less than a tolerance.

With regard to the thermomechanical tangent moduli $\boldsymbol{D}^{\varepsilon}$ and $\boldsymbol{D}^{\theta}$, the convex cutting plane approach can provide only the continuum forms of them. When plastic strain occurs, the assumption [3.29], the stress–strain relation [3.14] and the expanded yield criterion [3.15] provide the following linear system of equations (at the FE iteration increment)

$$\begin{aligned} \eth\boldsymbol{\sigma} &= \boldsymbol{C}: \left[\eth\boldsymbol{\varepsilon} - \eth\boldsymbol{\varepsilon}^{\mathrm{p}} - \boldsymbol{\alpha}\eth\theta\right], \\ \eth\boldsymbol{\varepsilon}^{\mathrm{p}} &= \boldsymbol{\Lambda}\eth p, \\ \eth\Phi &= 0 \quad \Rightarrow \quad \frac{\partial\Phi}{\partial\boldsymbol{\sigma}}:\eth\boldsymbol{\sigma} + \frac{\partial\Phi}{\partial p}\eth p + \frac{\partial\Phi}{\partial\theta}\eth\theta = 0. \end{aligned} \qquad [3.33]$$

This system leads to

$$\begin{aligned} \eth p &= \boldsymbol{\mathcal{P}}^{\varepsilon}:\eth\boldsymbol{\varepsilon} + \mathcal{P}^{\theta}\eth\theta, \qquad \eth\boldsymbol{\sigma} = \boldsymbol{D}^{\varepsilon}:\eth\boldsymbol{\varepsilon} + \boldsymbol{D}^{\theta}\eth\theta, \\ \boldsymbol{\mathcal{P}}^{\varepsilon} &= \frac{1}{A^{\mathrm{p}}}\frac{\partial\Phi}{\partial\boldsymbol{\sigma}}:\boldsymbol{C}, \quad \mathcal{P}^{\theta} = -\boldsymbol{\mathcal{P}}^{\varepsilon}:\boldsymbol{\alpha} + \frac{1}{A^{\mathrm{p}}}\frac{\partial\Phi}{\partial\theta}, \\ \boldsymbol{D}^{\varepsilon} &= \boldsymbol{C} - [\boldsymbol{C}:\boldsymbol{\Lambda}] \otimes \boldsymbol{\mathcal{P}}^{\varepsilon}, \quad \boldsymbol{D}^{\theta} = -\boldsymbol{C}:\boldsymbol{\alpha} - \boldsymbol{C}:\boldsymbol{\Lambda}\mathcal{P}^{\theta}, \\ A^{\mathrm{p}} &= \frac{\partial\Phi}{\partial\boldsymbol{\sigma}}:\boldsymbol{C}:\boldsymbol{\Lambda} - \frac{\partial\Phi}{\partial p}. \end{aligned} \qquad [3.34]$$

For the thermal tangent moduli, equation $[3.16]_1$ can be written as

$$
\begin{aligned}
\eth\gamma_{\text{loc}} &= \boldsymbol{\Gamma}^{\varepsilon}\!:\!\eth\boldsymbol{\varepsilon} + \Gamma^{\theta}\eth\theta,\\
\boldsymbol{\Gamma}^{\varepsilon} &= \frac{\Delta\boldsymbol{\varepsilon}^{\mathrm{p}}}{\Delta t}\!:\!\boldsymbol{D}^{\varepsilon} + \frac{1}{\Delta t}\left[\boldsymbol{\sigma}\!:\!\boldsymbol{\Lambda} - F_{pp}\Delta p - F_{p}\right]\boldsymbol{\mathcal{P}}^{\varepsilon},\\
\Gamma^{\theta} &= \frac{\Delta\boldsymbol{\varepsilon}^{\mathrm{p}}}{\Delta t}\!:\!\boldsymbol{D}^{\theta} + \frac{1}{\Delta t}\left[\boldsymbol{\sigma}\!:\!\boldsymbol{\Lambda} - F_{pp}\Delta p - F_{p}\right]\mathcal{P}^{\theta} - \frac{\Delta p}{\Delta t}F_{p\theta},
\end{aligned}
\qquad [3.35]
$$

which allows equation $[3.16]_2$ to be expressed as

$$
\begin{aligned}
\eth r &= \boldsymbol{R}^{\varepsilon}\!:\!\eth\boldsymbol{\varepsilon} + R^{\theta}\eth\theta,\\
\boldsymbol{R}^{\varepsilon} &= -\frac{\theta}{\Delta t}\left[\boldsymbol{\alpha}\!:\!\boldsymbol{D}^{\varepsilon} - F_{p\theta}\boldsymbol{\mathcal{P}}^{\varepsilon}\right] + \boldsymbol{\Gamma}^{\varepsilon},\\
R^{\theta} &= -\frac{1}{\Delta t}\left[\Delta\eta + c_0 + \theta\boldsymbol{\alpha}\!:\!\boldsymbol{D}^{\theta} - \theta F_{p\theta}\mathcal{P}^{\theta} - \theta F_{\theta\theta}\right] + \Gamma^{\theta}.
\end{aligned}
\qquad [3.36]
$$

As a remark, it should be mentioned that the convex cutting plane keeps the plastic strain direction constant during an iteration, making it less stable algorithm. However, it does not introduce additional fourth-order tensors, allowing faster computations per increment.

NOTE.– For both the closest point projection and the convex cutting plane algorithm, the tangent modulus $\boldsymbol{D}^{\varepsilon}$ presents minor and major symmetries as long as associated plasticity takes place. In non-associated plasticity mechanisms, where $\boldsymbol{\Lambda}(\boldsymbol{\sigma}) \neq \dfrac{\partial\Phi}{\partial\boldsymbol{\sigma}}$, $\boldsymbol{D}^{\varepsilon}$ is generally not fully symmetric, in the sense that $D^{\varepsilon}_{ijkl} = D^{\varepsilon}_{jikl} = D^{\varepsilon}_{ijlk}$ but $D^{\varepsilon}_{ijkl} \neq D^{\varepsilon}_{klij}$. Of course, the tangent modulus operator is utilized only for numerical purposes in the finite element calculations, thus a possible non-symmetric form does not cause violation of thermodynamic principles.

3.4.3. *Viscoplasticity and other phenomena*

Compared to the case of plasticity, viscoplasticity is characterized by the same set of equations except that no yield surface is defined. A viscoplastic material is rather characterized by a family of equipotential surfaces, where the pseudo-yield criterion is connected not only with the stress and the accumulated plastic strain p, but also with the rate $\dot{p}$, i.e. $\Phi := \Phi(\boldsymbol{\sigma}, p, \dot{p}, \theta)$. A typical viscoplastic behavior law is discussed in section 2.5.3. The algorithms presented previously are valid in the case of a viscoplastic material, with the only difference that in the expressions of $\delta\Phi$ and $\eth\Phi$ the term $\dfrac{\partial \Phi}{\partial p}$ should be substituted by $\dfrac{\partial \Phi}{\partial p} + \dfrac{1}{\Delta t}\dfrac{\partial \Phi}{\partial \dot{p}}$ (see section 3.2.1).

The numerical schemes describing elastoplastic materials are generally suitable (after proper modifications) to any type of rate-independent mechanisms, like the damage or the martensitic transformation in shape memory alloys [QID 00, HAR 10, CHE 11, LAG 12]. For rate-dependent mechanisms like viscoelasticity, these types of algorithms are applicable as long as explicit expressions for continuum or consistent tangent moduli can be obtained.

3.5. Special algorithms in viscoelasticity

A viscoelastic material of Poynting–Thomson type obeys a Helmholtz free energy potential of the form

$$\Psi\left(\boldsymbol{\varepsilon}, \theta, \boldsymbol{\varepsilon}^{\mathrm{v}}\right) = \frac{1}{2}\left[\boldsymbol{\varepsilon} - \boldsymbol{\varepsilon}^{\mathrm{v}} - \boldsymbol{\alpha}[\theta - \theta_0]\right] : \boldsymbol{C} : \left[\boldsymbol{\varepsilon} - \boldsymbol{\varepsilon}^{\mathrm{v}} - \boldsymbol{\alpha}[\theta - \theta_0]\right]$$
$$+ \frac{1}{2}\boldsymbol{\varepsilon}^{\mathrm{v}} : \boldsymbol{C}^{\mathrm{v}} : \boldsymbol{\varepsilon}^{\mathrm{v}} + c_0\left[[\theta - \theta_0] - \theta \ln\left(\frac{\theta}{\theta_0}\right)\right] - \eta_0\theta + \mathcal{E}_0.$$

Following the discussion of section 2.5.3 for the viscoelastic material, the evolution law for the viscoelastic strain can be expressed as

$$\boldsymbol{V} : \dot{\boldsymbol{\varepsilon}}^{\mathrm{v}} + \boldsymbol{C}^{\mathrm{v}} : \boldsymbol{\varepsilon}^{\mathrm{v}} = \boldsymbol{\sigma},$$

or equivalently

$$\dot{\boldsymbol{\varepsilon}}^{\mathrm{v}} = \boldsymbol{V}^{-1} : \left[\boldsymbol{\sigma} - \boldsymbol{C}^{\mathrm{v}} : \boldsymbol{\varepsilon}^{\mathrm{v}}\right].$$

The last expression permits us to re-express the evolution law in terms of the rate of a scalar v and a viscoelastic-type criterion Φ^{v}. Indeed, we can write

$$\dot{\boldsymbol{\varepsilon}}^{\mathrm{v}} = \boldsymbol{\Lambda}^{\mathrm{v}}\dot{v}, \quad \boldsymbol{\Lambda}^{\mathrm{v}} = \frac{\boldsymbol{\varepsilon}^{\mathrm{vis}}}{\|\boldsymbol{\varepsilon}^{\mathrm{vis}}\|},$$
$$\boldsymbol{\varepsilon}^{\mathrm{vis}} = \boldsymbol{V}^{-1}\colon [\boldsymbol{\sigma} - \boldsymbol{C}^{\mathrm{v}}\colon\boldsymbol{\varepsilon}^{\mathrm{v}}], \quad \Phi^{\mathrm{v}} = \|\boldsymbol{\varepsilon}^{\mathrm{vis}}\| - \dot{v} = 0,$$

where $\| \{\bullet\} \|$ expresses the usual norm, i.e. $\|\boldsymbol{\varepsilon}^{\mathrm{vis}}\| = \sqrt{\boldsymbol{\varepsilon}^{\mathrm{vis}}\colon\boldsymbol{\varepsilon}^{\mathrm{vis}}}$.

Using the above formalism, the discretized form of the viscoelastic behavior described in 2.5.3 can be written as

$$\boldsymbol{\sigma} = \boldsymbol{C}\colon [\boldsymbol{\varepsilon} - \boldsymbol{\varepsilon}^{\mathrm{v}} - \boldsymbol{\alpha}[\theta - \theta_0]],$$
$$\eta = \boldsymbol{\alpha}\colon\boldsymbol{\sigma} + c_0 \ln\left(\frac{\theta}{\theta_0}\right) + \eta_0,$$
$$\Delta\boldsymbol{\varepsilon}^{\mathrm{v}} = \boldsymbol{\Lambda}^{\mathrm{v}}\Delta v, \quad \boldsymbol{\Lambda}^{\mathrm{v}} = \frac{\boldsymbol{\varepsilon}^{\mathrm{vis}}}{\|\boldsymbol{\varepsilon}^{\mathrm{vis}}\|}, \quad \boldsymbol{\varepsilon}^{\mathrm{vis}} = \boldsymbol{V}^{-1}\colon [\boldsymbol{\sigma} - \boldsymbol{C}^{\mathrm{v}}\boldsymbol{\varepsilon}^{\mathrm{v}}],$$
$$\Phi^{\mathrm{v}}(\boldsymbol{\sigma}, \boldsymbol{\varepsilon}^{\mathrm{v}}, \dot{v}) = \|\boldsymbol{\varepsilon}^{\mathrm{vis}}\| - \frac{\Delta v}{\Delta t} = 0, \quad \Delta v \geq 0,$$
$$\gamma_{\mathrm{loc}} = [\boldsymbol{\sigma} - \boldsymbol{C}^{\mathrm{v}}\colon\boldsymbol{\varepsilon}^{\mathrm{v}}]\colon\frac{\Delta\boldsymbol{\varepsilon}^{\mathrm{v}}}{\Delta t},$$
$$r = -\theta\frac{\Delta\eta}{\Delta t} + \gamma_{\mathrm{loc}}. \qquad [3.37]$$

This system has similar structure with the discretized system of equations in elastoplasticity. Thus, we can apply the closest point projection or the convex cutting plane technique, with proper modifications, in order to identify numerically the viscoelastic response and the thermomechanical tangent moduli.

3.6. Numerical applications

3.6.1. *Thermoelastoplastic material*

Consider a thermoelastoplastic isotropic material with a Helmholtz free energy potential of the form

$$\Psi\left(\varepsilon,\theta,\varepsilon^{\mathrm{p}},p\right)=\frac{1}{2}\left[\varepsilon-\varepsilon^{\mathrm{p}}-\boldsymbol{\alpha}[\theta-\theta_0]\right]:\boldsymbol{C}:\left[\varepsilon-\varepsilon^{\mathrm{p}}-\boldsymbol{\alpha}[\theta-\theta_0]\right]$$
$$+c_0\left[[\theta-\theta_0]-\theta\ln\left(\frac{\theta}{\theta_0}\right)\right]+\frac{B}{n+1}p^{n+1}\left[1-\left[\frac{\theta-\theta_t}{\theta_m-\theta_t}\right]^m\right],$$

where B is a plastic coefficient, n is a plastic hardening exponent, m is a thermal softening exponent, θ_m is the melting temperature and θ_t is the transition temperature [RAN 13]. Moreover, the material is plastified according to the yield criterion

$$\Phi=\sigma^{\mathrm{vM}}-[A+Bp^n]\left[1-\left[\frac{\theta-\theta_t}{\theta_m-\theta_t}\right]^m\right].$$

In the above expression, A denotes the elastic limit at the transition temperature. The values of the material properties are summarized in Table 3.4.

It is noted that, while the expression of Ψ produces an inelastic part in the specific heat, the numerical calculations reveal that this part is negligible. Such result agrees with the experimental observations reported by Rosakis *et al.* [ROS 00].

The specimen initially is at temperature $\theta_0 = 673.15$ K and zero stress. At all times, the material is stress free in directions 2 and 3 and only the stress component σ_{11} is developed. The total loading path includes three steps:

1) Constant strain rate $\dot{\varepsilon}_{11} = 0.001$ 1/s for 300 s. Thermally, adiabatic conditions are considered.

2) Constant negative stress rate for 300 s until $\sigma_{11} = 0$. Thermally, adiabatic conditions are considered.

Property	TA6V
Young's modulus [MPa]	113800
Poisson's ratio	0.342
thermal expansion [1/K]	0.86E-5
density ρ [t/m^3]	4.4
specific heat capacity c_p [MJ/[t K]]	0.656
thermal conductivity [MN/[s K]]	6.7E-6
elastic limit A [MPa]	500
hardening parameter B [MPa]	160
plastic hardening exponent n	0.25
melting temperature θ_m [K]	1943.15
transition temperature θ_t [K]	293.15
thermal softening exponent m	1.055

Table 3.4. *Thermomechanical properties of the titanium alloy TA6V [RAN 13]*

3) Cooling with constant temperature rate for 300 s until $\theta = 673.15$. Mechanically, the stress is kept zero.

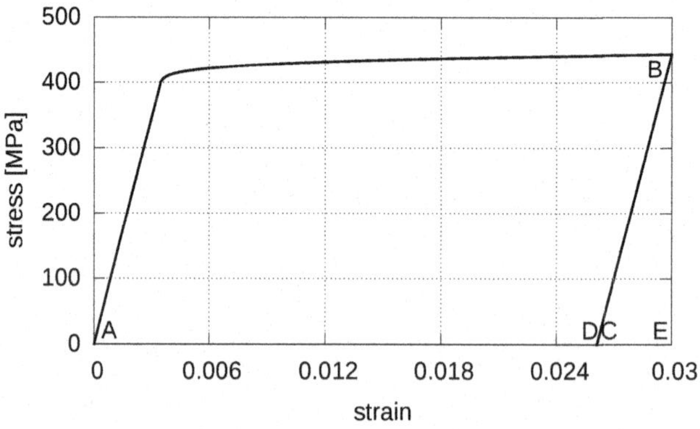

Figure 3.3. *Stress versus strain diagram*

Considering zero heat sources, the adiabaticity condition is expressed as $\mathcal{Q} = 0$. The aim of this example is to investigate the various energetic terms that appear due to the strong thermomechanical coupling (see the discussion in section 2.5.3). The computational scheme used here is the convex cutting plane. The stress–strain path is illustrated in Figure 3.3. From the numerical analysis, it is found that, at the end of the loading cycle,

$$\mathrm{W_m} = 11.33\,\mathrm{MJ/m^3}, \quad \mathrm{W^r_m} = 0.01\,\mathrm{MJ/m^3},$$
$$\mathrm{W^d_m} = 10.26\,\mathrm{MJ/m^3}, \quad \mathrm{W^{ir}_m} = 1.06\,\mathrm{MJ/m^3}.$$

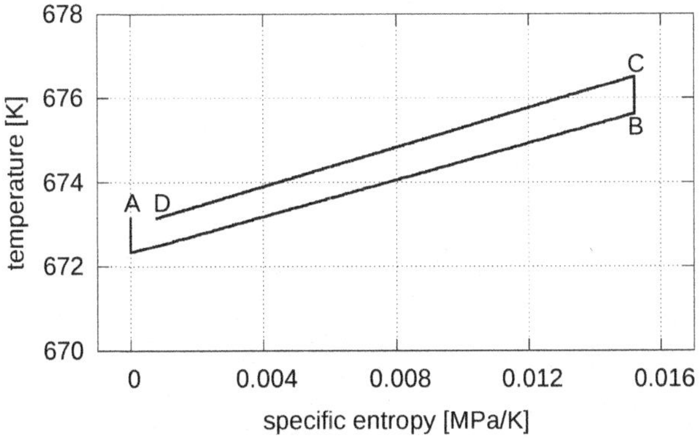

Figure 3.4. *Temperature versus specific entropy diagram*

The temperature-specific entropy diagram, presented in Figure 3.4, provides information about the distribution of the thermal work. At the end of the loading cycle, it is found that

$$\mathrm{W_t} = 0.52\,\mathrm{MJ/m^3}, \quad \mathrm{W^r_t} = -0.01\,\mathrm{MJ/m^3}, \quad \mathrm{W^{ir}_t} = 0.53\,\mathrm{MJ/m^3}.$$

Consequently, the recoverable and the irrecoverable internal energy are computed as

$$\mathcal{E}^{\mathrm{r}} = \mathrm{W^r_m} + \mathrm{W^r_t} = 0, \quad \mathcal{E}^{\mathrm{ir}} = \mathrm{W^{ir}_m} + \mathrm{W^{ir}_t} = 1.59\,\mathrm{MJ/m^3}.$$

This result indicates that the produced irrecoverable thermal work is approximately 33% of the stored internal energy. Finally, the energy balance dictates that the thermal energy loss at the end of the loading cycle is

$$\int \mathcal{Q}\mathrm{d}t = \mathcal{E} - \mathrm{W_m} = \mathrm{W_t} - \mathrm{W^d_m} = -9.74\,\mathrm{MJ/m^3}.$$

Figures 3.5 and 3.6 illustrate the distribution of the various mechanical and thermal works inside the material at the end of the first step. It is noted that the total recoverable mechanical work in this step (area of $\mathrm{W^r_m}$ in Figure 3.5) is somehow higher than the area of the triangle BCE of Figure 3.3 due to the thermal expansion, caused by the small temperature increase.

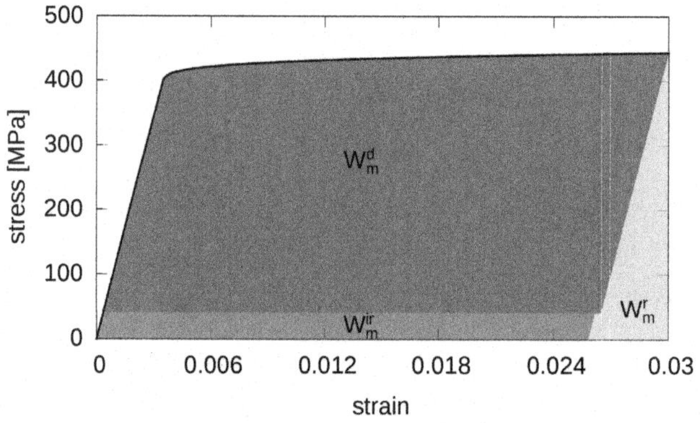

Figure 3.5. *Schematic distribution of mechanical work at the end of the first step: W^r_m=0.88 MJ/m^3, W^{ir}_m=1.06 MJ/m^3 and W^d_m=10.26 MJ/m^3. For a color version of this figure, see www.iste.co.uk/chatzigeorgiou/thermomechanical.zip*

3.6.2. *Viscoelastic material*

Consider a viscoelastic material of Poynting–Thomson type, as the one described in section 3.5. For isotropic viscoelastic response, the fourth-order tensors $\boldsymbol{C}$, $\boldsymbol{C}^{\mathrm{v}}$ and $\boldsymbol{V}$ can be written in the form

$$\begin{aligned}
\boldsymbol{C} &= 3K\boldsymbol{\mathcal{I}}^h + 2\mu\boldsymbol{\mathcal{I}}^d,\\
\boldsymbol{C}^{\mathrm{v}} &= 3K^{\mathrm{v}}\boldsymbol{\mathcal{I}}^h + 2\mu^{\mathrm{v}}\boldsymbol{\mathcal{I}}^d,\\
\boldsymbol{V} &= 3V^{\mathrm{b}}\boldsymbol{\mathcal{I}}^h + 2V^{\mathrm{s}}\boldsymbol{\mathcal{I}}^d.
\end{aligned}$$

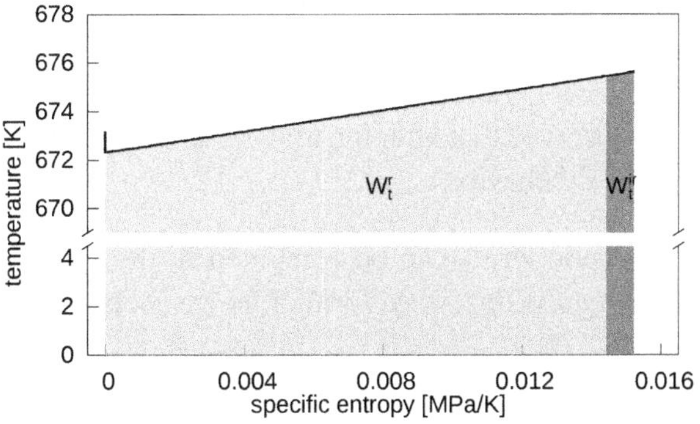

Figure 3.6. *Distribution of thermal work at the end of the first step: W_t^r=9.73 MJ/m^3 and W_t^{ir}=0.53 MJ/m^3. For a color version of this figure, see www.iste.co.uk/ chatzigeorgiou/thermomechanical.zip*

In the above expressions, $\mathcal{I}^h = \dfrac{1}{3}\boldsymbol{I} \otimes \boldsymbol{I}$ is the volumetric part of the symmetric fourth-order identity tensor and $\mathcal{I}^d = \mathcal{I} - \mathcal{I}^h$ is its deviatoric part. Isothermal conditions are considered during loading, thus the thermal response is not studied here. For the numerical example, it is assumed that:

$K = 6250$ MPa, $\mu = 1930.15$ MPa, $K^{\mathrm{v}} = 4761.9$ MPa, $\mu^{\mathrm{v}} = 1470.59$ MPa,

$V^{\mathrm{b}} = 107.14$ MPa/s, $V^{\mathrm{s}} = 33.09$ MPa/s.

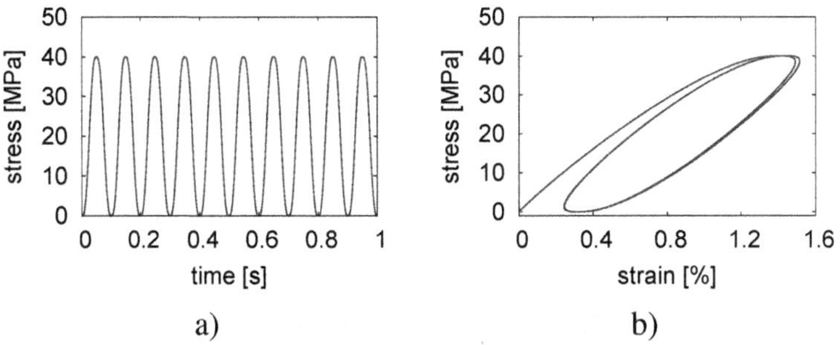

Figure 3.7. *a) Mechanical loading path (stress versus time) and b) stress–strain response of the viscoelastic material. For a color version of this figure, see www.iste.co.uk/ chatzigeorgiou/thermomechanical.zip*

The material is subjected to 10 uniaxial loading cycles, where the stress varies in a sinusoidal fashion with time between 0 and 40 MPa, as illustrated in Figure 3.7a. The stress–strain response during this loading path is presented in Figure 3.7b. It is observed that after the first cycle, the material tends to have a stabilized mechanical behavior.

The various energetic terms can be computed in the same way as in the previous example using the linearized form of the viscoelastic constitutive law equation (equation [3.37]).

4

Concepts for Heterogeneous Media

All real materials present an internal structure that often varies spatially and possesses significant complexity: this internal structure is generally referred as the material's microstructure. In the microstructure, local variations of the matter appear and these are defined as the heterogeneities. The presence of heterogeneities induces highly non-homogeneous physical fields like strain, temperature, etc., in the matter.

As an illustrative example of material heterogeneities, we can consider the case of metals. A metal is composed of crystals (called grains), each of them having a specific crystallographic orientation. Such aggregate forms a polycrystalline medium. In a single crystal, the components of the elastic stiffness tensor $\boldsymbol{C}$ depend on the interatomic distance. In the general case, the interatomic distance varies spatially inside the microstructure. A crystal with cubic structure has an elastic stiffness tensor of the form (in Voigt notation):

$$\boldsymbol{C} = \begin{bmatrix} C_{11} & C_{12} & C_{12} & 0 & 0 & 0 \\ C_{12} & C_{11} & C_{12} & 0 & 0 & 0 \\ C_{12} & C_{12} & C_{11} & 0 & 0 & 0 \\ 0 & 0 & 0 & C_{66} & 0 & 0 \\ 0 & 0 & 0 & 0 & C_{66} & 0 \\ 0 & 0 & 0 & 0 & 0 & C_{66} \end{bmatrix}.$$

This relation is established in the local coordinate of the crystal, which means that the axis 1 is in the [1 0 0] direction. If the coordinate systems of the crystal and the laboratory frame (where the boundary conditions are imposed) are not the same, then the crystal stiffness tensor must be properly rotated. This indicates that the elastic stiffness expressed in the laboratory coordinate system should vary from one grain to another as a function of the grain orientation. Thus, the grain orientation is already one source of heterogeneity in the microstructure and is referred as the texture.

There are of course other sources of heterogeneities. In steels, for example, a perlitic phase can be formed in the grain boundaries. This phase consists of ferrite lamellae (phase α) and cementite lamellae (Fe_3C phase). In this case, even if the crystals have the same orientation, the difference in composition induces a variation of the elastic constants between ferrite and perlite lamellae. Moreover, with regard to steel, there are additional phenomena that occur, e.g. the formation of martensitic phase, or the presence of residual austenite.

There are many ways to observe the materials' microstructure and especially the modern experimental methods now permit us to observe the microstructure in three dimensions. The combination of electron back scattered diffraction (EBSD) analysis with a focused ion beam (FIB) allows us to perform 3D texture analysis layer by layer. Another possibility is to analyze the material though an X-ray microtomograph. Many materials can be analyzed with this method (for instance polycrystalline materials, composite materials with random/continuous reinforcement, architectured materials, porous, etc.).

This chapter is devoted to the presentation of some important definitions and theorems which are quite general for materials with heterogeneities and they hold independently of the microstructure's complexity. It is divided into two parts: the first part presents the general principles of homogenization of composites from an engineering point of view, while the second part focuses on composites with periodic microstructure and the main concepts of the mathematical homogenization. Both frameworks are based on the assumption of scales separation and they introduce appropriate spaces for integrating the microstructural characteristics, either a representative volume element (for the classical homogenization approach) or a unit cell (for the mathematical homogenization).

The valuable mathematical tools discussed in the chapter are key factors in the development of homogenization methods for heterogeneous media. In what follows, all the operators are performed only in Cartesian coordinates. Homogenization in specific structures has also been investigated in curvilinear coordinate systems [CHA 11, TSU 12, CHA 12b, TSA 12, TSA 13a, KUN 14, DHA 15].

4.1. Preliminaries

Consider a continuum body occupying the space $\mathcal{B}$ and being bounded by the surface $\partial\mathcal{B}$. The volume occupied by the body is equal to V, while the vector normal to $\partial\mathcal{B}$ is denoted by $\boldsymbol{n}$ (Figure 4.1). The position of each material point is characterized by the vector $\boldsymbol{x}$ (in Cartesian coordinates). For the sake of simplicity, the discussion is based on the usual assumptions for small strains. In the sequel, the following simplified notation is utilized

$$\langle\{\bullet\}\rangle = \frac{1}{V}\int_{\mathcal{B}}\{\bullet\}\,\mathrm{d}v\,, \qquad \lceil\{\bullet\}\rfloor = \frac{1}{V}\int_{\partial\mathcal{B}}\{\bullet\}\,\mathrm{d}s\,.$$

In this formalism, the divergence theorem [2.10] for a tensor $\boldsymbol{A}$ of arbitrary order can be written in compact form as

$$\langle \mathrm{div}\boldsymbol{A}\rangle = \lceil \boldsymbol{A}\cdot\boldsymbol{n}\rfloor\,.$$

For a second-order tensor $\boldsymbol{A}$, the following properties hold

$$\left\langle \boldsymbol{A}^T\right\rangle = \langle\boldsymbol{A}\rangle^T\,, \qquad \left\lceil \boldsymbol{A}^T\right\rfloor = \lceil\boldsymbol{A}\rfloor^T\,.$$

Periodic tensors

Multiplication (single contraction, double contraction or dyadic product) of a periodic and an anti-periodic tensor produces an anti-periodic tensor.

For a periodic tensor $\boldsymbol{A}$ and an anti-periodic tensor $\boldsymbol{B}$ in a proper space $\mathcal{B}$ (i.e. space in which the boundaries are periodic and the vector $\boldsymbol{n}$ is anti-periodic) holds $\lceil \boldsymbol{A}\cdot\boldsymbol{B}\rfloor = 0$, $\lceil \boldsymbol{A}\otimes\boldsymbol{B}\rfloor = 0$. When a tensor $\boldsymbol{A}$ is periodic, then the tensors $\boldsymbol{A}\cdot\boldsymbol{n}$, $\boldsymbol{A}\otimes\boldsymbol{n}$ and $\boldsymbol{A}\times\boldsymbol{n}$ are anti-periodic. For a differentiable periodic tensor $\boldsymbol{A}$ in the space $\mathcal{B}$, we obtain

$$\begin{aligned}
\langle \mathrm{grad}\boldsymbol{A}\rangle &= \langle \mathrm{div}\left(\boldsymbol{A}\otimes\boldsymbol{I}\right)\rangle = \lceil \boldsymbol{A}\otimes\boldsymbol{n}\rfloor = \boldsymbol{0}\,,\\
\left\langle \mathrm{grad}_{\mathrm{sym}}\boldsymbol{A}\right\rangle &= \boldsymbol{0}\,,\\
\langle \mathrm{div}\boldsymbol{A}\rangle &= \lceil \boldsymbol{A}\cdot\boldsymbol{n}\rfloor = \boldsymbol{0}\,,\\
\langle \mathrm{curl}\boldsymbol{A}\rangle &= -\langle \mathrm{grad}\boldsymbol{A}\rangle : \boldsymbol{\epsilon} = \boldsymbol{0}\,.
\end{aligned}$$

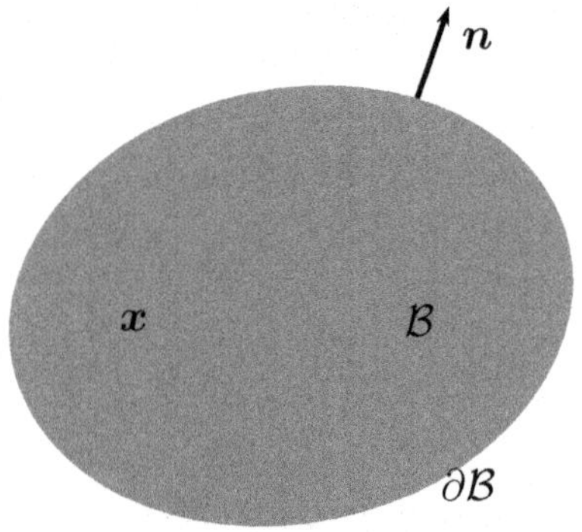

Figure 4.1. *Continuum body under small deformation*

4.2. Homogenization - engineering approach

The concepts of micromechanics, as identified by Hill [HIL 63], Hill and Rice [HIL 72b] and Hashin [HAS 83], define the main framework of almost all homogenization techniques in the literature. The approaches followed in these works provide the necessary tools to study composites with both random and periodic microstructures.

The general problem of the homogenization can be stated as:

Considering a given heterogeneous microstructure, how the overall properties of an equivalent homogeneous medium can be obtained?

In order to seek for an answer to the above question, we need first to specify the characteristic sizes of the heterogeneities with respect to the studied structure. In a metallic material, for instance, several characteristic scales exist: the scale of the structure, the scale of the crystalline structure, the scale of the intragranular structure (precipitates, lamellae perlite/martensite, etc.) and the scale of the crystallographic lattice.

As we consider finer scales, the behavior of the material is more accurately defined. But this also means that the representative volume of the medium becomes smaller, increasing tremendously the computational cost. Moreover, below a certain scale, continuum mechanics are unable to describe the material behavior, since the interatomic interactions must be accounted discretely. These methods are not presented here.

The classical homogenization approaches are valid under the following conditions:

– The characteristic size of the studied microstructure needs to be orders of magnitude smaller than the characteristic size of the structure. Thus, a typical volume that is large enough to be representative of the microstructure may be seen as a material point. This notion leads to the definition of a representative volume element (RVE), which plays a very important role in choosing an appropriate modeling strategy. The identification of the RVE in composite media is not always an easy task. In certain cases, advanced statistical tools are required in order to identify a proper RVE using experimentally obtained microstructural information [BAN 12, REM 16].

– The medium is statistically homogeneous. This statement expresses that, when the composite is subjected to uniform far field boundary conditions, the stress and strain averages over any RVE are essentially the same [NEM 99b]. For more in-depth discussion about statistical homogeneity, the interested reader is referred to the book of Qu and Cherkaoui [QU 06].

The scope of homogenization is to replace the real heterogeneous material of the RVE with a fictitious homogeneous material that is able to provide the same stress and strain fields, in an average sense, in the structure.

In the homogenization theory of random media, a composite is described through the introduction of two suitable scales. The first scale, the microscopic or RVE, represents the microstructure accounting for different material constituents and their geometry. The second scale, the macroscopic, considers the overall body as an imaginary homogeneous medium (Figure 4.2). At the microscale, the RVE occupies the space $\mathcal{B}$ with volume V and is bounded by the surface $\partial\mathcal{B}$ with normal vector $\boldsymbol{n}$. Each microscopic point is assigned with a position vector $\boldsymbol{x}$ in $\mathcal{B}$. However, at the macroscale, the continuum body occupies the space $\bar{\mathcal{B}}$ with volume $\overline{V}$ and bounded by a surface $\partial\bar{\mathcal{B}}$ with normal unit vector $\overline{\boldsymbol{n}}$. Each macroscopic point is assigned with a position vector $\overline{\boldsymbol{x}}$ in $\bar{\mathcal{B}}$ and is connected with an RVE. In a similar fashion with the previous section, a bar above a symbol denotes a macroscopic quantity or variable.

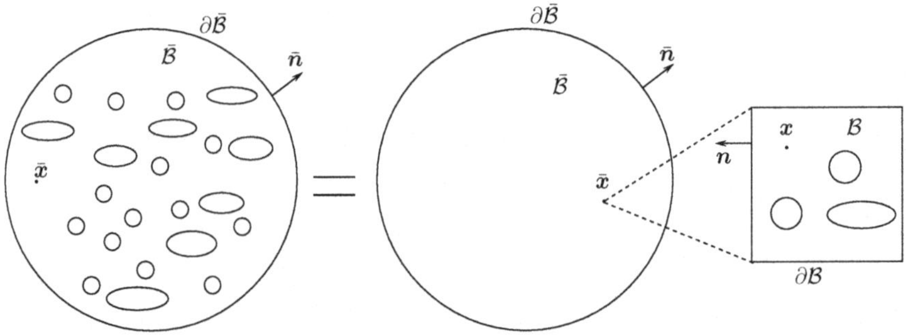

Figure 4.2. *Representation of a composite using two scales*

The standard methodology is to correlate macroscopic variables at a macroscopic position $\overline{\boldsymbol{x}}$ with the volume averaging of their microscopic counterparts over the RVE that corresponds to the position $\overline{\boldsymbol{x}}$ [HIL 67], i.e.

$$\bar{\boldsymbol{\varepsilon}} = \int_{\mathcal{B}} \boldsymbol{\varepsilon}\,\mathrm{d}v = \langle \boldsymbol{\varepsilon} \rangle\,, \quad \overline{\boldsymbol{\sigma}} = \frac{1}{V}\int_{\mathcal{B}} \boldsymbol{\sigma}\,\mathrm{d}v = \langle \boldsymbol{\sigma} \rangle\,,$$
$$\overline{\boldsymbol{\nabla}\boldsymbol{\theta}} = \frac{1}{V}\int_{\mathcal{B}} \boldsymbol{\nabla}\boldsymbol{\theta}\,\mathrm{d}v = \langle \boldsymbol{\nabla}\boldsymbol{\theta} \rangle\,, \quad \overline{\boldsymbol{q}} = \frac{1}{V}\int_{\mathcal{B}} \boldsymbol{q}\,\mathrm{d}v = \langle \boldsymbol{q} \rangle\,. \qquad [4.1]$$

In addition, the RVE is subjected to uniform temperature conditions, where the uniform temperature is equal to the macroscopic temperature $\overline{\theta}$. The latter is classical assumption in the micromechanics of random media [ROS 70, BEN 91]. Relations [4.1] are very popular, but they are not always convenient, especially when the RVE presents surfaces with displacement or traction jumps. Several articles in the literature suggest that macroscopic quantities should be instead defined through surface integrals of their microscopic counterparts over the boundary of the RVE [HIL 72a, BEN 01, COS 05]. This book does not include studies on special interfaces or surfaces, thus the definitions in [4.1] are considered sufficient for all the following discussions.

Another very useful concept in the micromechanics theories is the definition of the concentration tensors. Considering only mechanical fields, we can correlate microscopic strains or stresses with their corresponding macroscopic ones, at a specific RVE, through relations of the form

$$\boldsymbol{\varepsilon}(\boldsymbol{x}) = \boldsymbol{A}(\boldsymbol{x}):\bar{\boldsymbol{\varepsilon}}, \quad \boldsymbol{\sigma}(\boldsymbol{x}) = \boldsymbol{B}(\boldsymbol{x}):\bar{\boldsymbol{\sigma}}. \qquad [4.2]$$

In the above expressions, $\boldsymbol{A}(\boldsymbol{x})$ and $\boldsymbol{B}(\boldsymbol{x})$ are fourth-order tensors with minor symmetries, which depend on the position inside the RVE. The main advantage of these tensors is that their knowledge is sufficient to identify the correlation between the two scales, since the overall macroscopic properties can be directly computed with the help of $\boldsymbol{A}(\boldsymbol{x})$ or $\boldsymbol{B}(\boldsymbol{x})$ and the material properties of the microstructural constituents. In micromechanics of random media, the approaches that are called mean field theories provide analytical or semi-analytical expressions for the concentration tensors, which in turn are utilized to compute the average fields at each material constituent of the RVE using as information their macroscopic counterparts [MUR 87, NEM 99b, QU 06]. The advantage of the mean field theories is that they permit direct calculation of the tensors $\boldsymbol{A}$ or $\boldsymbol{B}$ using the Eshelby equivalence principle (see the discussion in Chapter 6).

Concentration tensors have also been introduced for the case of thermomechanical media [BEN 91] and media with other types of fields, like magnetomechanical or electromechanical [DUN 93]. A detailed discussion about the thermomechanical concentration tensors and their usefulness in the homogenization of random media is provided in Chapter 6.

Based on equations $[4.1]_{1,2}$, plenty of techniques have been developed to obtain the overall mechanical response of composites with random or repeated (periodic) microstructure. A very useful tool toward this goal is the well-known Hill–Mandel theorem.

4.2.1. *Hill's lemma and Hill–Mandel theorem: mechanical problem*

The discussion in this subsection is for operators related with small deformation processes, but it can by similarly extended to large deformations framework (see Appendix 1).

The following notion is adopted [NGU 88]:

DEFINITION 4.1.– *Kinematically admissible strain $\boldsymbol{\varepsilon}$ is called every symmetric second-order tensor that is related with a displacement vector $\boldsymbol{u}$ through the relation*

$$\boldsymbol{\varepsilon} = \text{grad}_{sym}\boldsymbol{u}\,.$$

DEFINITION 4.2.– *Statically admissible stress $\boldsymbol{\sigma}$ is called every symmetric second-order tensor that satisfies the equilibrium equation (ignoring body forces)*

$$\text{div}\boldsymbol{\sigma} = \boldsymbol{0}\,.$$

Using these definitions and the divergence theorem, the volume average for a kinematically admissible strain is given by

$$\begin{aligned}\langle\boldsymbol{\varepsilon}\rangle &= \frac{1}{2}\left\langle\text{grad}\boldsymbol{u} + [\text{grad}\boldsymbol{u}]^T\right\rangle = \frac{1}{2}\left\langle\text{div}\left(\boldsymbol{u}\otimes\boldsymbol{I}\right)\right\rangle \\ &+\frac{1}{2}\left\langle\text{div}\left(\boldsymbol{u}\otimes\boldsymbol{I}\right)\right\rangle^T = \frac{1}{2}\left\lceil\boldsymbol{u}\otimes\boldsymbol{n}+\boldsymbol{n}\otimes\boldsymbol{u}\right\rfloor\,, \end{aligned} \quad [4.3]$$

and for a statically admissible stress

$$\begin{aligned}\langle\boldsymbol{\sigma}\rangle &= \left\langle\text{grad}\boldsymbol{x} \,\tilde{\cdot}\, \boldsymbol{\sigma}\right\rangle^T = \left\langle\text{div}\left(\boldsymbol{x}\otimes\boldsymbol{\sigma}\right)\right\rangle^T - \left\langle\boldsymbol{x}\otimes\text{div}\boldsymbol{\sigma}\right\rangle^T \\ &= \left\lceil[\boldsymbol{\sigma}\cdot\boldsymbol{n}]\otimes\boldsymbol{x}\right\rfloor. \end{aligned} \quad [4.4]$$

The average mechanical work over the body, produced by a kinematically admissible strain $\boldsymbol{\varepsilon}$ and a statically admissible stress $\boldsymbol{\sigma}$, is half of the quantity

$$\langle\boldsymbol{\sigma}:\boldsymbol{\varepsilon}\rangle = \langle\boldsymbol{\sigma}:\text{grad}\boldsymbol{u}\rangle = \langle\text{div}(\boldsymbol{u}\cdot\boldsymbol{\sigma})\rangle - \langle\boldsymbol{u}\cdot\text{div}\boldsymbol{\sigma}\rangle = \left\lceil\boldsymbol{u}\cdot[\boldsymbol{\sigma}\cdot\boldsymbol{n}]\right\rfloor\,. \quad [4.5]$$

Considering the above definitions, the following theorems and lemmas can be proven:

THEOREM 4.1.– *Average strain theorem*: Assume that at each point of the boundary surface $\partial\mathcal{B}$, the displacement vector $\boldsymbol{u}_0 = \boldsymbol{\varepsilon}_0 \cdot \boldsymbol{x}$ with $\boldsymbol{\varepsilon}_0$ constant is applied. Then, the volume average of the strains inside the body $\mathcal{B}$ is equal to the boundary strain tensor, i.e.

$$\boldsymbol{\varepsilon}_0 = \langle \boldsymbol{\varepsilon} \rangle \,. \tag{4.6}$$

Indeed, using equation [4.3] and the symmetry of the strain tensor yields

$$\begin{aligned}
\langle \boldsymbol{\varepsilon} \rangle &= \frac{1}{2} \lceil \boldsymbol{u} \otimes \boldsymbol{n} + \boldsymbol{n} \otimes \boldsymbol{u} \rfloor = \frac{1}{2} \left[\boldsymbol{\epsilon}_0 \cdot \lceil \boldsymbol{x} \otimes \boldsymbol{n} \rfloor + \lceil \boldsymbol{n} \otimes \boldsymbol{x} \rfloor \cdot \boldsymbol{\epsilon}_0 \right] \\
&= \frac{1}{2} \left[\boldsymbol{\epsilon}_0 \cdot \langle \mathrm{grad} \boldsymbol{x} \rangle + \langle \mathrm{grad} \boldsymbol{x} \rangle^T \cdot \boldsymbol{\epsilon}_0 \right] = \boldsymbol{\epsilon}_0 \,.
\end{aligned}$$

THEOREM 4.2.– *Average stress theorem*: Assume that at each point of the boundary surface $\partial\mathcal{B}$, the traction vector $\boldsymbol{t}_0 = \boldsymbol{\sigma}_0 \cdot \boldsymbol{n}$ with $\boldsymbol{\sigma}_0$ constant is applied. If no body forces are present, the volume average of the stresses inside the body $\mathcal{B}$ is equal to the boundary stress tensor, i.e.

$$\boldsymbol{\sigma}_0 = \langle \boldsymbol{\sigma} \rangle \,. \tag{4.7}$$

Indeed, using equation [4.4] yields

$$\begin{aligned}
\langle \boldsymbol{\sigma} \rangle &= \lceil [\boldsymbol{\sigma} \cdot \boldsymbol{n}] \otimes \boldsymbol{x} \rfloor = \lceil [\boldsymbol{\sigma}_0 \cdot \boldsymbol{n}] \otimes \boldsymbol{x} \rfloor \\
&= \boldsymbol{\sigma}_0 \cdot \lceil \boldsymbol{n} \otimes \boldsymbol{x} \rfloor = \boldsymbol{\sigma}_0 \cdot \langle \mathrm{grad} \boldsymbol{x} \rangle^T = \boldsymbol{\sigma}_0 \,.
\end{aligned}$$

LEMMA 4.1.– *Hill's lemma*: Let $\boldsymbol{\varepsilon}$ be a kinematically admissible strain and $\boldsymbol{\sigma}$ be a statically admissible stress. Then, it holds

$$\langle \boldsymbol{\sigma} : \boldsymbol{\varepsilon} \rangle - \langle \boldsymbol{\sigma} \rangle : \langle \boldsymbol{\varepsilon} \rangle = \lceil [\boldsymbol{u} - \langle \boldsymbol{\varepsilon} \rangle \cdot \boldsymbol{x}] \cdot [\boldsymbol{\sigma} \cdot \boldsymbol{n} - \langle \boldsymbol{\sigma} \rangle \cdot \boldsymbol{n}] \rfloor \,. \tag{4.8}$$

Indeed, using the divergence theorem, equations [4.3]–[4.5] and the symmetry of the stress tensor, we obtain

$$\begin{aligned}\lceil [\boldsymbol{u} - \langle \boldsymbol{\varepsilon} \rangle \cdot \boldsymbol{x}] \cdot [\boldsymbol{\sigma} \cdot \boldsymbol{n} - \langle \boldsymbol{\sigma} \rangle \cdot \boldsymbol{n}] \rfloor &= \lceil \boldsymbol{u} \cdot [\boldsymbol{\sigma} \cdot \boldsymbol{n}] \rfloor - \langle \boldsymbol{\varepsilon} \rangle : \lceil [\boldsymbol{\sigma} \cdot \boldsymbol{n}] \otimes \boldsymbol{x} \rfloor \\ &\quad -\frac{1}{2} \langle \boldsymbol{\sigma} \rangle : \lceil \boldsymbol{u} \otimes \boldsymbol{n} + \boldsymbol{n} \otimes \boldsymbol{u} \rfloor \\ &\quad +[\langle \boldsymbol{\varepsilon} \rangle \underset{\sim}{:} \langle \boldsymbol{\sigma} \rangle] : \lceil \boldsymbol{x} \otimes \boldsymbol{n} \rfloor \\ &= \langle \boldsymbol{\sigma} : \boldsymbol{\varepsilon} \rangle - \langle \boldsymbol{\varepsilon} \rangle : \langle \boldsymbol{\sigma} \rangle \\ &\quad - \langle \boldsymbol{\sigma} \rangle : \langle \boldsymbol{\varepsilon} \rangle + \langle \boldsymbol{\varepsilon} \rangle : \langle \boldsymbol{\sigma} \rangle \\ &= \langle \boldsymbol{\sigma} : \boldsymbol{\varepsilon} \rangle - \langle \boldsymbol{\sigma} \rangle : \langle \boldsymbol{\varepsilon} \rangle \, .\end{aligned}$$

THEOREM 4.3.– *Hill–Mandel theorem*: Let $\boldsymbol{\varepsilon}$ be a kinematically admissible strain and $\boldsymbol{\sigma}$ be a statically admissible stress. Then, the three types of conditions:

1) $\boldsymbol{u} = \langle \boldsymbol{\varepsilon} \rangle \cdot \boldsymbol{x}$ on the boundary $\partial \mathcal{B}$;

2) $\boldsymbol{\sigma} \cdot \boldsymbol{n} = \langle \boldsymbol{\sigma} \rangle \cdot \boldsymbol{n}$ on the boundary $\partial \mathcal{B}$;

3) $\boldsymbol{\varepsilon} = \langle \boldsymbol{\varepsilon} \rangle + \text{grad}_{\text{sym}} \boldsymbol{z}$, with $\boldsymbol{z}$ periodic and $\boldsymbol{\sigma} \cdot \boldsymbol{n}$ anti-periodic;

satisfy the energy equivalence

$$\langle \boldsymbol{\sigma} : \boldsymbol{\varepsilon} \rangle = \langle \boldsymbol{\sigma} \rangle : \langle \boldsymbol{\varepsilon} \rangle \, . \qquad [4.9]$$

The proof of this theorem is as follows: the first two conditions ensure directly that the right-hand side of the Hill's lemma [4.8] is zero. For the third condition, using the properties of tensor derivatives discussed in section 1.1, the divergence theorem, the definition 4.2, the symmetry of the stress tensor and the properties of the products between periodic and anti-periodic vectors described in section 4.1, we obtain

$$\begin{aligned}\langle \boldsymbol{\sigma} : \boldsymbol{\varepsilon} \rangle &= \langle \boldsymbol{\sigma} \rangle : \langle \boldsymbol{\varepsilon} \rangle + \langle \boldsymbol{\sigma} : \text{grad} \boldsymbol{z} \rangle = \langle \boldsymbol{\sigma} \rangle : \langle \boldsymbol{\varepsilon} \rangle + \langle \boldsymbol{\sigma} : \text{grad} \boldsymbol{z} + \boldsymbol{z} \cdot \text{div} \boldsymbol{\sigma} \rangle \\ &= \langle \boldsymbol{\sigma} \rangle : \langle \boldsymbol{\varepsilon} \rangle + \langle \text{div} \, (\boldsymbol{z} \cdot \boldsymbol{\sigma}) \rangle = \langle \boldsymbol{\sigma} \rangle : \langle \boldsymbol{\varepsilon} \rangle + \lceil \boldsymbol{z} \cdot [\boldsymbol{\sigma} \cdot \boldsymbol{n}] \rfloor = \langle \boldsymbol{\sigma} \rangle : \langle \boldsymbol{\varepsilon} \rangle \, .\end{aligned}$$

It is noted that the Hill's lemma and the Hill–Mandel theorem are valid for the products $\boldsymbol{\sigma}:\dot{\boldsymbol{\varepsilon}}$ and $\dot{\boldsymbol{\sigma}}:\boldsymbol{\varepsilon}$. The proof is analogous for all the corresponding conditions.

Equations $[4.1]_{1,2}$ and the additional postulate of satisfying the energy equivalence of the Hill–Mandel theorem [4.9] under uniform tractions, linear displacements or periodicity conditions allow us to obtain useful information about the mechanical behavior of the composite. The energy equivalence, that the Hill–Mandel condition expresses, between the RVE and the macroscopic point of reference permits us to obtain the Reuss lower bound and the Voigt upper bound, or even the more strict Hashin and Shtrikman bounds [HAS 63], on the mechanical energy of composites. To obtain the overall elastic properties independently of the external boundary conditions, equation [4.9] requires to solve the equilibrium equations of the RVE, accompanied by the various phases constitutive laws, under appropriate conditions that provide consistency of the obtained results, i.e. the overall properties should be the same under uniform traction or linear displacement boundary conditions. For special microstructures, like those in unidirectional long fiber composites or spherical particle composites, exact solutions of the RVE boundary value problem are available and the macroscopic response can be obtained analytically through the composite cylinders and the composite spheres assemblage method, respectively [HAS 64, CHR 79]. In more complicated composites, no analytical solutions are available and thus numerical solution of the RVE problem is required. In the literature, there exist also methodologies that permit for specific type of composites to reduce the computational cost of FE calculations, like for instance the generalized method of cells [PAL 92].

4.2.2. *Hill's lemma and Hill–Mandel theorem: thermal problem*

The following notion is adopted:

DEFINITION 4.3.– *Admissible temperature gradient* $\boldsymbol{\nabla\theta}$ *is called every vector that is related with the temperature* θ *through the relation*

$$\boldsymbol{\nabla\theta} = \mathrm{grad}\theta\,.$$

DEFINITION 4.4.– *Admissible heat flux $\boldsymbol{q}$ is called every vector that satisfies the equation*

$$\text{div}\boldsymbol{q} = \mathbf{0}\,.$$

Using these definitions and the divergence theorem, the volume average for an admissible temperature gradient is given by

$$\langle \boldsymbol{\nabla\theta} \rangle = \langle \text{grad}\theta \rangle = \langle \text{div}\,(\theta \boldsymbol{I}) \rangle = \lceil \theta \boldsymbol{n} \rfloor\,, \qquad [4.10]$$

and for an admissible heat flux

$$\langle \boldsymbol{q} \rangle = \langle \text{grad}\boldsymbol{x}\cdot\boldsymbol{q} \rangle = \langle \text{div}\,(\boldsymbol{x}\otimes\boldsymbol{q}) \rangle - \langle \boldsymbol{x}\otimes\text{div}\boldsymbol{q} \rangle = \lceil [\boldsymbol{q}\cdot\boldsymbol{n}]\boldsymbol{x} \rfloor\,. \qquad [4.11]$$

The average thermal work over the body, produced by an admissible temperature gradient $\boldsymbol{\nabla\theta}$ and an admissible heat flux $\boldsymbol{q}$, is related with the quantity

$$\langle \boldsymbol{q}\cdot\boldsymbol{\nabla\theta} \rangle = \langle \boldsymbol{q}\cdot\text{grad}\theta \rangle = \langle \text{div}(\theta\boldsymbol{q}) \rangle - \langle \theta\text{div}\boldsymbol{q} \rangle = \lceil \theta[\boldsymbol{q}\cdot\boldsymbol{n}] \rfloor\,. \qquad [4.12]$$

Based on the above definitions and properties, the thermal Hill's lemma is expressed in the following way:

LEMMA 4.2.– Let $\boldsymbol{\nabla\theta}$ be an admissible temperature gradient and $\boldsymbol{q}$ be an admissible heat flux. Then, it holds

$$\langle \boldsymbol{q}\cdot\boldsymbol{\nabla\theta} \rangle - \langle \boldsymbol{q} \rangle \cdot \langle \boldsymbol{\nabla\theta} \rangle = \lceil [\theta - \langle \boldsymbol{\nabla\theta} \rangle \cdot\boldsymbol{x}]\,[\boldsymbol{q}\cdot\boldsymbol{n} - \langle \boldsymbol{q} \rangle \cdot\boldsymbol{n}] \rfloor\,. \qquad [4.13]$$

Indeed, using the divergence theorem and equations [4.10]–[4.12], we obtain

$$\begin{aligned}\lceil [\theta - \langle \boldsymbol{\nabla\theta} \rangle \cdot\boldsymbol{x}]\,[\boldsymbol{q}\cdot\boldsymbol{n} - \langle \boldsymbol{q} \rangle \cdot\boldsymbol{n}] \rfloor &= \lceil \theta[\boldsymbol{q}\cdot\boldsymbol{n}] \rfloor - \langle \boldsymbol{\nabla\theta} \rangle \cdot \lceil [\boldsymbol{q}\cdot\boldsymbol{n}]\boldsymbol{x} \rfloor \\ &\quad - \langle \boldsymbol{q} \rangle \cdot \lceil \theta\boldsymbol{n} \rfloor \\ &\quad + [\langle \boldsymbol{\nabla\theta} \rangle \otimes \langle \boldsymbol{q} \rangle] : \lceil \boldsymbol{x}\otimes\boldsymbol{n} \rfloor\end{aligned}$$

$$
\begin{aligned}
&= \langle \boldsymbol{q}\cdot\boldsymbol{\nabla\theta}\rangle - \langle \boldsymbol{q}\rangle \cdot \langle \boldsymbol{\nabla\theta}\rangle \\
&\quad - \langle \boldsymbol{q}\rangle \cdot \langle \boldsymbol{\nabla\theta}\rangle + \langle \boldsymbol{q}\rangle \cdot \langle \boldsymbol{\nabla\theta}\rangle \\
&= \langle \boldsymbol{q}\cdot\boldsymbol{\nabla\theta}\rangle - \langle \boldsymbol{q}\rangle \cdot \langle \boldsymbol{\nabla\theta}\rangle \ .
\end{aligned}
$$

Moreover, the thermal Hill–Mandel theorem can be stated in the following way:

THEOREM 4.4.– Let $\boldsymbol{\nabla\theta}$ be an admissible temperature gradient and $\boldsymbol{q}$ be an admissible heat flux. Then, the three types of conditions:

1) $\theta = \langle \boldsymbol{\nabla\theta}\rangle \cdot \boldsymbol{x}$ on the boundary $\partial\mathcal{B}$;

2) $\boldsymbol{q}\cdot\boldsymbol{n} = \langle \boldsymbol{q}\rangle \cdot \boldsymbol{n}$ on the boundary $\partial\mathcal{B}$;

3) $\boldsymbol{\nabla\theta} = \langle \boldsymbol{\nabla\theta}\rangle + \text{grad} z$, with z periodic and $\boldsymbol{q}\cdot\boldsymbol{n}$ anti-periodic;

satisfy the power equivalence

$$
\langle \boldsymbol{q}\cdot\boldsymbol{\nabla\theta}\rangle = \langle \boldsymbol{q}\rangle \cdot \langle \boldsymbol{\nabla\theta}\rangle . \qquad [4.14]
$$

The proof of this theorem for the first two conditions is derived directly from the Hill's lemma [4.13]. For the third condition, the properties of tensor derivatives discussed in section 1.1, the divergence theorem, the definition 4.4 and the properties of the products between periodic and anti-periodic vectors described in section 4.1, lead to

$$
\begin{aligned}
\langle \boldsymbol{q}\cdot\boldsymbol{\nabla\theta}\rangle &= \langle \boldsymbol{q}\rangle \cdot \langle \boldsymbol{\nabla\theta}\rangle + \langle \boldsymbol{q}\cdot \text{grad} z\rangle \\
&= \langle \boldsymbol{q}\rangle \cdot \langle \boldsymbol{\nabla\theta}\rangle + \langle \boldsymbol{q}\cdot \text{grad} z + z \,\text{div}\boldsymbol{q}\rangle \\
&= \langle \boldsymbol{q}\rangle \cdot \langle \boldsymbol{\nabla\theta}\rangle + \langle \text{div}\,(z\boldsymbol{q})\rangle = \langle \boldsymbol{q}\rangle \cdot \langle \boldsymbol{\nabla\theta}\rangle + \lceil z\,[\boldsymbol{q}\cdot\boldsymbol{n}]\rfloor \\
&= \langle \boldsymbol{q}\rangle \cdot \langle \boldsymbol{\nabla\theta}\rangle .
\end{aligned}
$$

4.2.3. *Divergence and curl properties for periodic fields*

A useful property in homogenization of composites with periodic microstructure is the following: consider the periodic vectors $\boldsymbol{A}$ and $\boldsymbol{B}$ for which

$$\mathrm{curl}\boldsymbol{A} = \boldsymbol{0}, \quad \mathrm{div}\boldsymbol{B} = 0.$$

Due to these expressions, we can identify a periodic scalar a and a periodic vector $\boldsymbol{b}$ for which

$$\boldsymbol{A} = \overline{\boldsymbol{A}} + \mathrm{grad}a, \quad \boldsymbol{B} = \overline{\boldsymbol{B}} + \mathrm{curl}\boldsymbol{b},$$

where $\overline{\boldsymbol{A}}$ and $\overline{\boldsymbol{B}}$ are constant vectors. Due to the periodicity of the fields, $\langle \boldsymbol{A} \rangle = \overline{\boldsymbol{A}}$ and $\langle \boldsymbol{B} \rangle = \overline{\boldsymbol{B}}$. Using the scalar a or the vector $\boldsymbol{b}$, we can show that the Hill–Mandel theorem is valid, i.e. $\langle \boldsymbol{A} \cdot \boldsymbol{B} \rangle = \langle \boldsymbol{A} \rangle \cdot \langle \boldsymbol{B} \rangle$.

Indeed, for the decomposition of $\boldsymbol{A}$, we have

$$\begin{aligned} \langle \boldsymbol{A} \cdot \boldsymbol{B} \rangle &= \overline{\boldsymbol{A}} \cdot \langle \boldsymbol{B} \rangle + \langle \boldsymbol{B} \cdot \mathrm{grad}a \rangle \\ &= \langle \boldsymbol{A} \rangle \cdot \langle \boldsymbol{B} \rangle + \langle \mathrm{div}(a\boldsymbol{B}) \rangle - \langle a\mathrm{div}\boldsymbol{B} \rangle = \langle \boldsymbol{A} \rangle \cdot \langle \boldsymbol{B} \rangle + \lceil a\boldsymbol{B} \cdot \boldsymbol{n} \rfloor \\ &= \langle \boldsymbol{A} \rangle \cdot \langle \boldsymbol{B} \rangle . \end{aligned}$$

Furthermore, for the decomposition of $\boldsymbol{B}$, we have

$$\begin{aligned} \langle \boldsymbol{A} \cdot \boldsymbol{B} \rangle &= \langle \boldsymbol{A} \rangle \cdot \overline{\boldsymbol{B}} + \langle \boldsymbol{A} \cdot \mathrm{curl}\boldsymbol{b} \rangle \\ &= \langle \boldsymbol{A} \rangle \cdot \langle \boldsymbol{B} \rangle + \langle \mathrm{div}(\boldsymbol{b} \times \boldsymbol{A}) \rangle + \langle \boldsymbol{b} \cdot \mathrm{curl}\boldsymbol{A} \rangle \\ &= \langle \boldsymbol{A} \rangle \cdot \langle \boldsymbol{B} \rangle + \lceil \boldsymbol{b} \cdot [\boldsymbol{A} \times \boldsymbol{n}] \rfloor = \langle \boldsymbol{A} \rangle \cdot \langle \boldsymbol{B} \rangle . \end{aligned}$$

It is noted that a generalization of this nice property between periodic fields has been established with the div-curl lemma developed by Murat and Tartar [MUR 78, TAR 79].

4.3. Mathematical homogenization of periodic media

Highly heterogeneous materials are very complicated and their behavior is difficult to be computed numerically. Mathematical homogenization is an efficient tool to overcome difficulties posed by the implementation of global finite element methods and bridge the gap between micromechanics and the overall behavior of complex structures.

A composite body occupies the space $\mathcal{B}^\epsilon$ with volume V^ϵ and has a periodic microstructure (Figure 4.3). The body is bounded by the surface $\partial\mathcal{B}^\epsilon$, with unit normal vector $\boldsymbol{n}^\epsilon$ associated at each point of the surface. The position vector on every point in $\mathcal{B}^\epsilon$ is denoted by $\overline{\boldsymbol{x}}$. Moreover, the characteristic length of the periodic microstructure is assigned with the scalar parameter ϵ.

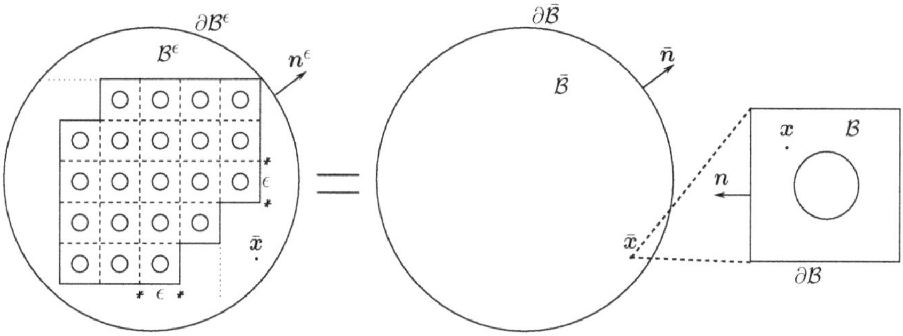

Figure 4.3. *Composite with periodic microstructure. The homogenization separates the problem into two scales, a macroscopic and a microscopic*

The mathematical (periodic) homogenization describes the composite using two scales. The first scale, the microscopic, takes into account the different material constituents and their geometry inside the microstructure. The second scale, the macroscopic, considers the overall body as an imaginary homogeneous medium. This homogenization method is in fact a convergence study, whose scope is to identify the connection between the scales, as the characteristic length ϵ tends to zero, in terms of: (i) the various fields (mechanical, thermal or other), and (ii) the conservation laws and the constitutive relations between the fields.

At the macroscale, the continuum body occupies the space $\bar{\mathcal{B}}$ with volume $\overline{V}$, bounded by the surface $\partial\bar{\mathcal{B}}$ with unit normal vector $\overline{\boldsymbol{n}}$. Each macroscopic point is assigned with a position vector $\overline{\boldsymbol{x}}$ in $\bar{\mathcal{B}}$. A periodic unit cell that describes the microscopic scale is attached to every position $\overline{\boldsymbol{x}}$. The unit cell occupies the space $\mathcal{B}$ with volume V, bounded by the surface $\partial\mathcal{B}$ with unit normal vector $\boldsymbol{n}$. Each microscopic point is assigned with a position vector $\boldsymbol{x}$

in $\mathcal{B}$ (Figure 4.3). The two scales $\overline{\boldsymbol{x}}$ and $\boldsymbol{x}$ are connected with the characteristic length ϵ through the relation $\boldsymbol{x} = \overline{\boldsymbol{x}}/\epsilon$. The periodic homogenization theory provides accurate results only when the characteristic length tends to zero, i.e. when the microstructure is extremely small compared to the actual size of the composite. Details about the mathematical framework in small deformation processes can be found in the literature [BEN 78, SAN 78, TAR 78, ALL 92, MUR 97]. In the following, all the variables that correspond to the general composite response are denoted with a superscript ϵ. Moreover, a bar above a symbol denotes a homogenized (macroscopic) quantity or variable, depending only on the vector $\overline{\boldsymbol{x}}$ and perhaps the time.

When considering the composite in a global sense, the gradient operator grad^ϵ is connected with the gradient operators of the macroscale ($\overline{\text{grad}}$) and microscale (grad) through the usual scale decomposition rule [SAN 78]

$$\text{grad}^\epsilon\{\bullet\} = \overline{\text{grad}}\{\bullet\} + \frac{1}{\epsilon}\text{grad}\{\bullet\}. \qquad [4.15]$$

Since all coordinate systems are Cartesian, the identity tensor $\boldsymbol{I}$ and the permutation tensor $\boldsymbol{\epsilon}$ are common for all of them and thus the other operators (grad_{sym}, div, curl) follow similar decomposition.

$\Re^N$ is the N_{th} dimensional Euclidean space, $C^\infty(\Omega)$ is the space of continuous functions with continuous derivatives of every order in the space Ω, $L^p(\Omega;\Re^N)$ denotes the Lebesgue space of all functions $f(\boldsymbol{x})$ from Ω to $\Re^N$ with

$$\left[\int_\Omega |f(\boldsymbol{x})|^p \text{d}\boldsymbol{x}\right]^{\frac{1}{p}} < \infty,$$

and $H^1(\Omega;\Re^N)$ denotes the Sobolev space of all functions $f(\boldsymbol{x})$ from Ω to $\Re^N$ with

$$\left[\int_\Omega \left[|f(\boldsymbol{x})|^2 + |\text{grad} f(\boldsymbol{x})|^2\right] \text{d}\boldsymbol{x}\right]^{\frac{1}{2}} < \infty.$$

In the above expressions, integration is performed on the space referred to Ω (for three-dimensional spaces $\mathrm{d}\boldsymbol{x}$ is simply $\mathrm{d}v$). A very useful tool in the mathematical homogenization theory is the notion of weak convergence.

DEFINITION 4.5.– *Let Ω be an open, bounded, smooth subset of $\Re^N$, $N \geq 2$. Moreover, assume $1 < p < \infty$ and q the conjugate exponent, $\frac{1}{p} + \frac{1}{q} = 1$. Defining $L^p(\Omega; \Re^N)$, a sequence $\{u_n\}_{n\geq 1} \subset L^p(\Omega)$ converges weakly to $u \in L^p(\Omega)$ if*

$$\int_\Omega u_n v \mathrm{d}x \to \int_\Omega u v \mathrm{d}x \quad \text{as } n \to \infty, \quad \forall v \in L^q(\Omega).$$

If u_n is periodic with respect to $\epsilon = \frac{1}{n}$, then it can be denoted as u^ϵ and its weak limit coincides with the mean value over a period.

EXAMPLE.– The sequence $u_n(x) = \sin(nx)$ converges weakly to zero. Indeed, as it can be seen in Figure 4.4, u_n is periodic with respect to nx with period $H = 2\pi$. While u_n does not have a limit as $n \to \infty$, the integral $\int_0^H u_n(x) g(x) \mathrm{d}x$ tends always to zero for any square-integrable function g on $[0, 2\pi]$. Thus, $U = \frac{1}{H} \int_0^H u_n(x) \mathrm{d}x = 0$ is considered to be the weak limit of the sequence u_n.

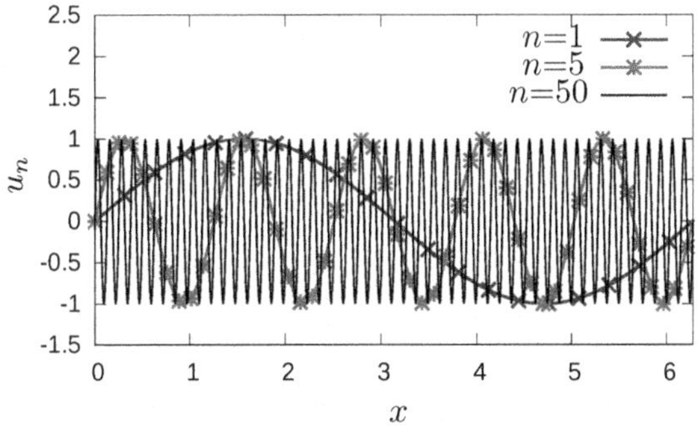

Figure 4.4. *Example of weakly convergent sequence. For a color version of this figure, see www.iste.co.uk/chatzigeorgiou/thermomechanical.zip*

4.3.1. *Approximating deformation functions*

One of the most striking features of the displacement, strain or stress distributions in heterogeneous materials is that the functions describing these quantities oscillate in space following the heterogeneity parameters $\epsilon_1, \epsilon_2, ...$ of the different scales (nano, micro, meso, etc.) that may exist in the composites' microstucture. For the sake of simplicity, we can assume that all the heterogeneities may be expressed in terms of a small characteristic length ϵ and that the body exhibits different periodicities $k\epsilon$, k taking positive values for a multiscale medium. Without loss of generality, it is considered that $k = 1$, i.e. that the studied problem corresponds to a two-scale composite with periodic microstructure, like the one in Figure 4.3.

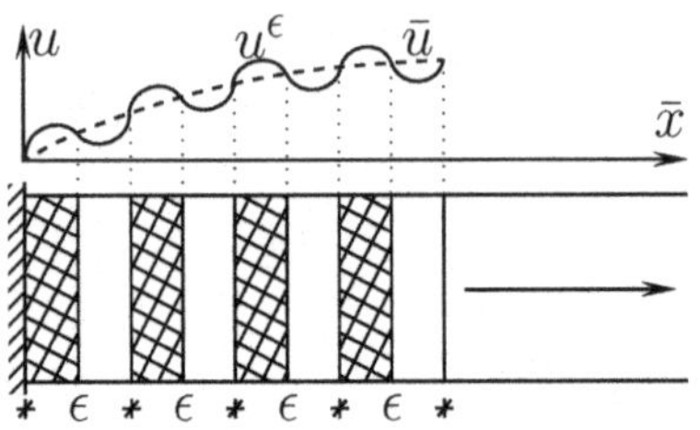

Figure 4.5. *Oscillating displacement in a stratified two-phase cantilever beam*

Figure 4.5 illustrates the displacement $u^{\epsilon}(\overline{x})$ of a stratified two-phase cantilever beam under longitudinal extension, made of alternate stiff and soft components. The problem for simplicity is treated as 1-D. For an oscillating function $u^{\epsilon}(\overline{x})$, the pointwise limit as $\epsilon \rightarrow 0$ cannot be defined and the function converges only weakly to the mean function $\langle u \rangle = \overline{u}$, as it is illustrated in the example of the weak convergence definition 4.5. However, even in the case where $u^{\epsilon}(\overline{x})$ presents small fluctuations (Figure 4.6), the gradient of $u^{\epsilon}(\overline{x})$ cannot be satisfactorily approximated by the gradient of the approximating function $\overline{u}(\overline{x})$.

For periodic microstructures in three-dimensional spaces, it is assumed that a generic function takes the form

$$\varphi^{\epsilon}(\overline{\boldsymbol{x}}) = \varphi\left(\overline{\boldsymbol{x}}, \frac{\overline{\boldsymbol{x}}}{\epsilon}\right), \text{ periodic with respect to } \boldsymbol{x} = \frac{\overline{\boldsymbol{x}}}{\epsilon}. \qquad [4.16]$$

sequences. This property causes the main difficulty in homogenization theory, since some fundamental quantities, such as the stress or the rate of energy or the dissipation, result from products of two weakly converging functions. Namely, the stress in elasticity is the product of the elasticity fourth-order tensor and the strain, the mechanical work is the product of stress and strain, and finally the dissipation in dissipative mechanisms is the product of the thermodynamic forces and thermodynamic fluxes. In the latest case, the situation is even worse in perfect elastoplasticity, where displacement exhibits discontinuities [FRA 12, FRA 14, FRA 15]. Searching for additional properties of weakly converging sequences, in order to find the limit of their product, covers a significant part of the research effort in this field. The most important theorem is the compensated compactness theorem by Murat and Tartar [TAR 77, MUR 78], which, applied in elasticity or conduction problems, allows for homogenizing the constitutive law and the mechanical work. The latter finding is a generalization of the Hill–Mandel theorem, independently of the type of boundary conditions.

Mathematical homogenization consists of setting the problem as a sequence of heterogeneous problems $\mathcal{P}^\epsilon$ in $\mathcal{B}^\epsilon$, in terms of the heterogeneity parameter ϵ. For instance, the boundary value problem in quasi-static elasticity takes the form (for fixed ϵ)

$$
\begin{aligned}
&\mathrm{div}^\epsilon \boldsymbol{\sigma}^\epsilon + \rho^\epsilon \boldsymbol{b}^\epsilon = \boldsymbol{0} \quad \text{in } \mathcal{B}^\epsilon,\\
&\boldsymbol{\sigma}^\epsilon = \boldsymbol{C}^\epsilon : \boldsymbol{\varepsilon}^\epsilon,\\
&\boldsymbol{\varepsilon}^\epsilon = \mathrm{grad}^\epsilon_{\mathrm{sym}} \boldsymbol{u}^\epsilon \quad \text{in } \mathcal{B}^\epsilon,\\
&\boldsymbol{\sigma}^\epsilon \cdot \boldsymbol{n}^\epsilon = \boldsymbol{t}^\epsilon \quad \text{on } \partial\mathcal{B}^{\mathrm{NB}\epsilon},\\
&\boldsymbol{u}^\epsilon = \boldsymbol{u}^{\mathrm{prescribed}} \quad \text{on } \partial\mathcal{B}^{\mathrm{EB}\epsilon},
\end{aligned}
\qquad [4.18]
$$

where $\partial\mathcal{B}^{\mathrm{EB}\epsilon}$ are the essential and $\partial\mathcal{B}^{\mathrm{NB}\epsilon}$ are the natural boundary conditions. Obviously, $\partial\mathcal{B}^{\mathrm{EB}\epsilon} \bigcup \partial\mathcal{B}^{\mathrm{NB}\epsilon} = \partial\mathcal{B}^\epsilon$. The above system of equations must be well posed and have at least one solution in a precise functional setting, which satisfies appropriate *a priori* estimates independently of ϵ. Then, the homogenized problem consists of passing to the limit for $\epsilon \to 0$ and the goal is to identify the form of the "homogenized" problem, the "homogenized" coefficients (called macroscopic coefficients if the homogenized material is homogeneous), as well as the homogenized solution. This task is not easy,

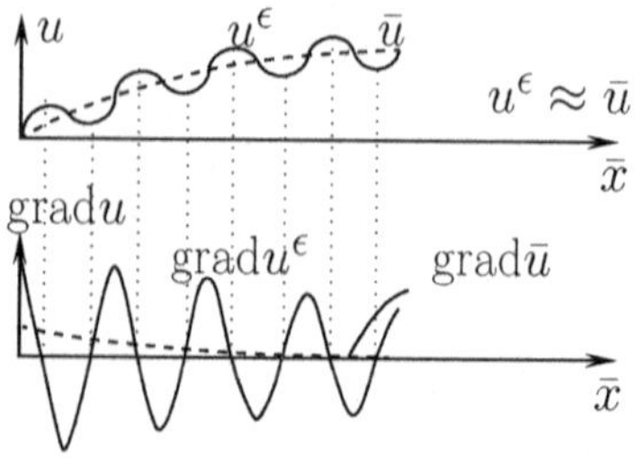

Figure 4.6. *Oscillating function and its gradient*

Contrarily to the strongly (i.e. in norm) converging sequences, like ϵ-independent functions, all oscillating sequences sufficiently regular (instance belonging to $L^2(\bar{\mathcal{B}})$) are only weakly convergent sequences in sense that

$$\lim_{\epsilon \to 0} \int_{\mathcal{B}^\epsilon} \varphi^\epsilon(\overline{\boldsymbol{x}})\psi(\overline{\boldsymbol{x}})\mathrm{d}\overline{\boldsymbol{x}} = \int_{\bar{\mathcal{B}}} \psi(\overline{\boldsymbol{x}})\frac{1}{V}\int_{\mathcal{B}} \varphi(\overline{\boldsymbol{x}}, \boldsymbol{x})\mathrm{d}\boldsymbol{x}\mathrm{d}\overline{\boldsymbol{x}},$$
$$\forall\, \psi(\overline{\boldsymbol{x}}) \in C^\infty(\bar{\mathcal{B}}), \qquad [4.1$$

where $\mathcal{B}^\epsilon$ is the approximating domain of the body, defined in Figure 4.7, wi $\mathcal{B}^\epsilon \to \bar{\mathcal{B}}$ as $\epsilon \to 0$.

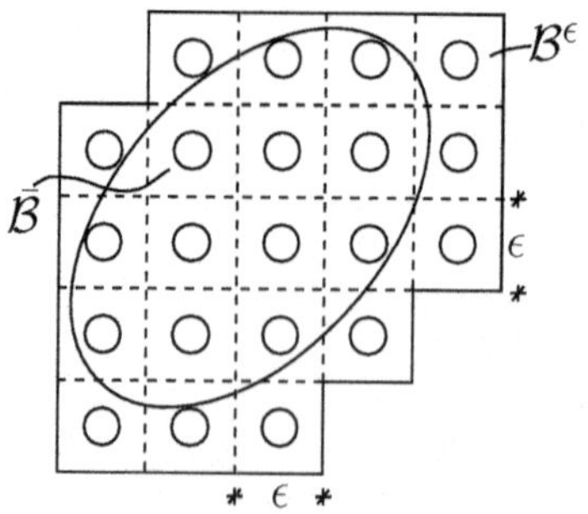

Figure 4.7. *Approximating domain occupied by the ϵ-periodic body*

It is worth noticing that the limit of the product of two weakly convergin sequences is not generally equal to the product of the weak limits of the two since energy equation, constitutive equation and, possibly, more general boundary conditions are non-linear.

Considering again the boundary value problem [4.18], we may observe that the meaning of the derivative in $[4.18]_1$ is not clear, since $\boldsymbol{C}^\epsilon$ in $[4.18]_2$ may be discontinuous, so the derivatives of $\boldsymbol{C}^\epsilon$ with respect to $\overline{\boldsymbol{x}}$ are not defined pointwise. Therefore, the heterogeneous problems must be put in weak form, which is appropriate for finite element implementation. In view of $[4.18]_1$, the equation of equilibrium after multiplication by a smooth vector function $\boldsymbol{\varphi}^\epsilon(\overline{\boldsymbol{x}})$ and integration over $\mathcal{B}^\epsilon$ gives

$$\int_{\mathcal{B}^\epsilon} \boldsymbol{\varphi}^\epsilon(\overline{\boldsymbol{x}}) \cdot \mathrm{div} \boldsymbol{\sigma}^\epsilon \mathrm{d}\overline{\boldsymbol{x}} + \int_{\mathcal{B}^\epsilon} \boldsymbol{\varphi}^\epsilon(\overline{\boldsymbol{x}}) \cdot \rho^\epsilon \boldsymbol{b}^\epsilon \mathrm{d}\overline{\boldsymbol{x}} = 0, \quad \forall \boldsymbol{\varphi}^\epsilon \in [C^\infty(\mathcal{B}^\epsilon)]^3, \quad [4.19]$$

from which, by divergence theorem and using the boundary conditions $[4.18]_4$ and $[4.18]_5$, and assuming for simplicity $\boldsymbol{u}^{\text{prescribed}} = \boldsymbol{0}$,

$$\int_{\mathcal{B}^\epsilon} \mathrm{grad}^\epsilon \boldsymbol{\varphi}^\epsilon : \boldsymbol{\sigma}^\epsilon \mathrm{d}\overline{\boldsymbol{x}} = \int_{\mathcal{B}^\epsilon} \boldsymbol{\varphi}^\epsilon \cdot \rho^\epsilon \boldsymbol{b}^\epsilon \mathrm{d}\overline{\boldsymbol{x}} + \int_{\partial \mathcal{B}^{\mathrm{NB}\epsilon}} \boldsymbol{\varphi}^\epsilon \cdot \boldsymbol{t}^\epsilon \mathrm{d}s,$$
$$\forall \boldsymbol{\varphi}^\epsilon \in [C^\infty(\mathcal{B}^\epsilon)]^3. \quad [4.20]$$

Under regularity assumptions on body forces and boundary conditions, elasticity theory provides that (i) a unique solution exists for fixed ϵ and (ii) the displacement $\boldsymbol{u}^\epsilon$ belongs to $H^1(\mathcal{B}^\epsilon)$, while the stress $\boldsymbol{\sigma}^\epsilon$ belongs to $L^2(\mathcal{B}^\epsilon)$. Now, we can pass to the limit in [4.20] for $\epsilon \to 0$. Assuming that $\rho^\epsilon \boldsymbol{b}^\epsilon \equiv \rho \boldsymbol{b}$, $\boldsymbol{t}^\epsilon \equiv \boldsymbol{t}$, $\boldsymbol{\varphi}^\epsilon \equiv \boldsymbol{\varphi}$ are non-oscillating vectors[1], and using the weak convergence limit definition [4.17] yields

$$\int_{\bar{\mathcal{B}}} \overline{\mathrm{grad}} \boldsymbol{\varphi} : \left[\frac{1}{V} \int_{\mathcal{B}} \boldsymbol{\sigma}(\overline{\boldsymbol{x}}, \boldsymbol{x}) \mathrm{d}\boldsymbol{x} \right] \mathrm{d}\overline{\boldsymbol{x}} = \int_{\bar{\mathcal{B}}} \boldsymbol{\varphi} \cdot \rho \boldsymbol{b} \mathrm{d}\overline{\boldsymbol{x}} + \int_{\partial \bar{\mathcal{B}}^{\mathrm{NB}}} \boldsymbol{\varphi} \cdot \boldsymbol{t} \mathrm{d}s,$$
$$\forall \boldsymbol{\varphi} \in \left[C^\infty(\bar{\mathcal{B}}) \right]^3. \quad [4.21]$$

1 In periodic media, the body forces can also be oscillating without affecting the main results, see, for instance, the work of Kikuchi and coworkers [GUE 90, TER 01].

The last equation states that the mean stress over the unit cell,

$$\overline{\boldsymbol{\sigma}}(\overline{\boldsymbol{x}}) = \frac{1}{V}\int_{\mathcal{B}} \boldsymbol{\sigma}(\overline{\boldsymbol{x}}, \boldsymbol{x})\mathrm{d}\boldsymbol{x} = \langle \boldsymbol{\sigma} \rangle, \qquad [4.22]$$

satisfies the homogenized equation of equilibrium [4.21] for the weak limits of the traction and body forces. However, we cannot yet pass to the limit in $[4.18]_2$ in order to obtain a homogenized equation of the form

$$\overline{\boldsymbol{\sigma}} = \overline{\boldsymbol{C}}:\overline{\boldsymbol{\varepsilon}}, \qquad [4.23]$$

for some homogenized elasticity tensor and the mean value of strain over the unit cell $\overline{\boldsymbol{\varepsilon}}$, and conclude that $\overline{\boldsymbol{\sigma}}$ is the homogenized stress. This is due to the oscillating character of both $\overline{\boldsymbol{C}}$ and $\overline{\boldsymbol{\varepsilon}}$. In other words, it is not only the weak convergence that matters. Another topology is needed, and this can be defined with the help of H-convergence, which will be presented in the sequel.

It is noted that, choosing a test function equal to a virtual displacement $\boldsymbol{u}^{*\epsilon}$ (i.e. a "small" displacement equal to zero on the boundary $\mathcal{B}^{\mathrm{EB}\epsilon}$), we obtain the heterogeneous virtual work equation

$$\int_{\mathcal{B}^\epsilon} \boldsymbol{\sigma}^\epsilon:\boldsymbol{\varepsilon}^{*\epsilon}\mathrm{d}\overline{\boldsymbol{x}} = \int_{\mathcal{B}^\epsilon} \rho\boldsymbol{b}\cdot\boldsymbol{u}^{*\epsilon}\mathrm{d}\overline{\boldsymbol{x}} + \int_{\mathcal{B}^{\mathrm{NB}\epsilon}} \boldsymbol{t}\cdot\boldsymbol{u}^{*\epsilon}\mathrm{d}s,$$
$$\forall \boldsymbol{u}^{*\epsilon} \in \left[C^\infty(\mathcal{B}^\epsilon)\right]^3. \qquad [4.24]$$

In order to pass to the limit in [4.24], we need an additional type of convergence, the two-scale convergence introduced by Nguetseng [NGU 89] and Allaire [ALL 92], which will be presented in the next subsection.

4.3.2. *Compensated compactness and two-scale convergence*

It is well known that the existence, the uniqueness and the finite element-based numerical treatment of boundary value problems depend crucially on the character of partial differential equations involved and on the functional setting of solutions of these problems. Usually, the equilibrium equations are of elliptic type, while the energy equation is of parabolic type, and the "nice" functional setting for applying finite element methods involves spatially distributed $L^2(\bar{\mathcal{B}})$-stress and strain, and $H^1(\bar{\mathcal{B}})$-displacements, smoothly varying with time. As it is stated in the weak convergence definition, if a function $\psi^\epsilon \in L^2(\bar{\mathcal{B}})$ satisfies

$$\int_{\bar{\mathcal{B}}} \psi^\epsilon \varphi \mathrm{d}\bar{\boldsymbol{x}} \to \int_{\bar{\mathcal{B}}} \psi \varphi \mathrm{d}\bar{\boldsymbol{x}}, \quad \forall \varphi \in L^2(\bar{\mathcal{B}}),$$

then ψ^ϵ converges weakly in $L^2(\bar{\mathcal{B}})$ to ψ. The application of H^1-convergence in linear elasticity states that, if the strain $\boldsymbol{\varepsilon}^\epsilon$ converges weakly to $\bar{\boldsymbol{\varepsilon}}$ in $L^2(\bar{\mathcal{B}})$ and div $(\boldsymbol{C}^\epsilon : \boldsymbol{\varepsilon}^\epsilon)$ converges strongly to $\boldsymbol{b}$ in $L^2(\bar{\mathcal{B}})$, then the stress $\boldsymbol{\sigma}^\epsilon = \boldsymbol{C}^\epsilon : \boldsymbol{\varepsilon}^\epsilon$ converges weakly to $\bar{\boldsymbol{\sigma}} = \overline{\boldsymbol{C}} : \bar{\boldsymbol{\varepsilon}}$ in $L^2(\bar{\mathcal{B}})$ and the energy $\boldsymbol{\sigma}^\epsilon : \boldsymbol{\varepsilon}^\epsilon$ converges in the sense of distributions

$$\int_{\bar{\mathcal{B}}} \varphi \boldsymbol{\sigma}^\epsilon : \boldsymbol{\varepsilon}^\epsilon \mathrm{d}\bar{\boldsymbol{x}} \to \int_{\bar{\mathcal{B}}} \varphi \bar{\boldsymbol{\sigma}} : \bar{\boldsymbol{\varepsilon}} \mathrm{d}\bar{\boldsymbol{x}}, \quad \forall \varphi \in C^\infty(\bar{\mathcal{B}}).$$

The above finding is a consequence of the theorem of the compensated compactness [TAR 77, MUR 78] and represents a generalization of the Hill–Mandel theorem for the homogenized mechanical work as the product of the homogenized stress and homogenized strain. It is worth noticing that the macroscopic elastic tensor $\overline{\boldsymbol{C}}$ is defined only in terms of the $\boldsymbol{C}^\epsilon$'s and the volume fractions of the phases.

Another more "detailed" weak convergence, using oscillating test functions $\varphi^\epsilon(\bar{\boldsymbol{x}})$, was introduced by Nguetseng [NGU 89] and Allaire [ALL 92]. According to this definition, a function $\psi^\epsilon(\bar{\boldsymbol{x}})$ two-scale converges to $\psi^{(0)}(\bar{\boldsymbol{x}}, \boldsymbol{x})$ if, for every $\varphi^\epsilon(\bar{\boldsymbol{x}}) = \varphi\left(\bar{\boldsymbol{x}}, \frac{\bar{\boldsymbol{x}}}{\epsilon}\right) \in L^2(\bar{\mathcal{B}} \times \mathcal{B})$,

$$\lim_{\epsilon \to 0} \int_{\bar{\mathcal{B}}} \psi^\epsilon(\bar{\boldsymbol{x}}) \varphi^\epsilon(\bar{\boldsymbol{x}}) \mathrm{d}\bar{\boldsymbol{x}} = \int_{\bar{\mathcal{B}}} \frac{1}{V} \int_{\mathcal{B}} \psi^{(0)}(\bar{\boldsymbol{x}}, \boldsymbol{x}) \varphi(\bar{\boldsymbol{x}}, \boldsymbol{x}) \mathrm{d}\boldsymbol{x} \mathrm{d}\bar{\boldsymbol{x}}. \qquad [4.25]$$

An important example of two-scale converging functions is given by the functions having an asymptotic expansion of the form

$$\psi^\epsilon(\bar{\boldsymbol{x}}) = \psi(\bar{\boldsymbol{x}}, \boldsymbol{x}) = \psi^{(0)}(\bar{\boldsymbol{x}}, \boldsymbol{x}) + \epsilon \psi^{(1)}(\bar{\boldsymbol{x}}, \boldsymbol{x}) + \epsilon^2 \psi^{(2)}(\bar{\boldsymbol{x}}, \boldsymbol{x}) + \ldots, \qquad [4.26]$$

where $\psi^{(0)}$, $\psi^{(1)}$, $\psi^{(2)}$ etc., are periodic in $\boldsymbol{x} = \dfrac{\overline{\boldsymbol{x}}}{\epsilon}$. More details about the asymptotic expansion method are provided in Chapter 5. Note that the two-scale limit $\psi^{(0)}$ carries more information than the weak limit

$$< \psi >= \frac{1}{V} \int_{\mathcal{B}} \psi^{(0)}(\overline{\boldsymbol{x}}, \boldsymbol{x}) \mathrm{d}\boldsymbol{x}. \qquad [4.27]$$

An equally important lemma in Nguetseng [NGU 89] and Allaire [ALL 92] is the two-scale convergence of derivatives of H^1-functions. Application of this lemma to the displacement gradient yields to the following assertion, revealing the existence of the micro-displacement $\boldsymbol{u}^{(1)}$: if the displacement $\boldsymbol{u}^{\epsilon}(\overline{\boldsymbol{x}})$ converges weakly in $H^1(\bar{\mathcal{B}})$ to $\boldsymbol{u}(\overline{\boldsymbol{x}})$, then $\boldsymbol{u}^{\epsilon}(\overline{\boldsymbol{x}})$ two-scale converges to $\boldsymbol{u}^{(0)} \equiv \boldsymbol{u}(\overline{\boldsymbol{x}})$ and there exists a micro-displacement $\boldsymbol{u}^{(1)}$ such that

$$\lim_{\epsilon \to 0} \int_{\bar{\mathcal{B}}} \mathrm{grad}^{\epsilon} \boldsymbol{u}^{\epsilon}(\overline{\boldsymbol{x}}) \varphi^{\epsilon}(\overline{\boldsymbol{x}}) \mathrm{d}\overline{\boldsymbol{x}}$$
$$= \int_{\bar{\mathcal{B}}} \frac{1}{V} \int_{\mathcal{B}} \left[\overline{\mathrm{grad}} \boldsymbol{u}^{(0)}(\overline{\boldsymbol{x}}) + \mathrm{grad} \boldsymbol{u}^{(1)}(\overline{\boldsymbol{x}}, \boldsymbol{x}) \right] \varphi(\overline{\boldsymbol{x}}, \boldsymbol{x}) \mathrm{d}\boldsymbol{x} \mathrm{d}\overline{\boldsymbol{x}}, \qquad [4.28]$$

$$\lim_{\epsilon \to 0} \int_{\bar{\mathcal{B}}} \boldsymbol{\varepsilon}^{\epsilon}(\overline{\boldsymbol{x}}) \varphi^{\epsilon}(\overline{\boldsymbol{x}}) \mathrm{d}\overline{\boldsymbol{x}}$$
$$= \int_{\bar{\mathcal{B}}} \frac{1}{V} \int_{\mathcal{B}} \left[\overline{\boldsymbol{\varepsilon}}(\overline{\boldsymbol{x}}) + \mathrm{grad}_{\mathrm{sym}} \boldsymbol{u}^{(1)}(\overline{\boldsymbol{x}}, \boldsymbol{x}) \right] \varphi(\overline{\boldsymbol{x}}, \boldsymbol{x}) \mathrm{d}\boldsymbol{x} \mathrm{d}\overline{\boldsymbol{x}}, \qquad [4.29]$$

for every $\varphi(\overline{\boldsymbol{x}}, \boldsymbol{x}) \in C^{\infty}(\bar{\mathcal{B}} \times \mathcal{B})$. The fact that the displacement two-scale converges to the first term of its asymptotic expansion $\boldsymbol{u}^{(0)}$ which is independent of the micro-coordinate can be proven rigorously in the case of elasticity [KAL 97]. In the case of nonlinear materials, the same conclusion can also be derived through an approach that is discussed in Chapter 5. Therefore, the first term of the strain in the right-hand side of equation [4.29] coincides with the macroscopic strain $\overline{\boldsymbol{\varepsilon}} = \overline{\mathrm{grad}}_{\mathrm{sym}} \boldsymbol{u}^{(0)}$. It is noted that the appearance of $\mathrm{grad}_{\mathrm{sym}} \boldsymbol{u}^{(1)}(\overline{\boldsymbol{x}}, \boldsymbol{x})$ fits with the introduction of the actual displacement within the cell defined by Suquet [SUQ 87] as

$$\boldsymbol{w}(\overline{\boldsymbol{x}}, \boldsymbol{x}) = \overline{\boldsymbol{\varepsilon}} \cdot \boldsymbol{x} + \boldsymbol{u}^{(1)}(\overline{\boldsymbol{x}}, \boldsymbol{x}).$$

At this stage, we can pass to the limit in [4.24] for ε^* equal to the actual strain ε and obtain with the help of [4.29]

$$\int_{\bar{\mathcal{B}}}\frac{1}{V}\int_{\mathcal{B}}\boldsymbol{\sigma}^{(0)}:\bar{\varepsilon}\mathrm{d}\boldsymbol{x}\mathrm{d}\bar{\boldsymbol{x}}+\int_{\bar{\mathcal{B}}}\frac{1}{V}\int_{\mathcal{B}}\boldsymbol{\sigma}^{(0)}:\mathrm{grad}\boldsymbol{u}^{(1)}\mathrm{d}\boldsymbol{x}\mathrm{d}\bar{\boldsymbol{x}}$$
$$=\int_{\bar{\mathcal{B}}}\rho\boldsymbol{b}\cdot\boldsymbol{u}^{(0)}\mathrm{d}\bar{\boldsymbol{x}}+\int_{\partial\bar{\mathcal{B}}^{\mathrm{NB}}}\boldsymbol{t}\cdot\boldsymbol{u}^{(0)}\mathrm{d}s. \quad [4.30]$$

Using that $\bar{\varepsilon}$ is independent of $\boldsymbol{x}$ and that $\bar{\boldsymbol{\sigma}}=\dfrac{1}{V}\displaystyle\int_{\mathcal{B}}\boldsymbol{\sigma}^{(0)}\mathrm{d}\boldsymbol{x}$, the above relation leads to the satisfaction of the following equations,

$$\int_{\bar{\mathcal{B}}}\bar{\boldsymbol{\sigma}}:\bar{\varepsilon}\mathrm{d}\bar{\boldsymbol{x}}=\int_{\bar{\mathcal{B}}}\rho\boldsymbol{b}\cdot\boldsymbol{u}^{(0)}\mathrm{d}\bar{\boldsymbol{x}}+\int_{\partial\bar{\mathcal{B}}^{\mathrm{NB}}}\boldsymbol{t}\cdot\boldsymbol{u}^{(0)}\mathrm{d}s, \quad [4.31]$$

$$\int_{\mathcal{B}}\boldsymbol{\sigma}^{(0)}:\mathrm{grad}\boldsymbol{u}^{(1)}\mathrm{d}\boldsymbol{x}=0. \quad [4.32]$$

Equation [4.31] is the macro-equilibrium equation and [4.32] is the cell problem in the microscale. An analytical presentation of the asymptotic expansion homogenization and the solution of the cell problem will be presented in Chapter 5. It is worth noting that the virtual work equation and the convergence results presented above are independent of the constitutive law of the materials.

As a remark, it should be mentioned that the steady state heat conduction problem

$$\mathrm{div}^{\epsilon}\boldsymbol{q}^{\epsilon}-\rho^{\epsilon}\mathcal{R}^{\epsilon}=0 \quad \text{in } \mathcal{B}^{\epsilon},$$
$$\boldsymbol{q}^{\epsilon}=-\boldsymbol{k}^{\epsilon}\cdot\boldsymbol{\nabla}\boldsymbol{\theta}^{\epsilon},$$
$$\boldsymbol{\nabla}\boldsymbol{\theta}^{\epsilon}=\mathrm{grad}^{\epsilon}\theta^{\epsilon} \quad \text{in } \mathcal{B}^{\epsilon},$$
$$-\boldsymbol{q}^{\epsilon}\cdot\boldsymbol{n}^{\epsilon}=q^{s\ \epsilon} \quad \text{on } \partial\mathcal{B}^{\mathrm{QB}\epsilon},$$
$$\theta^{\epsilon}=\theta^{\mathrm{prescribed}} \quad \text{on } \partial\mathcal{B}^{\mathrm{TB}\epsilon},$$

with $\partial\mathcal{B}^{\mathrm{TB}\epsilon}\bigcup\partial\mathcal{B}^{\mathrm{QB}\epsilon} = \partial\mathcal{B}^{\epsilon}$, can be treated with the same mathematical tools, since it has the same structure with the mechanical problem. However, the coupled thermoelastic problem accounting for inertia effects presents certain theoretical difficulties and special treatment is required, as well as proper choice of the initial conditions [BRA 88, BRA 92] (further discussion about this point is provided is Chapter 5).

4.3.3. *General results from mathematical homogenization*

Homogenization techniques, combined with numerical analyses of large engineering structures subjected to loadings leading to inelastic behavior or highly localized strain, still remain a challenge. However, there exists a compact support of knowledge due to the work of mathematicians, which paved the way for the development of consistent numerical methods. Specifically, optimal bounds in linear elasticity [FRA 86b] and level set methods [ALL 04] promote the design of complex structures. However, the knowledge of general homogenization-induced properties, such as the loss of isotropy, the deviation of the macroscopic constitutive law from the corresponding laws of the composite constituents and the dependence of macroscopic properties on initial and/or boundary conditions in inelastic problems, permits the correct characterization of modern composites. Simple inelastic model cases help to understand the behavior of complex problems under different loading paths. For instance, the analysis of 1-D viscoplastic problems has shown that the models exhibiting power law strain rate sensitivity are not stable by homogenization and that a memory effect appears under time dependent boundary and body forces [CHA 10b, CHA 10c]. Additionally, the homogenized strain hardening (or softening) of thermoviscoplastic power law materials is no more given by a power law. One-dimensional stratified materials, very useful in treating sequentially laminated multiscale complex structures with local periodicity, exhibit explicit formulas for the macroscopic parameters in elasticity and even in some inelastic problems. The fact that the unit cell can be expressed as a periodic 1-D problem leads to analytical and semi-analytical methods that are very efficient when combined with micromechanics-based methods, as in the case of stratified materials made of composite sheets.

In the homogenization framework of dissipative generalized standard materials, there exist two categories of micro-functions: the first category comprises micro-displacement, micro-stress and micro-strain, which appear exactly as the two-scale convergence limits from the heterogeneous equations and are controlled directly *via* the macro-strain path. However, the second set of micro-functions comprises the internal variables (as the plastic strain), which depend on both micro- and macro-dissipation inequality. The mathematical homogenization of elastic-plastic hardening materials and elastoviscoplastic materials has been fully studied in the literature. The evolution of their solution in time is smooth. For fixed time interval, their solution can be easier computed numerically, since they belong to H^1 for displacements and L^2 for strain and stress. This is essentially due to the absence of plasticity generated discontinuities in displacement, leading to a dissipation without concentrations. The numerical investigation does not need any loading safety control and is based on a complete computational scheme, a return mapping algorithm (RMA) [SIM 98] (see Chapter 3 for details). On the contrary, composites with elastic-perfectly plastic components are not handled easily by homogenization methods. In elastic-perfectly plastic components, although the evolution is smooth in time for quasi-static evolutions, the dissipation product is the product of a stress and a plastic strain that is a measure [FRA 12, FRA 14, FRA 15]. Elastic-plastic evolutions may exhibit strain localization, shear banding or failure and need loading control.

5

Composites with Periodic Structure

Material science has grown tremendously in the recent years to meet the extensive needs of several engineering applications. The automotive and aerospace industries require novel, innovative and multifunctional composite materials that can be utilized in complicated structures with high demands in strength, durability and long lifetime during repeated loading cycles.

To match those high requirements, composites are frequently exposed in regimes where dissipative phenomena like viscoelasticity, viscoplasticity or plasticity occur. Such deformation mechanisms are often accompanied by significant temperature increases, which influences in return the material's response. The thermomechanical couplings play a key role in thermoplastics operating in temperature ranges close to the glass transition zones, as well as in metals during phase transformation. In addition, thermal and mechanical fields are strongly connected in the case of fatigue. Generally, fatigue is a complicated phenomenon that depends not only on the stress state, but also on the energy dissipation that occurs during inelastic mechanisms [BEN 15]. A proper study of the fatigue of a structure requires to account for the energy exchanges during thermomechanical loading cycles. In the case of homogeneous materials, strong thermomechanical couplings are examined through appropriate thermodynamics frameworks, like for instance the theory of the generalized standard materials [GER 73, HAL 75, GER 82, GER 83]. For composites, however, such studies are very limited.

Composites with periodic microstructure appear very often in engineering applications. Laminate structures with repeated layers and woven composites are typical examples of periodic media. Additionally, particulate and unidirectional fiber composites where the particles or fibers are evenly distributed into the matrix material can be seen as periodic. The identification of the overall response of a composite with periodic microstructure often requires efficient computational tools like finite element or finite volume methods. Multiscale approaches have been developed in order to examine the overall behavior of viscoelastic, elastoplastic or viscoplastic composites [SUQ 87, TER 01, YU 02, ABO 03, ABO 04, ASA 07, KHA 09, CAV 09, CAV 11, KRU 11, CAV 16]. Most methodologies for nonlinear composites utilize appropriate tangent-stiffness methods, but there also exist frameworks based on secant approximations [GON 04]. To reduce the computational cost of FE calculations, a method that utilizes fast Fourier transforms has been developed [MOU 98]. Analytical or semi-analytical homogenization solutions exist for elastic multilayered composites [KAL 97] and for elastic fiber composites with square or hexagonal symmetry [ROD 01, GUI 01, WAN 16a, WAN 16b].

The asymptotic expansion homogenization (AEH) method was originally developed in the late 1970s [BEN 78, SAN 78]. It is a homogenization framework for periodic media with the advantage that it does not require initial assumption about the form of the microscopic and macroscopic equations which describe the problem under investigation. AEH initially considers a general form of the system of equations, and through appropriate asymptotic expansions on the field variables (displacement, temperature, strain, stress, etc.) identifies in a rigorous manner the representation of the various equations in microscale and macroscale. The method is classical and fairly straightforward regarding the case of thermoelastic media. The AEH has been implemented successfully for composites with elastoplastic components [FIS 97], as well as for shape memory alloy composites [HER 07, CHA 15], but these works focus exclusively on the mechanical composite response. A limited number of studies combine the conservations of linear momentum and energy in fully coupled thermomechanical processes [ENE 83, FRA 86a, YU 02, TEM 12]. Yu and Fish [YU 02] use the AEH in order to perform a fully coupled thermomechanical analysis for viscoelastic composites with periodic microstructure. Their analysis is based on specific form for the materials constitutive law. Sengupta *et al.* [SEN 12] have proposed a fully coupled thermomechanical framework for composites under large deformation processes. This study is based on certain hypotheses on how the conservation laws are expressed in both scales and which is the correlation between microscopic and macroscopic variables.

This chapter follows the framework proposed by Chatzigeorgiou *et al.* [CHA 16] and presents a fully coupled thermomechanical homogenization framework, applicable to any type of constitutive law that is based on thermodynamic principles. Thus, it allows different types of inelastic material behaviors for the various constituents of the composite. The developed methodology identifies initially the micro- and macro-conservation laws and investigates afterward how a general energy potential (for instance, Helmholtz or Gibbs), or more accurately its rate, is properly formulated in both scales. The outcome of this study is that it proves in a systematic and consistent manner which variables and equations can be rigorously identified at the macroscopic scale and which variables require either additional assumptions or numerical treatment.

As already explained in section 4.3, the limit of product of two oscillating functions does not converge to the product of their limits, unless additional conditions hold, as in the case of H-convergence and its application to the energy product (Hill–Mandel theorem). In the homogenization of fully coupled thermomechanical processes, products of oscillating functions appear necessarily after time-differentiation of functions depending on the internal variables $\boldsymbol{\xi}$, such as the internal energy and the free energies (Helmholtz or Gibbs). These products are composed by the rates of the internal variables and the partial derivatives of the considered energies with respect to the internal variables (see section 2.5.2). The above rates of energy enter the energy equation and the dissipation function, prohibiting the derivation of homogenized expressions in a form similar to the heterogeneous and microscale expressions. This fact suggests a theoretical setting and a subsequent computational framework based on the division of the global process time into small intervals, transforming the thermomechanical evolution to a succession of slow changes of variables and applying an incremental linearized procedure with small increments of all functions, where all unknowns, including internal variables, are expressed in terms of the macro-strain and macro-temperature. Such a computational procedure is utilized here by introducing the "linearized incremental formulation".

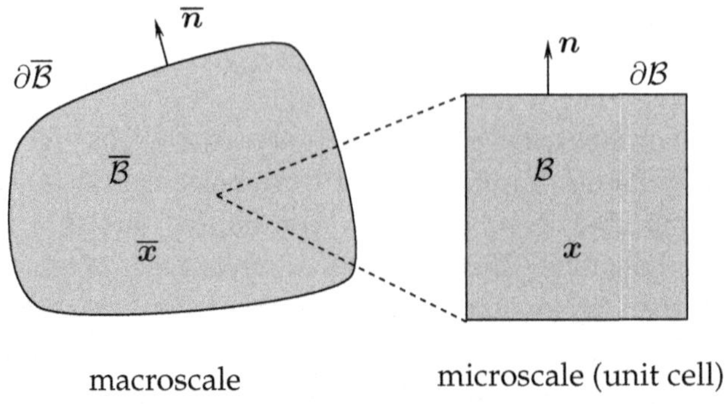

Figure 5.1. *Macroscale and microscale of composite. For a color version of this figure, see www.iste.co.uk/chatzigeorgiou/thermomechanical.zip*

5.1. Thermomechanical processes

As already explained in section 4.3, periodic homogenization describes the composite using two scales. At the macroscale, the continuum body occupies the space $\overline{\mathcal{B}}$ with volume $\overline{V}$ and is bounded by the surface $\partial\overline{\mathcal{B}}$ with normal unit vector $\overline{\boldsymbol{n}}$. Each macroscopic point is assigned with a position vector $\overline{\boldsymbol{x}}$ in $\overline{\mathcal{B}}$. However, at the microscale level, there is a periodic unit cell that occupies the space $\mathcal{B}$ with volume V and is bounded by the surface $\partial\mathcal{B}$ with normal unit vector $\boldsymbol{n}$. Each microscopic point is assigned with a position vector $\boldsymbol{x}$ in $\mathcal{B}$ (Figure 5.1). The two scales $\overline{\boldsymbol{x}}$ and $\boldsymbol{x}$ are connected with a characteristic length ϵ through the relation $\boldsymbol{x} = \overline{\boldsymbol{x}}/\epsilon$.

It is recalled from Chapter 4 that all the variables corresponding to the general composite response are denoted with a superscript ϵ. Moreover, a bar above a symbol denotes a macroscopic quantity or variable, depending only on the vector $\overline{\boldsymbol{x}}$ and perhaps the time. Additionally, the mean operators $\langle\{\bullet\}\rangle$ and $\lceil\{\bullet\}\rfloor$ introduced in section 4.1 refer to volume averaging over the unit cell. Moreover, the operations with derivatives, like the grad^ϵ, and the div^ϵ, follow the decomposition rule [4.15].

5.1.1. *Kinematics*

As described in Chapter 2, the strain tensor $\boldsymbol{\varepsilon}^\epsilon$ can be expressed as a function of the displacement vector $\boldsymbol{u}^\epsilon$ through the relation

$$\boldsymbol{\varepsilon}^\epsilon = \mathrm{grad}^\epsilon_{\mathrm{sym}} \boldsymbol{u}^\epsilon. \qquad [5.1]$$

The periodicity of the composite microstructure allows us to assume the following asymptotic series expansion for the displacement [SAN 78],

$$\boldsymbol{u}^\epsilon(\overline{\boldsymbol{x}}, t) = \boldsymbol{u}^{(0)}(\overline{\boldsymbol{x}}, \boldsymbol{x}, t) + \epsilon\, \boldsymbol{u}^{(1)}(\overline{\boldsymbol{x}}, \boldsymbol{x}, t) + \epsilon^2 \boldsymbol{u}^{(2)}(\overline{\boldsymbol{x}}, \boldsymbol{x}, t) + ... \qquad [5.2]$$

The main assumption in the asymptotic expansion homogenization framework is that all two-scale functions $\{\bullet\}^{(0)}$, $\{\bullet\}^{(1)}$ etc., are considered periodic in $\boldsymbol{x}$. Using equation [5.2] in [5.1] and taking into account [4.15] yields

$$\boldsymbol{\varepsilon}^\epsilon = \frac{1}{\epsilon} \mathrm{grad}_{\mathrm{sym}} \boldsymbol{u}^{(0)} + \overline{\mathrm{grad}}_{\mathrm{sym}} \boldsymbol{u}^{(0)} + \mathrm{grad}_{\mathrm{sym}} \boldsymbol{u}^{(1)} + \epsilon ....$$

For $\epsilon \rightarrow 0$, the deformation gradient is bounded only when the ϵ^{-1} term vanishes [TEM 12]. Taking into account that rigid body rotations are not periodic motions, we conclude that

$$\mathrm{grad}_{\mathrm{sym}} \boldsymbol{u}^{(0)} = \boldsymbol{0} \quad \rightarrow \quad \boldsymbol{u}^{(0)} := \overline{\boldsymbol{u}}(\overline{\boldsymbol{x}}, t). \qquad [5.3]$$

The fact that the first expanding term of the displacement depends only on the macroscopic vector $\overline{\boldsymbol{x}}$ is very common in the periodic homogenization theory (see, for instance, in the cases of thermoelasticity [ENE 83, KAL 97] and thermoviscoelasticity [YU 02]). Using the final result, we can write the strain tensor in the general expanded form

$$\boldsymbol{\varepsilon}^\epsilon(\overline{\boldsymbol{x}}, t) = \boldsymbol{\varepsilon}^{(0)}(\overline{\boldsymbol{x}}, \boldsymbol{x}, t) + \epsilon\, \boldsymbol{\varepsilon}^{(1)}(\overline{\boldsymbol{x}}, \boldsymbol{x}, t) + \epsilon^2 \boldsymbol{\varepsilon}^{(2)}(\overline{\boldsymbol{x}}, \boldsymbol{x}, t) + ..., \qquad [5.4]$$

with

$$\boldsymbol{\varepsilon}^{(0)} = \overline{\mathrm{grad}}_{\mathrm{sym}} \overline{\boldsymbol{u}} + \mathrm{grad}_{\mathrm{sym}} \boldsymbol{u}^{(1)}. \qquad [5.5]$$

Defining $\overline{\boldsymbol{\varepsilon}} := \overline{\text{grad}}_{\text{sym}}\overline{\boldsymbol{u}}$ and taking into account the relation (see section 4.1)

$$\left\langle \text{grad}_{\text{sym}}\boldsymbol{u}^{(1)} \right\rangle = \mathbf{0},$$

yields

$$\overline{\boldsymbol{\varepsilon}} = \left\langle \boldsymbol{\varepsilon}^{(0)} \right\rangle. \qquad [5.6]$$

The tensor $\boldsymbol{\varepsilon}^{(0)}$ is defined as the microscopic strain, while the tensor $\overline{\boldsymbol{\varepsilon}}$, depending exclusively on $\overline{\boldsymbol{x}}$ and the time, is defined as the macroscopic strain.

5.1.2. *Kinetics*

The stress is assumed to have a similar asymptotic series expansion with the strain, i.e.

$$\boldsymbol{\sigma}^{\epsilon}(\overline{\boldsymbol{x}}, t) = \boldsymbol{\sigma}^{(0)}(\overline{\boldsymbol{x}}, \boldsymbol{x}, t) + \epsilon\, \boldsymbol{\sigma}^{(1)}(\overline{\boldsymbol{x}}, \boldsymbol{x}, t) + \epsilon^2 \boldsymbol{\sigma}^{(2)}(\overline{\boldsymbol{x}}, \boldsymbol{x}, t) + ... \qquad [5.7]$$

Similarly to the strain, the tensor $\boldsymbol{\sigma}^{(0)}$ is defined as the microscopic stress, while its weak limit, i.e. its average over the unit cell $\overline{\boldsymbol{\sigma}} := \left\langle \boldsymbol{\sigma}^{(0)} \right\rangle$, is defined as the macroscopic stress. Obviously, $\overline{\boldsymbol{\sigma}}$ does not depend on $\boldsymbol{x}$.

5.1.3. *Conservation laws*

5.1.3.1. *Conservation of mass*

The following asymptotic series expansion for the density is assumed:

$$\rho^{\epsilon}(\overline{\boldsymbol{x}}, t) = \rho^{(0)}(\overline{\boldsymbol{x}}, \boldsymbol{x}, t) + \epsilon\, \rho^{(1)}(\overline{\boldsymbol{x}}, \boldsymbol{x}, t) + \epsilon^2 \rho^{(2)}(\overline{\boldsymbol{x}}, \boldsymbol{x}, t) + ... \qquad [5.8]$$

According to Table 2.1, the conservation of mass is written as

$$\frac{D\rho^{\epsilon}}{Dt} \simeq 0. \qquad [5.9]$$

Inserting [5.8] in [5.9] yields

$$\frac{D\rho^{(0)}}{Dt} + \epsilon ... = 0. \qquad [5.10]$$

Since the scope of homogenization is to study the system of equations when ϵ tends to zero, the following equation leads to the microscale conservation of mass

$$\frac{D\rho^{(0)}}{Dt} = 0 \quad \Rightarrow \quad \rho^{(0)} := \rho^{(0)}(\overline{\boldsymbol{x}}, \boldsymbol{x}). \qquad [5.11]$$

The macroscale conservation of mass is identified by obtaining the weak limit of equation [5.11], i.e. averaging it over the unit cell,

$$\left\langle \frac{D\rho_0^{(0)}}{Dt} \right\rangle = 0,$$

or, by defining $\overline{\rho} := \left\langle \rho^{(0)} \right\rangle$,

$$\frac{D\overline{\rho}}{Dt} = 0 \quad \Rightarrow \quad \overline{\rho} := \overline{\rho}(\overline{\boldsymbol{x}}). \qquad [5.12]$$

From these results, it becomes clear that $\rho^{(0)}$ represents the microscopic density and $\overline{\rho}$ represents the macroscopic density.

5.1.3.2. *Conservation of linear momentum*

According to Table 2.1, the conservation of linear momentum is written as

$$\rho^{\epsilon} \frac{D\dot{\boldsymbol{u}}^{\epsilon}}{Dt} = \rho^{\epsilon} \boldsymbol{b}^{\epsilon} + \operatorname{div}^{\epsilon} \boldsymbol{\sigma}^{\epsilon}. \qquad [5.13]$$

Considering an asymptotic series expansion for the body forces

$$\boldsymbol{b}^{\epsilon}(\overline{\boldsymbol{x}}, t) = \boldsymbol{b}^{(0)}(\overline{\boldsymbol{x}}, \boldsymbol{x}, t) + \epsilon\, \boldsymbol{b}^{(1)}(\overline{\boldsymbol{x}}, \boldsymbol{x}, t) + \epsilon^2 \boldsymbol{b}^{(2)}(\overline{\boldsymbol{x}}, \boldsymbol{x}, t) + ..., \qquad [5.14]$$

and with the help of equations [4.15], [5.2], [5.7] and [5.8], we obtain

$$\rho^{(0)}\frac{D\dot{\bar{\boldsymbol{u}}}}{Dt} + \epsilon ... = \rho^{(0)}\boldsymbol{b}^{(0)} + \overline{\text{div}}\boldsymbol{\sigma}^{(0)} + \text{div}\boldsymbol{\sigma}^{(1)} + \frac{1}{\epsilon}\text{div}\boldsymbol{\sigma}^{(0)} + \epsilon ... \quad [5.15]$$

For $\epsilon \to 0$, the term ϵ^{-1} should vanish, leading to the conservation of linear momentum in the microscale

$$\text{div}\boldsymbol{\sigma}^{(0)} = \boldsymbol{0}. \quad [5.16]$$

The conservation of linear momentum in the macroscale is obtained by averaging over the unit cell the ϵ^0 term of equation [5.15]. Noting that $\left\langle \text{div}\boldsymbol{\sigma}^{(1)} \right\rangle = \boldsymbol{0}$ (see section 4.1) and defining $\bar{\boldsymbol{b}} := \left\langle \rho^{(0)}\boldsymbol{b}^{(0)} \right\rangle / \bar{\rho}$, the average of the ϵ^0 term takes the form

$$\bar{\rho}\frac{D\dot{\bar{\boldsymbol{u}}}}{Dt} = \bar{\rho}\bar{\boldsymbol{b}} + \overline{\text{div}}\bar{\boldsymbol{\sigma}}. \quad [5.17]$$

As it can be seen from [5.16], no inertia term is included in the microscopic linear momentum. This result is a consequence of the asymptotic expansion approach and the zeroth order form of homogenization (i.e. the higher order terms are neglected). These assumptions though are not always valid, since the actual composite under dynamic effects may behave in a non-periodic manner [BRA 92, NEM 11]. The obtained microscopic and macroscopic linear momentum equations can be considered reliable as long as the unit cell size is much smaller than the wavelength of the dynamic loading [CHA 10a]. When the wavelength is comparable with the unit cell size, then either higher order terms should be taken into account in the homogenization procedure [CHE 01, HUI 14], or alternative homogenization techniques should be utilized [YAN 94, AGG 04, NEM 11]. Since the present book focuses on quasi-static processes, in later chapters, the inertia terms are omitted.

NOTE.– The microscopic strain is expressed by equations [5.5] and [5.6]. However the microscopic stress is statically admissible due to equation [5.16]. Thus, according to the discussion in section 4.2.1, $\boldsymbol{\sigma}^{(0)}$ and $\boldsymbol{\varepsilon}^{(0)}$ satisfy the Hill–Mandel theorem, i.e.

$$\left\langle \boldsymbol{\sigma}^{(0)}:\dot{\boldsymbol{\varepsilon}}^{(0)}\right\rangle = \left\langle \boldsymbol{\sigma}^{(0)}\right\rangle : \left\langle \dot{\boldsymbol{\varepsilon}}^{(0)}\right\rangle = \overline{\boldsymbol{\sigma}}:\dot{\overline{\boldsymbol{\varepsilon}}},$$

$$\left\langle \dot{\boldsymbol{\sigma}}^{(0)}:\boldsymbol{\varepsilon}^{(0)}\right\rangle = \left\langle \dot{\boldsymbol{\sigma}}^{(0)}\right\rangle : \left\langle \boldsymbol{\varepsilon}^{(0)}\right\rangle = \dot{\overline{\boldsymbol{\sigma}}}:\overline{\boldsymbol{\varepsilon}}. \qquad [5.18]$$

5.1.3.3. *Conservation of angular momentum*

According to Table 2.1, the conservation of angular momentum is simply expressed as

$$\boldsymbol{\sigma}^{\epsilon} = [\boldsymbol{\sigma}^{\epsilon}]^{T}. \qquad [5.19]$$

With the help of equation [5.7] and setting $\epsilon \to 0$, the angular momentum in the microscale is written as

$$\boldsymbol{\sigma}^{(0)} = \left[\boldsymbol{\sigma}^{(0)}\right]^{T}. \qquad [5.20]$$

The volume average over the unit cell of the above expression provides the conservation of angular momentum in the macroscale. Taking into account the properties presented in section 4.1, the final expression for the macroscale takes the form

$$\overline{\boldsymbol{\sigma}} = \overline{\boldsymbol{\sigma}}^{T}. \qquad [5.21]$$

5.1.3.4. *Conservation of energy*

According to equations [2.30] and [2.33], the conservation of energy is written as

$$\dot{\mathcal{E}}^{\epsilon} = \boldsymbol{\sigma}^{\epsilon}:\dot{\boldsymbol{\varepsilon}}^{\epsilon} - \mathrm{div}\boldsymbol{q}^{\epsilon} + \rho^{\epsilon}\mathcal{R}^{\epsilon}, \quad \text{or} \quad r^{\epsilon} - \mathrm{div}\boldsymbol{q}^{\epsilon} + \rho^{\epsilon}\mathcal{R}^{\epsilon} = 0. \qquad [5.22]$$

Considering an asymptotic series expansion for the internal energy, the heat fluxes and the heat sources in the reference configuration

$$\mathcal{E}^{\epsilon}(\overline{\boldsymbol{x}},t) = \mathcal{E}^{(0)}(\overline{\boldsymbol{x}},\boldsymbol{x},t) + \epsilon\,\mathcal{E}^{(1)}(\overline{\boldsymbol{x}},\boldsymbol{x},t) + \epsilon^{2}\mathcal{E}^{(2)}(\overline{\boldsymbol{x}},\boldsymbol{x},t) + ...,$$

$$\begin{aligned}\boldsymbol{q}^{\epsilon}(\overline{\boldsymbol{x}},t) &= \boldsymbol{q}^{(0)}(\overline{\boldsymbol{x}},\boldsymbol{x},t)+\epsilon\,\boldsymbol{q}^{(1)}(\overline{\boldsymbol{x}},\boldsymbol{x},t)+\epsilon^{2}\boldsymbol{q}^{(2)}(\overline{\boldsymbol{x}},\boldsymbol{x},t)+...,\\ \mathcal{R}^{\epsilon}(\overline{\boldsymbol{x}},t) &= \mathcal{R}^{(0)}(\overline{\boldsymbol{x}},\boldsymbol{x},t)+\epsilon\,\mathcal{R}^{(1)}(\overline{\boldsymbol{x}},\boldsymbol{x},t)+\epsilon^{2}\mathcal{R}^{(2)}(\overline{\boldsymbol{x}},\boldsymbol{x},t)+...,\\ r^{\epsilon}(\overline{\boldsymbol{x}},t) &= r^{(0)}(\overline{\boldsymbol{x}},\boldsymbol{x},t)+\epsilon\,r^{(1)}(\overline{\boldsymbol{x}},\boldsymbol{x},t)+\epsilon^{2}\,r^{(2)}(\overline{\boldsymbol{x}},\boldsymbol{x},t)+...,\end{aligned} \quad [5.23]$$

and with the help of equation [4.15] and the expansions [5.4], [5.7] and [5.8], we obtain

$$\begin{aligned}\dot{\mathcal{E}}^{(0)}+\epsilon... &= \boldsymbol{\sigma}^{(0)}:\dot{\boldsymbol{\varepsilon}}^{(0)}-\overline{\text{div}}\boldsymbol{q}^{(0)}-\text{div}\boldsymbol{q}^{(1)}+\rho^{(0)}\mathcal{R}^{(0)}\\ &\quad -\frac{1}{\epsilon}\text{div}\boldsymbol{q}^{(0)}+\epsilon....\end{aligned} \quad [5.24]$$

For $\epsilon \rightarrow 0$, the term ϵ^{-1} should vanish, leading to the conservation of energy in the microscale

$$\text{div}\boldsymbol{q}^{(0)}=\boldsymbol{0}. \quad [5.25]$$

The conservation of energy in the macroscale is obtained by volume averaging over the unit cell the ϵ^{0} term of equation [5.24]. Noting that (see section 4.1)

$$\left\langle \text{div}\boldsymbol{q}^{(1)}\right\rangle = \boldsymbol{0},$$

and defining $\overline{\mathcal{E}} := \left\langle \mathcal{E}^{(0)}\right\rangle$, $\overline{\mathcal{R}} := \left\langle \rho^{(0)}\mathcal{R}^{(0)}\right\rangle/\overline{\rho}$ and $\overline{\boldsymbol{q}} := \left\langle \boldsymbol{q}^{(0)}\right\rangle$, the average of the ϵ^{0} term, with the help of relation [5.18], takes the form

$$\dot{\overline{\mathcal{E}}} = \overline{\boldsymbol{\sigma}}:\dot{\overline{\boldsymbol{\varepsilon}}}-\overline{\text{div}}\overline{\boldsymbol{q}}+\overline{\rho}\overline{\mathcal{R}}. \quad [5.26]$$

Finally, when ϵ tends to 0, the energy rate term r (the difference between the rates of the mechanical work and the internal energy) obtains its microscopic representation

$$r^{(0)} = \boldsymbol{\sigma}^{(0)}:\dot{\boldsymbol{\varepsilon}}^{(0)}-\dot{\mathcal{E}}^{(0)}, \quad [5.27]$$

and the volume average of the last expression, combined with the Hill–Mandel theorem provides the macroscopic $\bar{r}$ as

$$\bar{r} = \left\langle r^{(0)} \right\rangle = \bar{\boldsymbol{\sigma}}:\dot{\bar{\boldsymbol{\varepsilon}}} - \dot{\bar{\mathcal{E}}}. \qquad [5.28]$$

5.1.3.5. *Entropy inequality*

According to equation [2.32], the entropy inequality is written as

$$\theta^\epsilon \dot{\eta}^\epsilon - \frac{1}{\theta^\epsilon}\boldsymbol{q}^\epsilon \cdot \boldsymbol{\nabla\theta}^\epsilon + \boldsymbol{\sigma}^\epsilon:\dot{\boldsymbol{\varepsilon}}^\epsilon - \dot{\mathcal{E}}^\epsilon \geq 0, \quad \boldsymbol{\nabla\theta}^\epsilon = \text{grad}^\epsilon\theta^\epsilon. \qquad [5.29]$$

The periodicity of the composite microstructure allows us to assume the following asymptotic series expansion for the absolute temperature and the entropy:

$$\begin{aligned} \theta^\epsilon(\bar{\boldsymbol{x}}, t) &= \theta^{(0)}(\bar{\boldsymbol{x}}, \boldsymbol{x}, t) + \epsilon\, \theta^{(1)}(\bar{\boldsymbol{x}}, \boldsymbol{x}, t) + \epsilon^2\theta^{(2)}(\bar{\boldsymbol{x}}, \boldsymbol{x}, t) + ..., \\ \eta^\epsilon(\bar{\boldsymbol{x}}, t) &= \eta^{(0)}(\bar{\boldsymbol{x}}, \boldsymbol{x}, t) + \epsilon\, \eta^{(1)}(\bar{\boldsymbol{x}}, \boldsymbol{x}, t) + \epsilon^2\eta^{(2)}(\bar{\boldsymbol{x}}, \boldsymbol{x}, t) + ... \end{aligned} \qquad [5.30]$$

Combining $[5.30]_1$ with $[5.29]_2$ and using [4.15] yields

$$\boldsymbol{\nabla\theta}^\epsilon = \frac{1}{\epsilon}\text{grad}\theta^{(0)} + \overline{\text{grad}}\theta^{(0)} + \text{grad}\theta^{(1)} + \epsilon....$$

For $\epsilon \to 0$, the temperature gradient is bounded only when the ϵ^{-1} term vanish [TEM 12], leading to

$$\text{grad}\theta^{(0)} = \boldsymbol{0} \quad \rightarrow \quad \theta^{(0)} := \bar{\theta}(\bar{\boldsymbol{x}}, t). \qquad [5.31]$$

The conclusion that the first expanding term of the absolute temperature depends only on the macroscopic vector $\bar{\boldsymbol{x}}$ has been also verified in the cases of thermoelasticity [ENE 83, KAL 97] and thermoviscoelasticity [YU 02]. Using the final result, we can write the temperature gradient as

$$\begin{aligned} \boldsymbol{\nabla\theta}^\epsilon(\bar{\boldsymbol{x}}, t) &= \boldsymbol{\nabla\theta}^{(0)}(\bar{\boldsymbol{x}}, \boldsymbol{x}, t) + \epsilon\, \boldsymbol{\nabla\theta}^{(1)}(\bar{\boldsymbol{x}}, \boldsymbol{x}, t) \\ &\quad + \epsilon^2\boldsymbol{\nabla\theta}^{(2)}(\bar{\boldsymbol{x}}, \boldsymbol{x}, t) + ..., \end{aligned} \qquad [5.32]$$

with

$$\boldsymbol{\nabla}\boldsymbol{\theta}^{(0)} = \overline{\text{grad}\bar{\theta}} + \text{grad}\theta^{(1)}. \qquad [5.33]$$

Defining $\overline{\boldsymbol{\nabla}\boldsymbol{\theta}} := \overline{\text{grad}\bar{\theta}}$ and recalling from section 4.1 that

$$\left\langle \text{grad}\theta^{(1)} \right\rangle = \mathbf{0},$$

we obtain that

$$\overline{\boldsymbol{\nabla}\boldsymbol{\theta}} = \left\langle \boldsymbol{\nabla}\boldsymbol{\theta}^{(0)} \right\rangle. \qquad [5.34]$$

It is noted that equations [5.33] and [5.34] in conjunction with [5.25] lead to the conclusion that the Hill–Mandel thermal power equivalence holds, i.e.

$$\left\langle \boldsymbol{q}^{(0)} \cdot \boldsymbol{\nabla}\boldsymbol{\theta}^{(0)} \right\rangle = \left\langle \boldsymbol{q}^{(0)} \right\rangle \cdot \left\langle \boldsymbol{\nabla}\boldsymbol{\theta}^{(0)} \right\rangle = \bar{\boldsymbol{q}} \cdot \overline{\boldsymbol{\nabla}\boldsymbol{\theta}}, \qquad [5.35]$$

(see the discussion in section 4.2.2).

The expanded form of equation [5.29] is written as

$$\bar{\theta}\dot{\eta}^{(0)} - \frac{1}{\bar{\theta} + \epsilon ...}\boldsymbol{q}^{(0)} \cdot \boldsymbol{\nabla}\boldsymbol{\theta}^{(0)} + \boldsymbol{\sigma}^{(0)} : \dot{\boldsymbol{\varepsilon}}^{(0)} - \dot{\mathcal{E}}^{(0)} + \epsilon ... \geq 0. \qquad [5.36]$$

For $\epsilon \to 0$, the entropy inequality in the microscale is obtained,

$$\bar{\theta}\dot{\eta}^{(0)} - \frac{1}{\bar{\theta}}\boldsymbol{q}^{(0)} \cdot \boldsymbol{\nabla}\boldsymbol{\theta}^{(0)} + \boldsymbol{\sigma}^{(0)} : \dot{\boldsymbol{\varepsilon}}^{(0)} - \dot{\mathcal{E}}^{(0)} \geq 0. \qquad [5.37]$$

Defining $\bar{\eta} := \left\langle \eta^{(0)} \right\rangle$, the macroscale entropy inequality takes the form

$$\bar{\theta}\dot{\bar{\eta}} - \frac{1}{\bar{\theta}}\bar{\boldsymbol{q}} \cdot \overline{\boldsymbol{\nabla}\boldsymbol{\theta}} + \bar{\boldsymbol{\sigma}} : \dot{\bar{\boldsymbol{\varepsilon}}} - \dot{\bar{\mathcal{E}}} \geq 0, \qquad [5.38]$$

where equations [5.18] and [5.35] have been utilized.

Table 5.1 summarizes the various variables and forms of conservation laws at both the microscale and macroscale. In the table and for the remaining part of this chapter, the superscript (0) of the first asymptotic expansion terms is omitted and the superscript (1) of the second terms is substituted with a tilde above the variable. A similar analysis for large deformation processes is provided in Appendix 2. Finally, it is worth mentioning that the results obtained are consistent with those already established in the cases of thermoelasticity [ENE 83] and thermoviscoelasticity [FRA 86a, YU 02], the discussed framework though is general and does not require a specific constitutive law.

5.2. Constitutive law

As discussed extensively in section 2.5, the internal energy $\mathcal{E}^{\epsilon}$ can be expressed in terms of certain variables that dictate the status of the material at each time instant. Apart from the strain and entropy, a set of internal variables could be necessary to describe the material state of a dissipative material,

$$\mathcal{E}^{\epsilon} := \mathcal{E}^{\epsilon}(\boldsymbol{\varepsilon}^{\epsilon}, \eta^{\epsilon}, \boldsymbol{\xi}^{\epsilon}). \qquad [5.39]$$

In the above expression, $\boldsymbol{\xi}$ denotes the list of all internal variables of the composite material (i.e. of all constituents that appear inside the composite). Of course, identifying constitutive laws in terms of entropy is impractical, thus it is more convenient to introduce either a Helmholtz or a Gibbs free energy potential

$$\Psi^{\epsilon} := \Psi^{\epsilon}(\boldsymbol{\varepsilon}^{\epsilon}, \theta^{\epsilon}, \boldsymbol{\xi}^{\epsilon}), \quad G^{\epsilon} := G^{\epsilon}(\boldsymbol{\sigma}^{\epsilon}, \theta^{\epsilon}, \boldsymbol{\xi}^{\epsilon}), \qquad [5.40]$$

which are connected with the internal energy through the relations [2.37], i.e.

$$\Psi^{\epsilon} = \mathcal{E}^{\epsilon} - \eta^{\epsilon}\theta^{\epsilon}, \quad G^{\epsilon} = \mathcal{E}^{\epsilon} - \eta^{\epsilon}\theta^{\epsilon} - \boldsymbol{\sigma}^{\epsilon}:\boldsymbol{\varepsilon}^{\epsilon}. \qquad [5.41]$$

Variable/equation	Microscale	Macroscale
displacement	$\boldsymbol{u} = \overline{\boldsymbol{u}}$	$\overline{\boldsymbol{u}}$
strain	$\boldsymbol{\varepsilon} = \overline{\boldsymbol{\varepsilon}} + \text{grad}_{\text{sym}}\widetilde{\boldsymbol{u}}$	$\overline{\boldsymbol{\varepsilon}} = \overline{\text{grad}}_{\text{sym}}\overline{\boldsymbol{u}} = \langle \boldsymbol{\varepsilon} \rangle$
stress	$\boldsymbol{\sigma}$	$\overline{\boldsymbol{\sigma}} = \langle \boldsymbol{\sigma} \rangle$
density	ρ	$\overline{\rho} = \langle \rho \rangle$
body forces	$\boldsymbol{b}$	$\overline{\boldsymbol{b}} = \langle \rho \boldsymbol{b} \rangle / \overline{\rho}$
internal energy	$\mathcal{E}$	$\overline{\mathcal{E}} = \langle \mathcal{E} \rangle$
energy rate term r	$r = \boldsymbol{\sigma} : \dot{\boldsymbol{\varepsilon}} - \dot{\mathcal{E}}$	$\overline{r} = \overline{\boldsymbol{\sigma}} : \dot{\overline{\boldsymbol{\varepsilon}}} - \dot{\overline{\mathcal{E}}} = \langle r \rangle$
heat flux	$\boldsymbol{q}$	$\overline{\boldsymbol{q}} = \langle \boldsymbol{q} \rangle$
heat sources	$\mathcal{R}$	$\overline{\mathcal{R}} = \langle \rho \mathcal{R} \rangle / \overline{\rho}$
temperature	$\theta = \overline{\theta}$	$\overline{\theta}$
temperature gradient	$\boldsymbol{\nabla\theta} = \overline{\boldsymbol{\nabla\theta}} + \text{grad}\widetilde{\theta}$	$\overline{\boldsymbol{\nabla\theta}} = \overline{\text{grad}}\overline{\theta} = \langle \boldsymbol{\nabla\theta} \rangle$
specific entropy	η	$\overline{\eta} = \langle \eta \rangle$
equilibrium	$\text{div}\boldsymbol{\sigma} = \boldsymbol{0}$	$\overline{\rho}\overline{\boldsymbol{b}} + \overline{\text{div}}\overline{\boldsymbol{\sigma}} = \boldsymbol{0}$
angular momentum	$\boldsymbol{\sigma} = \boldsymbol{\sigma}^T$	$\overline{\boldsymbol{\sigma}} = \overline{\boldsymbol{\sigma}}^T$
energy balance	$\text{div}\boldsymbol{q} = 0$	$\overline{r} - \overline{\text{div}}\overline{\boldsymbol{q}} + \overline{\rho}\overline{\mathcal{R}} = 0$
entropy inequality	$\overline{\theta}\dot{\eta} + r - \dfrac{\boldsymbol{q}}{\overline{\theta}} \cdot \boldsymbol{\nabla\theta} \geq 0$	$\overline{\theta}\dot{\overline{\eta}} + \overline{r} - \dfrac{\overline{\boldsymbol{q}}}{\overline{\theta}} \cdot \overline{\boldsymbol{\nabla\theta}} \geq 0$

Table 5.1. *Variables and conservation laws in micro/macroscale*

Based on the discussion in section 2.5.1, the following relations are obtained:

$$\boldsymbol{\sigma}^\epsilon = \frac{\partial \Psi^\epsilon}{\partial \boldsymbol{\varepsilon}^\epsilon}, \quad \eta^\epsilon = -\frac{\partial \Psi^\epsilon}{\partial \theta^\epsilon}, \quad \text{or} \quad \boldsymbol{\varepsilon}^\epsilon = -\frac{\partial G^\epsilon}{\partial \boldsymbol{\sigma}^\epsilon}, \quad \eta^\epsilon = -\frac{\partial G^\epsilon}{\partial \theta^\epsilon}, \qquad [5.42]$$

while the intrinsic dissipation takes the form

$$\gamma_{\text{loc}}^{\epsilon} = -\frac{\partial \Psi^{\epsilon}}{\partial \boldsymbol{\xi}^{\epsilon}} : \dot{\boldsymbol{\xi}}^{\epsilon}, \quad \text{or} \quad \gamma_{\text{loc}}^{\epsilon} = -\frac{\partial G^{\epsilon}}{\partial \boldsymbol{\xi}^{\epsilon}} : \dot{\boldsymbol{\xi}}^{\epsilon}. \qquad [5.43]$$

The rate of the Helmholtz free energy potential is expressed as

$$\dot{\Psi}^{\epsilon} = \boldsymbol{\sigma}^{\epsilon} : \dot{\boldsymbol{\varepsilon}}^{\epsilon} - \eta^{\epsilon}\dot{\theta}^{\epsilon} + \frac{\partial \Psi^{\epsilon}}{\partial \boldsymbol{\xi}^{\epsilon}} : \dot{\boldsymbol{\xi}}^{\epsilon} = \boldsymbol{\sigma}^{\epsilon} : \dot{\boldsymbol{\varepsilon}}^{\epsilon} - \eta^{\epsilon}\dot{\theta}^{\epsilon} - \gamma_{\text{loc}}^{\epsilon}, \qquad [5.44]$$

and the rate of the Gibbs free energy potential is written as

$$\dot{G}^{\epsilon} = -\boldsymbol{\varepsilon}^{\epsilon} : \dot{\boldsymbol{\sigma}}^{\epsilon} - \eta^{\epsilon}\dot{\theta}^{\epsilon} + \frac{\partial G^{\epsilon}}{\partial \boldsymbol{\xi}^{\epsilon}} : \dot{\boldsymbol{\xi}}^{\epsilon} = -\boldsymbol{\varepsilon}^{\epsilon} : \dot{\boldsymbol{\sigma}}^{\epsilon} - \eta^{\epsilon}\dot{\theta}^{\epsilon} - \gamma_{\text{loc}}^{\epsilon}. \qquad [5.45]$$

To proceed further and identify microscopic and macroscopic counterparts of the various thermodynamic quantities, the following variables are considered to follow an asymptotic series expansion:

$$\begin{aligned}
\Psi^{\epsilon}(\overline{\boldsymbol{x}}, t) &= \Psi(\overline{\boldsymbol{x}}, \boldsymbol{x}, t) + \epsilon\, \widetilde{\Psi}(\overline{\boldsymbol{x}}, \boldsymbol{x}, t) + ..., \\
G^{\epsilon}(\overline{\boldsymbol{x}}, t) &= G(\overline{\boldsymbol{x}}, \boldsymbol{x}, t) + \epsilon\, \widetilde{G}(\overline{\boldsymbol{x}}, \boldsymbol{x}, t) + ..., \\
\boldsymbol{\xi}^{\epsilon}(\overline{\boldsymbol{x}}, t) &= \boldsymbol{\xi}(\overline{\boldsymbol{x}}, \boldsymbol{x}, t) + \epsilon\, \widetilde{\boldsymbol{\xi}}(\overline{\boldsymbol{x}}, \boldsymbol{x}, t) + ..., \\
\gamma_{\text{loc}}^{\epsilon}(\overline{\boldsymbol{x}}, t) &= \gamma_{\text{loc}}(\overline{\boldsymbol{x}}, \boldsymbol{x}, t) + \epsilon\, \widetilde{\gamma}_{\text{loc}}(\overline{\boldsymbol{x}}, \boldsymbol{x}, t) + ..., \\
\boldsymbol{\Xi}_{\Psi}^{\epsilon}(\overline{\boldsymbol{x}}, t) &:= -\frac{\partial \Psi^{\epsilon}}{\partial \boldsymbol{\xi}^{\epsilon}}(\overline{\boldsymbol{x}}, t) = \boldsymbol{\Xi}_{\Psi}(\overline{\boldsymbol{x}}, \boldsymbol{x}, t) + \epsilon\, \widetilde{\boldsymbol{\Xi}}_{\Psi}(\overline{\boldsymbol{x}}, \boldsymbol{x}, t) + ..., \\
\boldsymbol{\Xi}_{G}^{\epsilon}(\overline{\boldsymbol{x}}, t) &:= -\frac{\partial G^{\epsilon}}{\partial \boldsymbol{\xi}^{\epsilon}}(\overline{\boldsymbol{x}}, t) = \boldsymbol{\Xi}_{G}(\overline{\boldsymbol{x}}, \boldsymbol{x}, t) + \epsilon\, \widetilde{\boldsymbol{\Xi}}_{G}(\overline{\boldsymbol{x}}, \boldsymbol{x}, t) + ..., \qquad [5.46]
\end{aligned}$$

where, as usual, all two-scale functions are periodic in $\boldsymbol{x}$. Then, for $\epsilon \to 0$, equations [5.41], [5.44] and [5.45], combined with [5.46] and the asymptotic expansions presented in section 5.1, provide the following expressions in the microscale:

$$\Psi = \mathcal{E} - \eta\bar{\theta},$$

$$
\begin{aligned}
\dot{\Psi} &= \boldsymbol{\sigma}:\dot{\boldsymbol{\varepsilon}} - \eta\dot{\bar{\theta}} - \boldsymbol{\Xi}_{\Psi}:\dot{\boldsymbol{\xi}} = \boldsymbol{\sigma}:\dot{\boldsymbol{\varepsilon}} - \eta\dot{\bar{\theta}} - \gamma_{\text{loc}}, \\
G &= \mathcal{E} - \eta\bar{\theta} - \boldsymbol{\sigma}:\boldsymbol{\varepsilon}, \\
\dot{G} &= -\boldsymbol{\varepsilon}:\dot{\boldsymbol{\sigma}} - \eta\dot{\bar{\theta}} - \boldsymbol{\Xi}_{G}:\dot{\boldsymbol{\xi}} = -\boldsymbol{\varepsilon}:\dot{\boldsymbol{\sigma}} - \eta\dot{\bar{\theta}} - \gamma_{\text{loc}}.
\end{aligned}
\qquad [5.47]
$$

From these expressions, it becomes clear that

$$
\begin{aligned}
\Psi &:= \Psi\left(\boldsymbol{\varepsilon}, \bar{\theta}, \boldsymbol{\xi}\right), \quad \boldsymbol{\sigma} = \frac{\partial \Psi}{\partial \boldsymbol{\varepsilon}}, \quad \eta = -\frac{\partial \Psi}{\partial \bar{\theta}}, \\
\boldsymbol{\Xi}_{\Psi} &= -\frac{\partial \Psi}{\partial \boldsymbol{\xi}}, \quad \gamma_{\text{loc}} = -\frac{\partial \Psi}{\partial \boldsymbol{\xi}}:\dot{\boldsymbol{\xi}},
\end{aligned}
\qquad [5.48]
$$

and

$$
\begin{aligned}
G &:= G\left(\boldsymbol{\sigma}, \bar{\theta}, \boldsymbol{\xi}\right), \quad \boldsymbol{\varepsilon} = -\frac{\partial G}{\partial \boldsymbol{\sigma}}, \quad \eta = -\frac{\partial G}{\partial \bar{\theta}}, \\
\boldsymbol{\Xi}_{G} &= -\frac{\partial G}{\partial \boldsymbol{\xi}}, \quad \gamma_{\text{loc}} = -\frac{\partial G}{\partial \boldsymbol{\xi}}:\dot{\boldsymbol{\xi}}.
\end{aligned}
\qquad [5.49]
$$

As it can be observed from the expression $[5.48]_2$, the micro-stress depends on the macroscopic temperature, since Ψ is a function of $\bar{\theta}$. This result is compatible with the one obtained in linear thermoelasticity [ENE 83, KAL 97, CHA 12a]. Averaging equations [5.47] over the unit cell and considering the Hill–Mandel theorem described in section 4.2.1, yields

$$
\begin{aligned}
\overline{\Psi} &= \langle\Psi\rangle = \overline{\mathcal{E}} - \overline{\eta}\bar{\theta}, \quad \dot{\overline{\Psi}} = \overline{\boldsymbol{\sigma}}:\dot{\overline{\boldsymbol{\varepsilon}}} - \overline{\eta}\dot{\bar{\theta}} - \overline{\gamma}_{\text{loc}}, \\
\overline{G} &= \langle G\rangle = \overline{\mathcal{E}} - \overline{\eta}\bar{\theta} - \overline{\boldsymbol{\sigma}}:\overline{\boldsymbol{\varepsilon}}, \quad \dot{\overline{G}} = -\overline{\boldsymbol{\varepsilon}}:\dot{\overline{\boldsymbol{\sigma}}} - \overline{\eta}\dot{\bar{\theta}} - \overline{\gamma}_{\text{loc}}, \\
\overline{\gamma}_{\text{loc}} &= \langle\gamma_{\text{loc}}\rangle.
\end{aligned}
\qquad [5.50]
$$

The obtained results are summarized in Tables 5.2 and 5.3.

From the above analysis, two crucial points need to be discussed:

1) As already mentioned in Chapter 2, it is customary in the second law of thermodynamics to consider two separate dissipative mechanisms (due to mechanical work and due to heat conduction) which are both non-negative.

For the actual composite, the dissipation due to heat conduction is expressed by

$$-\frac{1}{\theta^\epsilon}\boldsymbol{q}^\epsilon\cdot\boldsymbol{\nabla}\boldsymbol{\theta}^\epsilon \geq 0\,, \qquad [5.51]$$

while the dissipation due to mechanical work is written as

$$\boldsymbol{\sigma}^\epsilon:\dot{\boldsymbol{\varepsilon}}^\epsilon - \dot{\theta}^\epsilon\eta^\epsilon - \dot{\Psi}^\epsilon \geq 0\,, \qquad [5.52]$$

Expression	Microscale
Helmholtz free energy	$\Psi = \mathcal{E} - \eta\bar{\theta} := \Psi(\boldsymbol{\varepsilon}, \bar{\theta}, \boldsymbol{\xi})$
Gibbs free energy	$G = \mathcal{E} - \eta\bar{\theta} - \boldsymbol{\sigma}:\boldsymbol{\varepsilon} := G(\boldsymbol{\sigma}, \bar{\theta}, \boldsymbol{\xi})$
intrinsic dissipation	$\gamma_{\text{loc}} = -\dfrac{\partial\Psi}{\partial\boldsymbol{\xi}}:\dot{\boldsymbol{\xi}}$ or $\gamma_{\text{loc}} = -\dfrac{\partial G}{\partial\boldsymbol{\xi}}:\dot{\boldsymbol{\xi}}$
rate of Helmholtz free energy	$\dot{\Psi} = \boldsymbol{\sigma}:\dot{\boldsymbol{\varepsilon}} - \eta\dot{\bar{\theta}} + \dfrac{\partial\Psi}{\partial\boldsymbol{\xi}}:\dot{\boldsymbol{\xi}}$
rate of Gibbs free energy	$\dot{G} = -\boldsymbol{\varepsilon}:\dot{\boldsymbol{\sigma}} - \eta\dot{\bar{\theta}} + \dfrac{\partial G}{\partial\boldsymbol{\xi}}:\dot{\boldsymbol{\xi}}$

Table 5.2. *Microscale expressions related with the material's law*

Expression	Macroscale
Helmholtz free energy	$\bar{\Psi} = \langle\Psi\rangle = \bar{\mathcal{E}} - \bar{\eta}\bar{\theta}$
Gibbs free energy	$\bar{G} = \langle G\rangle = \bar{\mathcal{E}} - \bar{\eta}\bar{\theta} - \bar{\boldsymbol{\sigma}}:\bar{\boldsymbol{\varepsilon}}$
intrinsic dissipation	$\bar{\gamma}_{\text{loc}} = \langle\gamma_{\text{loc}}\rangle$
rate of Helmholtz free energy	$\dot{\bar{\Psi}} = \bar{\boldsymbol{\sigma}}:\dot{\bar{\boldsymbol{\varepsilon}}} - \bar{\eta}\dot{\bar{\theta}} - \bar{\gamma}_{\text{loc}}$
rate of Gibbs free energy	$\dot{\bar{G}} = -\bar{\boldsymbol{\varepsilon}}:\dot{\bar{\boldsymbol{\sigma}}} - \bar{\eta}\dot{\bar{\theta}} - \bar{\gamma}_{\text{loc}}$

Table 5.3. *Macroscale expressions related with the material's law*

when Helmholtz free energy potential is considered, or

$$-\boldsymbol{\varepsilon}^{\epsilon}:\dot{\boldsymbol{\sigma}}^{\epsilon} - \dot{\theta}^{\epsilon}\eta^{\epsilon} - \dot{G}^{\epsilon} \geq 0\,, \qquad [5.53]$$

when Gibbs free energy potential is considered. Combining the results presented in Table 5.1, inequality [5.51] is written for both microscale and macroscale as

$$-\frac{1}{\bar{\bar{\theta}}}\boldsymbol{q}\cdot\boldsymbol{\nabla\theta} \geq 0\,, \quad -\frac{1}{\bar{\bar{\theta}}}\overline{\boldsymbol{q}}\cdot\overline{\boldsymbol{\nabla\theta}} \geq 0\,. \qquad [5.54]$$

The inequality $[5.54]_1$ is always true once the Fourier law for the heat flux vector is considered,

$$\boldsymbol{q} = -\boldsymbol{\kappa}\cdot\boldsymbol{\nabla\theta}, \qquad [5.55]$$

with positive definite thermal conductivity tensor $\boldsymbol{\kappa}$. Since the macroscopic expression is derived from its microscopic counterpart by volume averaging over the unit cell, $[5.54]_2$ is also true when [5.55] holds. With regard to the dissipation due to mechanical work, Tables 5.1–5.3 allow us to write for both microscale and macroscale

$$\boldsymbol{\sigma}:\dot{\boldsymbol{\varepsilon}} - \dot{\bar{\theta}}\eta - \dot{\Psi} \geq 0\,, \quad \overline{\boldsymbol{\sigma}}:\dot{\overline{\boldsymbol{\varepsilon}}} - \dot{\bar{\theta}}\bar{\eta} - \dot{\overline{\Psi}} \geq 0\,, \qquad [5.56]$$

when Helmholtz free energy potential is considered, or

$$-\boldsymbol{\varepsilon}:\dot{\boldsymbol{\sigma}} - \dot{\bar{\theta}}\eta - \dot{G} \geq 0\,, \quad -\overline{\boldsymbol{\varepsilon}}:\dot{\overline{\boldsymbol{\sigma}}} - \dot{\bar{\theta}}\bar{\eta} - \dot{\overline{G}} \geq 0\,, \qquad [5.57]$$

when Gibbs free energy potential is considered. Moreover, according to $[5.48]_1$ and $[5.49]_1$, the microscopic Helmholtz and Gibbs free energy potentials are functions of the macroscopic temperature. These observations indicate that the microscopic stress depends on the macroscopic temperature, and consequently the microscale equilibrium equation, $\mathrm{div}\boldsymbol{\sigma} = \mathbf{0}$, needs to be solved under uniform temperature conditions, i.e. constant temperature equal to the macro-temperature. It also becomes evident that constitutive laws satisfying the micro-dissipation inequality $[5.57]_1$ lead to automatic satisfaction of the the macro-dissipation inequality $[5.57]_2$, which is derived from its microscopic counterpart by volume averaging.

2) While the internal variables and their conjugate thermodynamic forces are well defined at the microscale, this is not true for the macroscale problem. The macroscopic intrinsic dissipation (equation [5.50]$_5$) cannot be decomposed into a macroscopic variables product. The only available information is that the average of γ_{loc} is identified as the macroscopic $\overline{\gamma}_{\text{loc}}$. This is a well-known issue in the case of plasticity [SUQ 87], and it is due to the fact that, in contrast to the total strains and stresses, plastic strains do not always converge rigorously to a limit value in a weak sense (i.e. a macroscopic counterpart). When multiple types of nonlinear mechanisms exist in the composite, the macroscopic intrinsic dissipation can be expressed in terms of the complete set of the internal variables linked with the various dissipative responses [TSA 15]. Analytical homogenized constitutive laws can be obtained only for composites with special microstructure and whose constituents are described by simple inelastic constitutive laws (see, for instance, the case studied by Chatzigeorgiou *et al.* [CHA 09]). One way to overcome the theoretical issues and to provide homogenized mechanical constitutive laws for generally nonlinear composites is through the identification of macroscopic inelastic stresses [PIN 09].

Considering a fully coupled thermomechanical framework, the implications arising from the intrinsic dissipation influence also the energy equation, which cannot be expressed analytically in the macroscale. The linearized, incremental formulation presented in the next section overpasses these difficulties.

5.2.1. *Linearized, incremental formulation*

The relationship between stress, strain and temperature is often very complicated in nonlinear materials. For dissipative mechanisms where internal variables are involved, such a relationship is expressed only in an implicit way (see section 2.5). As was already discussed in Chapter 3, in computational formulations of constitutive laws, it is customary to first linearize the actual problem in time and then proceed to a second linearization of the nonlinear expressions. Here, an approach based on the well-known return mapping algorithm scheme is utilized [SIM 98]. Details about this approach are provided in Chapter 3. Similar numerical homogenization techniques have been proposed in the literature for elastoplastic [TER 01, ASA 07, TSA 13a], magnetoelastic [JAV 13] and shape memory alloy composites [CHA 15]. In brief, the two problems (macroscale and unit cell) are solved simultaneously, using an iterative scheme (Figure 5.2): from the macroscale analysis, the macroscopic strains, temperatures and temperature gradients are obtained, which are used in the unit cell problem to compute the microscale variables and the rest of the unknown macroscopic variables (stresses, heat fluxes, etc.). Moreover, the unit cell problem is utilized to compute the macroscopic tangent moduli, which are required for the macroscale analysis.

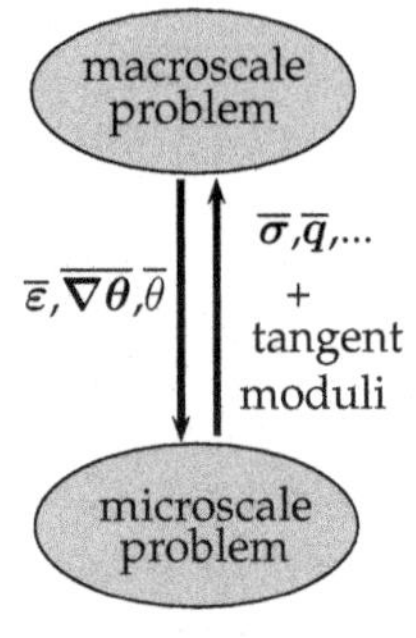

Figure 5.2. *Computational homogenization scheme*

The discussed framework can be easily extended to account for more complicated microstructures exhibiting generalized periodicity [TSA 13a, TSA 13b].

This section follows the methodology and definitions described in Chapter 3 (see the preliminary part of section 3.2). For the macroscale analysis, there are two types of increments for a variable $\bar{b}$:

– The time increment between the steps n and $n+1$,

$$\Delta\bar{b}^{(n+1)} = \bar{b}^{(n+1)} - \bar{b}^{(n)}.$$

– The FE iterative increment between the steps m and $m+1$,

$$\begin{aligned}\eth\bar{b}^{(n+1)(m)} &= \bar{b}^{(n+1)(m+1)} - \bar{b}^{(n+1)(m)}\\ &= \Delta\bar{b}^{(n+1)(m+1)} - \Delta\bar{b}^{(n+1)(m)}.\end{aligned}$$

For the microscale analysis, there are three types of increments for a variable b:

– The time increment between the steps n and $n+1$,

$$\Delta b^{n+1} = b^{n+1} - b^n.$$

– The macroscopic tangent moduli increment between the steps m and $m+1$,

$$\begin{aligned} \eth b^{(n+1)(m)} &= b^{(n+1)(m+1)} - b^{(n+1)(m)} \\ &= \Delta b^{(n+1)(m+1)} - \Delta b^{(n+1)(m)}. \end{aligned}$$

– The unit cell iterative increment between the steps m^* and m^*+1,

$$\begin{aligned} \delta b^{(n+1)(m+1)(m^*)} &= b^{(n+1)(m+1)(m^*+1)} - b^{(n+1)(m+1)(m^*)} \\ &= \Delta b^{(n+1)(m+1)(m^*+1)} - \Delta b^{(n+1)(m+1)(m^*)}. \end{aligned}$$

In order to avoid lengthy expressions, every quantity that refers to the time step $n+1$ will be presented without index. The type of increment and the scale (microscopic or macroscopic) dictates the appropriate step. Only quantities that refer to the previous time steps are presented with the exponent (n).

5.2.1.1. *Unit cell problem*

As illustrated in Figure 5.3, the unit cell problem is split in two parts: the first part uses as input the macroscopic strain, temperature and temperature gradient, it is iterative and it calculates the macroscopic stress, heat flux and the energy rate term r. The second part uses as input the obtained microscopic thermomechanical tangent moduli at the end of the first part and calculates their macroscopic counterparts through the solution of a linear problem.

1) In the first part, the macro-strain tensor, macro-temperature and macro-temperature gradient are provided exclusively by the analysis in the macroscale (presented later in this section), which means that they do not iterate during the calculations, and thus there is no δ increment in $\bar{\varepsilon}$, $\bar{\theta}$ and $\overline{\boldsymbol{\nabla\theta}}$. Since the micro-equilibrium is solved under uniform temperature conditions, the microscopic

stress tensor at the time step $n+1$ can be expressed in an iterative linearized fashion as

$$\boldsymbol{\sigma}^{\text{updated}} = \boldsymbol{\sigma} + \boldsymbol{D}^{\varepsilon}:\delta\boldsymbol{\varepsilon}, \quad \delta\boldsymbol{\varepsilon} = \text{grad}_{\text{sym}}\delta\widetilde{\boldsymbol{u}},$$

where $\boldsymbol{D}^{\varepsilon}$ denotes the mechanical tangent modulus. Similarly, the microscopic heat flux can be written as

$$\boldsymbol{q}^{\text{updated}} = \boldsymbol{q} - \boldsymbol{\kappa}\cdot\delta\boldsymbol{\nabla\theta}, \quad \delta\boldsymbol{\nabla\theta} = \text{grad}\delta\widetilde{\theta}.$$

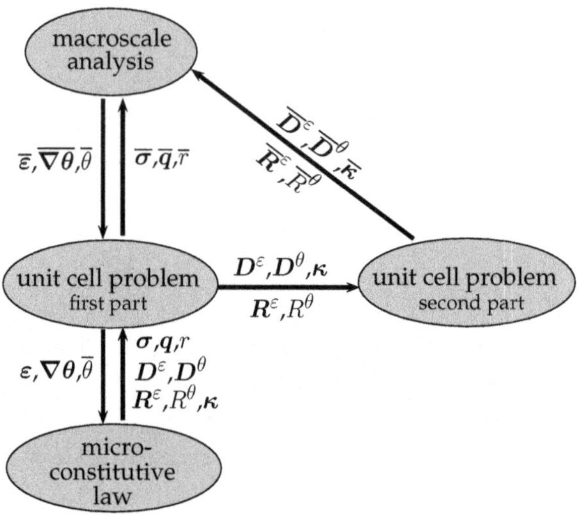

Figure 5.3. *Detailed scheme of the homogenization problem*

To ensure the existence and uniqueness of solution of the microscale problem, the fourth-order tensor $\boldsymbol{D}^{\varepsilon}$ and the second-order thermal conductivity tensor $\boldsymbol{\kappa}$ need to be positive definite at all times. Using the above linearized expressions, the micro-equilibrium and micro-energy balance are written as

$$\text{div}\left(\boldsymbol{\sigma} + \boldsymbol{D}^{\varepsilon}:\text{grad}\delta\widetilde{\boldsymbol{u}}\right) = 0, \qquad \text{div}\left(\boldsymbol{q} - \boldsymbol{\kappa}\cdot\text{grad}\delta\widetilde{\theta}\right) = 0. \qquad [5.58]$$

These conservation laws are accompanied by appropriate microscopic constitutive laws to update the current stress $\boldsymbol{\sigma}$ and the energy rate term

$r = \sigma : \dot{\varepsilon} - \dot{\mathcal{E}}$. With regard to boundary conditions, periodicity of $\delta\widetilde{\boldsymbol{u}}$, $\delta\widetilde{\boldsymbol{\theta}}$ and antiperiodicity of $\boldsymbol{\sigma} \cdot \boldsymbol{n}$ and $\boldsymbol{q} \cdot \boldsymbol{n}$ are imposed. The global system of equations is solved iteratively using a procedure analogous to the one described in Chapter 3. When the convergence is achieved (i.e. when $\delta\widetilde{\boldsymbol{u}}$ and $\delta\widetilde{\theta}$ are almost zero), the time increments and the actual values of the fluctuating terms $\widetilde{\boldsymbol{u}}$ and $\widetilde{\theta}$ are evaluated, allowing all the microscopic quantities and their averaged (macroscopic) counterparts of Table 5.1 to be computed. Moreover, the thermomechanical tangent moduli $\boldsymbol{D}^{\varepsilon}$, $\boldsymbol{D}^{\theta}$, $\boldsymbol{R}^{\varepsilon}$, R^{θ} and $\boldsymbol{\kappa}$ at all microscopic points can be calculated when the convergence is achieved (see Chapter 3 for details). These moduli are essential to obtain their macroscopic counterparts in the second part of the unit cell problem. Table 5.4 illustrates the algorithmic scheme for the first part of the unit cell problem. Table 5.5 provides the necessary constitutive law algorithm at each microscopic point in the unit cell.

2) Once the first part is completed, the residual terms of the microscopic stress and heat flux are assumed exactly zero. The strain, the temperature gradient, the equilibrium equation and the conservation of energy at the microscale presented in Table 5.1 are written in the linearized, incremental form as

$$\eth\boldsymbol{\varepsilon} = \eth\overline{\boldsymbol{\varepsilon}} + \text{grad}_{\text{sym}}\eth\widetilde{\boldsymbol{u}}, \quad \text{div}\eth\boldsymbol{\sigma} = \boldsymbol{0},$$

$$\eth\boldsymbol{\nabla\theta} = \eth\overline{\boldsymbol{\nabla\theta}} + \text{grad}\eth\widetilde{\theta}, \text{div}\eth\boldsymbol{q} = \boldsymbol{0}. \qquad [5.59]$$

With regard to boundary conditions, periodicity of $\eth\widetilde{\boldsymbol{u}}$, $\eth\widetilde{\boldsymbol{\theta}}$ and antiperiodicity of $\eth\boldsymbol{\sigma}\cdot\boldsymbol{n}$, $\eth\boldsymbol{q}\cdot\boldsymbol{n}$ are imposed. Moreover, with the help of the thermomechanical tangent moduli, the increments of the stress and the scalar r can be expressed as

$$\eth\boldsymbol{\sigma} = \boldsymbol{D}^{\varepsilon}{:}\eth\boldsymbol{\varepsilon} + \boldsymbol{D}^{\theta}\eth\overline{\theta}, \qquad \eth r = \boldsymbol{R}^{\varepsilon}{:}\eth\boldsymbol{\varepsilon} + R^{\theta}\eth\overline{\theta}. \qquad [5.60]$$

Note that the microscopic tensors $\boldsymbol{D}^{\varepsilon}$, $\boldsymbol{D}^{\theta}$, $\boldsymbol{R}^{\varepsilon}$ and R^{θ} are obtained at the end of the first part of the unit cell problem. In sections 3.3 and 3.4, the analytical expressions of these tensors per incremental step are presented for elastic and elastoplastic materials, respectively. Finally, the increment of the heat flux vector can be described in terms of the temperature gradient vector by a linear law of the form

$$\eth\boldsymbol{q} = -\boldsymbol{\kappa}\cdot\eth\boldsymbol{\nabla\theta}, \qquad [5.61]$$

where the tensor $\boldsymbol{\kappa}$ is the one obtained at the end of the first part of the unit cell problem. Combining equations [5.59], [5.60] and [5.61] yields

$$\text{div}\left(\boldsymbol{D}^{\varepsilon}:\eth\bar{\boldsymbol{\varepsilon}} + \boldsymbol{D}^{\theta}\eth\bar{\theta} + \boldsymbol{D}^{\varepsilon}:\text{grad}\eth\widetilde{\boldsymbol{u}}\right) = \boldsymbol{0},$$

$$\text{div}\left(\boldsymbol{\kappa}\cdot\eth\overline{\boldsymbol{\nabla\theta}} + \boldsymbol{\kappa}\cdot\text{grad}\eth\widetilde{\theta}\right) = 0. \qquad [5.62]$$

The solution of the above homogeneous system (up to an arbitrary macroscopic constant) is written in the form [ENE 83]

$$\eth\widetilde{\boldsymbol{u}} = \eth\bar{\boldsymbol{\varepsilon}}:\boldsymbol{\chi}^{\varepsilon} + \eth\bar{\theta}\,\boldsymbol{\chi}^{\theta}, \qquad \eth\widetilde{\theta} = \eth\overline{\boldsymbol{\nabla\theta}}\cdot\boldsymbol{\psi}^{\theta}, \qquad [5.63]$$

where the third-order tensor $\boldsymbol{\chi}^{\varepsilon}$ and the vectors $\boldsymbol{\chi}^{\theta}$, $\boldsymbol{\psi}^{\theta}$ are periodic and they are called corrector terms. Substituting [5.63] in [5.62] yields

$$\eth\bar{\boldsymbol{\varepsilon}}:\text{div}\left([\boldsymbol{D}^{\varepsilon} + \boldsymbol{D}^{\varepsilon}\,\tilde{:}\,\text{grad}\boldsymbol{\chi}^{\varepsilon}]^{T}\right) + \eth\bar{\theta}\,\text{div}\left(\boldsymbol{D}^{\theta} + \boldsymbol{D}^{\varepsilon}:\text{grad}\boldsymbol{\chi}^{\theta}\right) = \boldsymbol{0},$$

$$\eth\overline{\boldsymbol{\nabla\theta}}\cdot\text{div}\left(\left[\boldsymbol{\kappa} + \boldsymbol{\kappa}\,\tilde{\cdot}\,\text{grad}\boldsymbol{\psi}^{\theta}\right]^{T}\right) = \boldsymbol{0}.$$

The last expressions are valid for arbitrary values of the macroscale variables $\eth\bar{\boldsymbol{\varepsilon}}$, $\eth\bar{\theta}$ and $\eth\overline{\boldsymbol{\nabla\theta}}$ only if the corrector terms satisfy the linear equations[1]

$$\text{div}\left([\boldsymbol{D}^{\varepsilon} + \boldsymbol{D}^{\varepsilon}\,\tilde{:}\,\text{grad}\boldsymbol{\chi}^{\varepsilon}]^{T}\right) = \boldsymbol{0}, \quad \text{div}\left(\boldsymbol{D}^{\theta} + \boldsymbol{D}^{\varepsilon}:\text{grad}\boldsymbol{\chi}^{\theta}\right) = \boldsymbol{0},$$

$$\text{div}\left(\left[\boldsymbol{\kappa} + \boldsymbol{\kappa}\,\tilde{\cdot}\,\text{grad}\boldsymbol{\psi}^{\theta}\right]^{T}\right) = \boldsymbol{0}. \qquad [5.64]$$

1 In indicial notation, these expressions are written as:

$$\frac{\partial}{\partial x_j}\left(D^{\varepsilon}_{ijkl} + D^{\varepsilon}_{ijmn}\frac{\partial \chi^{\varepsilon}_{klm}}{\partial x_n}\right) = 0, \quad \frac{\partial}{\partial x_j}\left(D^{\theta}_{ij} + D^{\varepsilon}_{ijkl}\frac{\partial \chi^{\theta}_{k}}{\partial x_l}\right) = 0,$$

$$\frac{\partial}{\partial x_i}\left(\kappa_{ij} + \kappa_{ik}\frac{\partial \psi^{\theta}_{j}}{\partial x_k}\right) = 0.$$

Using the solution [5.63], the increments of the other microscopic fields are written as

$$\begin{aligned}
\eth\varepsilon &= \boldsymbol{A}^{\varepsilon}:\eth\bar{\varepsilon} + \boldsymbol{A}^{\theta}\eth\bar{\theta}, \\
\eth\boldsymbol{\nabla\theta} &= \boldsymbol{A}^{\kappa}\cdot\eth\overline{\boldsymbol{\nabla\theta}}, \\
\eth\boldsymbol{\sigma} &= \boldsymbol{D}^{\varepsilon}:\boldsymbol{A}^{\varepsilon}:\eth\bar{\varepsilon} + \left[\boldsymbol{D}^{\theta} + \boldsymbol{D}^{\varepsilon}:\boldsymbol{A}^{\theta}\right]\eth\bar{\theta}, \\
\eth\boldsymbol{q} &= -\boldsymbol{\kappa}\cdot\boldsymbol{A}^{\kappa}\cdot\eth\overline{\boldsymbol{\nabla\theta}}, \\
\eth r &= \boldsymbol{R}^{\varepsilon}:\boldsymbol{A}^{\varepsilon}:\eth\bar{\varepsilon} + \left[R^{\theta} + \boldsymbol{R}^{\varepsilon}:\boldsymbol{A}^{\theta}\right]\eth\bar{\theta},
\end{aligned} \qquad [5.65]$$

where

$$\boldsymbol{A}^{\varepsilon} = \boldsymbol{\mathcal{I}} + \boldsymbol{\mathcal{I}} \,\tilde{:}\, \mathrm{grad}\boldsymbol{\chi}^{\varepsilon}, \quad \boldsymbol{A}^{\theta} = \mathrm{grad}_{\mathrm{sym}}\boldsymbol{\chi}^{\theta}, \quad \boldsymbol{A}^{\kappa} = \boldsymbol{I} + \left[\mathrm{grad}\boldsymbol{\psi}^{\theta}\right]^{T}, \qquad [5.66]$$

are strain and temperature gradient related concentration tensors. It is recalled that $\boldsymbol{\mathcal{I}}$ denotes the fourth-order symmetric identity tensor and $\boldsymbol{I}$ is the second-order identity tensor. Averaging of these fields over the unit cell yields the macroscopic tensors

$$\begin{aligned}
\overline{\boldsymbol{D}}^{\varepsilon} &= \langle \boldsymbol{D}^{\varepsilon}:\boldsymbol{A}^{\varepsilon}\rangle, \overline{\boldsymbol{D}}^{\theta} = \left\langle \boldsymbol{D}^{\theta} + \boldsymbol{D}^{\varepsilon}:\boldsymbol{A}^{\theta}\right\rangle, \overline{\boldsymbol{\kappa}} = \langle \boldsymbol{\kappa}\cdot\boldsymbol{A}^{\kappa}\rangle, \\
\overline{\boldsymbol{R}}^{\varepsilon} &= \langle \boldsymbol{R}^{\varepsilon}:\boldsymbol{A}^{\varepsilon}\rangle, \overline{R}^{\theta} = \left\langle R^{\theta} + \boldsymbol{R}^{\varepsilon}:\boldsymbol{A}^{\theta}\right\rangle.
\end{aligned} \qquad [5.67]$$

5.2.1.2. *Macroscale problem*

The macroscale problem is solved using a numerical scheme analogous to the one presented in Table 3.1 of Chapter 3. Each macroscopic point is linked with a unit cell that allows us to compute macroscopic variables and tangent moduli. Ignoring body forces, inertia forces and heat sources (radiation), the macroscopic equations of the equilibrium and the energy balance are written in the linearized, incremental form as

$$\begin{aligned}
\eth\bar{\varepsilon} &= \overline{\mathrm{grad}}_{\mathrm{sym}}\eth\bar{\boldsymbol{u}}, \overline{\mathrm{div}}\left(\bar{\boldsymbol{\sigma}} + \eth\bar{\boldsymbol{\sigma}}\right) = \boldsymbol{0}, \\
\eth\overline{\boldsymbol{\nabla\theta}} &= \overline{\mathrm{grad}}\eth\bar{\theta}, \overline{\mathrm{div}}\left(\bar{\boldsymbol{q}} + \eth\bar{\boldsymbol{q}}\right) = \bar{r} + \eth\bar{r},
\end{aligned} \qquad [5.68]$$

where $\overline{r} = \overline{\boldsymbol{\sigma}} : \dot{\overline{\varepsilon}} - \dot{\overline{\mathcal{E}}} = \langle r \rangle$. The values of the macroscopic stress, the macroscopic heat flux and the $\overline{r}$ are computed in the first part of the unit cell problem. The above equations are accompanied by appropriate, incremental, constitutive relations of the form

$$\eth\overline{\boldsymbol{\sigma}} = \overline{\boldsymbol{D}}^{\varepsilon}:\eth\overline{\varepsilon} + \overline{\boldsymbol{D}}^{\theta}\eth\overline{\theta}, \quad \eth\overline{r} = \overline{\boldsymbol{R}}^{\varepsilon}:\eth\overline{\varepsilon} + \overline{R}^{\theta}\eth\overline{\theta}, \quad \eth\overline{\boldsymbol{q}} = -\overline{\boldsymbol{\kappa}}\cdot\eth\overline{\boldsymbol{\nabla}\boldsymbol{\theta}}. \qquad [5.69]$$

1. At time step n everything is known in both scales.

2. At time step $n+1$ and at a specific macro-iteration, $\Delta\overline{\theta}$, $\Delta\overline{\varepsilon}$ and $\Delta\overline{\boldsymbol{\nabla}\boldsymbol{\theta}}$ are provided by the **macroscale analysis**. At the beginning of the micro-iterations ($m^* = 0$) set at every microscopic point $\Delta\boldsymbol{\varepsilon} = \Delta\overline{\varepsilon}$, $\Delta\boldsymbol{\nabla}\boldsymbol{\theta} = \Delta\overline{\boldsymbol{\nabla}\boldsymbol{\theta}}$, $\Delta\widetilde{\boldsymbol{u}} = 0$, $\Delta\widetilde{\theta} = 0$.

3. At every microscopic point evaluate $\boldsymbol{\sigma}$, r, $\boldsymbol{q}$ and the micro-tangent thermomechanical moduli $\boldsymbol{D}^{\varepsilon}$, $\boldsymbol{D}^{\theta}$, $\boldsymbol{R}^{\varepsilon}$, R^{θ}, $\boldsymbol{\kappa}$ using the **micro-constitutive law**.

4. Compute the virtual increments $\delta\widetilde{\boldsymbol{u}}$ and $\delta\widetilde{\theta}$ from the micro-equilibrium and micro-energy equations

$$\operatorname{div}\left(\boldsymbol{D}^{\varepsilon} : \operatorname{grad}\delta\widetilde{\boldsymbol{u}} + \boldsymbol{\sigma}\right) = 0, \qquad \operatorname{div}\left(\boldsymbol{q} - \boldsymbol{\kappa}\cdot\operatorname{grad}\delta\widetilde{\theta}\right) = 0\,,$$

   with periodic microscopic boundary conditions.

5. At every microscopic point update the micro-quantities

$$\Delta\widetilde{\boldsymbol{u}} = \Delta\widetilde{\boldsymbol{u}} + \delta\widetilde{\boldsymbol{u}}, \quad \delta\boldsymbol{\varepsilon} = \operatorname{grad}_{\text{sym}}\delta\widetilde{\boldsymbol{u}}, \quad \Delta\boldsymbol{\varepsilon} = \Delta\boldsymbol{\varepsilon} + \delta\boldsymbol{\varepsilon},$$
$$\Delta\widetilde{\theta} = \Delta\widetilde{\theta} + \delta\widetilde{\theta}, \quad \delta\boldsymbol{\nabla}\boldsymbol{\theta} = \operatorname{grad}\delta\widetilde{\theta}, \quad \Delta\boldsymbol{\nabla}\boldsymbol{\theta} = \Delta\boldsymbol{\nabla}\boldsymbol{\theta} + \delta\boldsymbol{\nabla}\boldsymbol{\theta}.$$

6. At every microscopic point evaluate $\boldsymbol{\sigma}$, r, $\boldsymbol{q}$ and the micro-tangent thermomechanical moduli $\boldsymbol{D}^{\varepsilon}$, $\boldsymbol{D}^{\theta}$, $\boldsymbol{R}^{\varepsilon}$, R^{θ}, $\boldsymbol{\kappa}$ using the **micro-constitutive law**.

7. If the convergence criterion is satisfied then continue with step 8, else return to step 4.

8. Compute the macro-stress $\overline{\boldsymbol{\sigma}} = \langle\boldsymbol{\sigma}\rangle$, the macro-heat flux $\overline{\boldsymbol{q}} = \langle\boldsymbol{q}\rangle$ and the macro-scalar $\overline{r} = \langle r\rangle$.

Table 5.4. *Algorithm for the first part of the unit cell problem*

1. At time step n everything is known in both scales.
2. At time step $n+1$ the time increments $\Delta\varepsilon$ and $\Delta\nabla\theta$ are provided by the **first part of the unit cell problem**. Moreover $\Delta\bar{\theta}$ is known from the **macroscale analysis**.
3. The stress tensor $\boldsymbol{\sigma}$, the internal variables $\boldsymbol{\xi}$, the scalar r and the heat flux $\boldsymbol{q}$ are computed through a **constitutive law algorithm**, described in subsection 3.2.3.
4. The thermomechanical moduli $\boldsymbol{D}^{\varepsilon}$, $\boldsymbol{D}^{\theta}$, $\boldsymbol{R}^{\varepsilon}$, R^{θ} and $\boldsymbol{\kappa}$ are computed from a **tangent moduli algorithm**, described in subsection 3.2.4.

Table 5.5. *Algorithm for the micro-constitutive law*

1. At every microscopic point in the unit cell the tensors $\boldsymbol{D}^{\varepsilon}$, $\boldsymbol{D}^{\theta}$, $\boldsymbol{R}^{\varepsilon}$, R^{θ} and $\boldsymbol{\kappa}$ are provided by the **first part of the unit cell problem**.
2. Compute the correctors $\boldsymbol{\chi}^{\varepsilon}$, $\boldsymbol{\chi}^{\theta}$ and ψ^{θ} from the equations

$$
\begin{aligned}
\operatorname{div}\left([\boldsymbol{D}^{\varepsilon} + \boldsymbol{D}^{\varepsilon} \,\tilde{\vdots}\, \operatorname{grad}\boldsymbol{\chi}^{\varepsilon}]^{T}\right) &= \mathbf{0},\\
\operatorname{div}\left(\boldsymbol{D}^{\theta} + \boldsymbol{D}^{\varepsilon} : \operatorname{grad}\boldsymbol{\chi}^{\theta}\right) &= \mathbf{0},\\
\operatorname{div}\left(\left[\boldsymbol{\kappa} + \boldsymbol{\kappa} \,\tilde{\cdot}\, \operatorname{grad}\psi^{\theta}\right]^{T}\right) &= \mathbf{0}.
\end{aligned}
$$

3. Evaluate the effective tangent moduli

$$
\overline{\boldsymbol{D}}^{\varepsilon} = \langle \boldsymbol{D}^{\varepsilon} : \boldsymbol{A}^{\varepsilon} \rangle, \quad \overline{\boldsymbol{D}}^{\theta} = \left\langle \boldsymbol{D}^{\theta} + \boldsymbol{D}^{\varepsilon} : \boldsymbol{A}^{\theta} \right\rangle, \quad \overline{\boldsymbol{\kappa}} = \langle \boldsymbol{\kappa} \cdot \boldsymbol{A}^{\kappa} \rangle,
$$

$$
\overline{\boldsymbol{R}}^{\varepsilon} = \langle \boldsymbol{R}^{\varepsilon} : \boldsymbol{A}^{\varepsilon} \rangle, \quad \overline{R}^{\theta} = \left\langle R^{\theta} + \boldsymbol{R}^{\varepsilon} : \boldsymbol{A}^{\theta} \right\rangle,
$$

where $\boldsymbol{A}^{\varepsilon} = \boldsymbol{\mathcal{I}} + \boldsymbol{\mathcal{I}} \,\tilde{\vdots}\, \operatorname{grad}\boldsymbol{\chi}^{\varepsilon}$, $\boldsymbol{A}^{\theta} = \operatorname{grad}_{\mathrm{sym}}\boldsymbol{\chi}^{\theta}$, $\boldsymbol{A}^{\kappa} = \boldsymbol{I} + \left[\operatorname{grad}\psi^{\theta}\right]^{T}$.

Table 5.6. *Algorithm for the second part of the unit cell problem*

The moduli $\overline{\boldsymbol{D}}^{\varepsilon}$, $\overline{\boldsymbol{D}}^{\theta}$, $\overline{\boldsymbol{R}}^{\varepsilon}$, $\overline{R}^{\theta}$ and $\overline{\boldsymbol{\kappa}}$ are provided by the second part of the unit cell problem. The macroscale and microscale analyses are solved simultaneously, according to the iterative framework demonstrated in Figure 5.3. Table 5.7 illustrates the algorithmic scheme for the macroscale analysis.

5.3. Discussion

Based on the analysis presented in this chapter, the following points are worthy of discussion:

– According to the asymptotic expansion homogenization framework, in the case of smooth media where no displacement or temperature jumps due to cracks are present and no strain or temperature gradient singularities appear due to instability phenomena (i.e. $\boldsymbol{u}^{\epsilon}$, θ^{ϵ}, $\boldsymbol{\varepsilon}^{\epsilon}$ and $\boldsymbol{\nabla\theta}^{\epsilon}$ are bounded), the displacement vector $\boldsymbol{u}^{\epsilon}$ and the temperature θ^{ϵ} coincide asymptotically with their macroscopic counterparts $\overline{\boldsymbol{u}}$ and $\overline{\theta}$, respectively (Table 5.1). The fluctuating parts $\widetilde{\boldsymbol{u}}$ and $\widetilde{\theta}$ are influencing only the microscopic strain tensor and the microscopic temperature gradient vector, respectively, which can vary significantly inside the unit cell. As a result, material constants that may depend on the temperature (for instance, the thermal conductivity $\boldsymbol{\kappa}$) should be considered as functions of the macro-temperature even in the microscale.

– Equivalent results with the asymptotic analysis approach can be obtained by using the Suquet's methodology [SUQ 87]. In the zeroth order periodic homogenization, we can alternatively consider that the displacement and the temperature in the microscale are expressed by two terms, a macroscopic and a fluctuating,

$$\boldsymbol{u}^{\text{micro}} = \overline{\boldsymbol{\varepsilon}}\cdot\boldsymbol{x} + \widetilde{\boldsymbol{u}}, \quad \theta^{\text{micro}} = \overline{\boldsymbol{\nabla\theta}}\cdot\boldsymbol{x} + \widetilde{\theta}, \qquad [5.70]$$

where $\widetilde{\boldsymbol{u}}$, $\widetilde{\theta}$ are periodic functions (see also the discussion in sections 4.2.1 and 4.2.2). The gradients of these expressions provide the strain and the temperature gradient in the microscale. Even though equations [5.70] are very efficient in expressing the unit cell problem, they cannot be used to determine the macroscopic displacement and temperature, since $\widetilde{\boldsymbol{u}}$ and $\widetilde{\theta}$ depend not only on the microscopic position, but also on the size of the unit cell (i.e. the characteristic length ϵ). Thus, we need to postulate the constitutive law of a thermomechanical medium as function of the macro-temperature [MAG 03].

The same hypothesis is also considered in the case of micromechanics of general composite media [ROS 70].

1. At time step n everything is known. At time step $n+1$ begin the macro-iterations by setting the values of all quantities and tangent moduli in both scales equal to the previous time step values. Also set $\Delta\overline{\boldsymbol{u}} = 0$, $\Delta\overline{\theta} = 0$.

2. Compute the virtual increments of $\overline{\boldsymbol{u}}$ and $\overline{\theta}$ from the macroscopic equilibrium and macroscopic energy balance

$$\begin{aligned} \operatorname{div}\left(\overline{\boldsymbol{\sigma}} + \overline{\boldsymbol{D}}^{\varepsilon} : \overline{\operatorname{grad}}\eth\overline{\boldsymbol{u}} + \overline{\boldsymbol{D}}^{\theta}\eth\overline{\theta}\right) &= \boldsymbol{0}, \\ \overline{\operatorname{div}}\left(\overline{\boldsymbol{q}} - \overline{\boldsymbol{\kappa}} \cdot \overline{\operatorname{grad}}\eth\overline{\theta}\right) &= \overline{r} + \overline{\boldsymbol{R}}^{\varepsilon} : \overline{\operatorname{grad}}\eth\overline{\boldsymbol{u}} + \overline{R}^{\theta}\eth\overline{\theta}, \end{aligned}$$

with appropriate macroscopic boundary conditions.

3. Update the macro-quantities

$$\begin{gathered} \Delta\overline{\boldsymbol{u}} = \Delta\overline{\boldsymbol{u}} + \eth\overline{\boldsymbol{u}}, \quad \eth\overline{\varepsilon} = \overline{\operatorname{grad}}_{\text{sym}}\eth\overline{\boldsymbol{u}}, \quad \Delta\overline{\varepsilon} = \Delta\overline{\varepsilon} + \eth\overline{\varepsilon}, \\ \Delta\overline{\theta} = \Delta\overline{\theta} + \eth\overline{\theta}, \quad \eth\overline{\boldsymbol{\nabla\theta}} = \overline{\operatorname{grad}}\eth\overline{\theta}, \quad \Delta\overline{\boldsymbol{\nabla\theta}} = \Delta\overline{\boldsymbol{\nabla\theta}} + \eth\overline{\boldsymbol{\nabla\theta}}. \end{gathered}$$

4. At each macroscopic point compute $\overline{\boldsymbol{\sigma}}$, $\overline{\boldsymbol{q}}$, $\overline{r}$, through its corresponding unit cell, using the **first part of the unit cell problem**. Also compute the macroscopic moduli $\overline{\boldsymbol{D}}^{\varepsilon}$, $\overline{\boldsymbol{D}}^{\theta}$, $\overline{\boldsymbol{R}}^{\varepsilon}$, $\overline{R}^{\theta}$, $\overline{\boldsymbol{\kappa}}$ using the **second part of the unit cell problem.**

5. If the convergence criterion is satisfied
   then update the values of all the microscopic variables in the unit cells linked with the macroscopic points,
   set $n = n+1$ and return to step 1 for the next time increment,
   else return to step 2.

Table 5.7. *Algorithm for the macroscale analysis*

– The developed framework reveals that a fully coupled thermomechanical analysis for periodic composites through homogenization has two significant differences compared to a corresponding analysis for homogeneous materials:

i) The micro-equilibrium is solved at each macroscopic point under uniform macroscopic temperature (which of course can vary from one unit cell to another), while the balance of energy in the microscale includes only the heat conduction part. Thus, the microscale equilibrium and energy equation are practically uncoupled and they are solved separately, independently of how complicated the expressions of the micro-energy potentials are. The coupling arising from the thermomechanical energetic terms and the mechanical dissipation is taken into account only in the macroscale energy equation (Tables 5.1–5.3).

ii) The linearized incremental formulation yields that the macroscopic tangent moduli $\overline{\boldsymbol{R}}^{\varepsilon}$ and $\overline{R}^{\theta}$, given by the linearized equations $[5.67]_{4,5}$, depend on (a) their microscopic counterparts (which are not involved in the solution of the microscale problem since it is decoupled) and (b) the concentration tensors obtained exclusively from the solution of the micro-equations $[5.64]_{1,2}$. This formalism is useful in the case of random media, in which the Eshelby-based approaches provide analytical or semi-analytical expressions for the concentration tensors $\boldsymbol{A}^{\varepsilon}$ and $\boldsymbol{A}^{\theta}$ [BEN 91].

– Combining the expression $r = \boldsymbol{\sigma}:\dot{\boldsymbol{\varepsilon}} - \dot{\mathcal{E}}$ with $[5.47]_{1,3}$ or $[5.47]_{2,4}$ yields

$$r = -\bar{\theta}\dot{\eta} + \gamma_{\mathrm{loc}}. \qquad [5.71]$$

In this representation, the r can be considered as the difference between the intrinsic dissipation and the rate of thermal work. In the linearized, incremental formulation, we can identify relations for the entropy and the dissipation increments of the form

$$\eth\eta = \boldsymbol{H}^{\varepsilon}:\eth\boldsymbol{\varepsilon} + H^{\theta}\eth\bar{\theta}, \quad \eth\gamma_{\mathrm{loc}} = \boldsymbol{\Gamma}^{\varepsilon}:\eth\boldsymbol{\varepsilon} + \Gamma^{\theta}\eth\bar{\theta}. \qquad [5.72]$$

The first expression of [5.72] provides a constitutive relation between the entropy, temperature and strain. Such type of constitutive law can be very useful (in extended form) in the case of multiphysics homogenization problems with additional fields like magnetic or electric [BRA 09]. The second expression of [5.72] is considered as a linearized representation of the intrinsic dissipation. Combining [5.71] and [5.72], we obtain

$$\eth r = \left[-\frac{\bar{\theta}}{\Delta t}\boldsymbol{H}^{\varepsilon} + \boldsymbol{\Gamma}^{\varepsilon}\right]:\eth\boldsymbol{\varepsilon} + \left[-\frac{\Delta\eta}{\Delta t} - \frac{\bar{\theta}}{\Delta t}H^{\theta} + \Gamma^{\theta}\right]\eth\bar{\theta}. \qquad [5.73]$$

Comparing the last expression with $[5.60]_2$ yields

$$\boldsymbol{R}^{\varepsilon} = -\frac{\bar{\theta}}{\Delta t}\boldsymbol{H}^{\varepsilon} + \boldsymbol{\Gamma}^{\varepsilon}, \quad R^{\theta} = -\frac{\Delta\eta}{\Delta t} - \frac{\bar{\theta}}{\Delta t}H^{\theta} + \Gamma^{\theta}. \qquad [5.74]$$

– With regard to the unit cell problem (Tables 5.4 and 5.6), it is customary in the literature to work with a displacement field of Suquet-type (equation $[5.70]_1$) and not only with the periodic part $\widetilde{\boldsymbol{u}}$. The periodicity condition at the boundaries is imposed by relations of the form

$$\Delta\boldsymbol{u}^{+} - \Delta\boldsymbol{u}^{-} = \Delta\bar{\boldsymbol{\varepsilon}}\cdot[\boldsymbol{x}^{+} - \boldsymbol{x}^{-}],$$

where the superscripts $^+$ and $^-$ denote opposite faces of the unit cell. From computational point of view, the macroscopic strain increment $\Delta\bar{\boldsymbol{\varepsilon}}$ is imposed at "dummy" nodes, which are connected with the nodes of the unit cell external boundaries [MIC 99] (see Figure 5.4 for points at two opposite boundaries). In finite element calculations, the dummy nodes develop reaction forces that are equal to the macroscopic stress increment $\Delta\bar{\boldsymbol{\sigma}}$ multiplied by the volume of the unit cell.

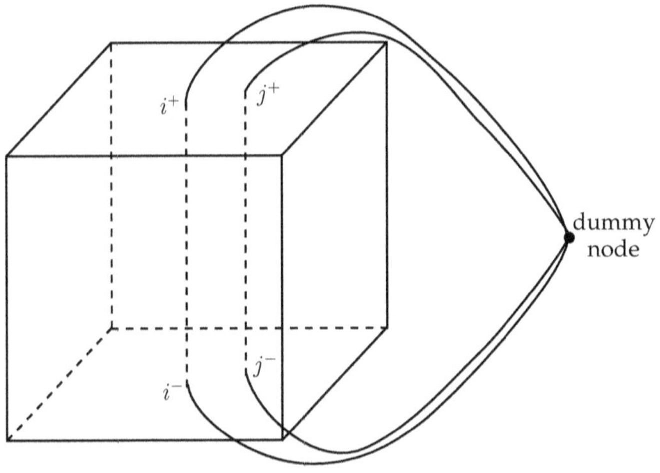

Figure 5.4. *Opposite sides of a unit cell are connected with a dummy node*

Such an approach is equivalent to the numerical scheme described here and provides exactly the same results. This equivalence is explained in the following manner: mechanically speaking, the asymptotic expansion homogenization method treats the unit cell problem in two steps: first, the uniform strain increment $\Delta\bar{\varepsilon}$ is imposed everywhere in the unit cell. Thus, an inconsistency due to the traction continuity between the different material phases appears. In the second step, this inconsistency is taken care of by allowing the unit cell to "relax", leading to the development of the periodic displacement increment $\Delta\widetilde{\boldsymbol{u}}$. Due to the linearized character of the iterative scheme, the principle of superposition holds, permitting to solve the unit cell problem directly in one step, by applying the Suquet-type boundary conditions.

5.4. Example: multilayered composite

The simplest problem in periodic homogenization with analytical (or semi-analytical, for constituents that exhibit a non-linear response) solution is the multilayered composite, i.e. a material whose unit cell consists of N stacked constituents, as demonstrated in Figure 5.5. Each constituent has volume fraction f_r and thermomechanical tangent moduli $\boldsymbol{D}^{\varepsilon(r)}$, $\boldsymbol{D}^{\theta(r)}$ $(r = 1, 2, ..., N)$. Obviously, $\sum_{r=1}^{N} f_r = 1$.

5.4.1. *Unit cell problem: microscopic fields*

In the unit cell of the multilayered material and under periodicity conditions, the periodic part of the displacement, $\delta\widetilde{\boldsymbol{u}}$, and the temperature, $\delta\widetilde{\theta}$, are uniform along the x_2 and x_3 axes, presenting non-uniformity only on the x_1 axis. Thus, equations [5.58] for each material r can be expressed in the 1-D differential form

$$\frac{\mathrm{d}}{\mathrm{d}x_1}\left(\boldsymbol{D}^{\varepsilon(r)}_{nn} \cdot \frac{\mathrm{d}\delta\widetilde{\boldsymbol{u}}^{(r)}}{\mathrm{d}x_1} + \boldsymbol{\sigma}^{(r)}_n\right) = \boldsymbol{0}\,, \quad \frac{\mathrm{d}}{\mathrm{d}x_1}\left(q^{(r)}_1 - \kappa^{(r)}_{11}\frac{\mathrm{d}\delta\widetilde{\theta}^{(r)}}{\mathrm{d}x_1}\right) = 0\,, \quad [5.75]$$

with

$$\boldsymbol{D}_{nn}^{\varepsilon(r)} = \begin{bmatrix} D_{1111}^{\varepsilon(r)} & D_{1121}^{\varepsilon(r)} & D_{1131}^{\varepsilon(r)} \\ D_{2111}^{\varepsilon(r)} & D_{2121}^{\varepsilon(r)} & D_{2131}^{\varepsilon(r)} \\ D_{3111}^{\varepsilon(r)} & D_{3121}^{\varepsilon(r)} & D_{3131}^{\varepsilon(r)} \end{bmatrix},$$

$$\widetilde{\boldsymbol{u}}^{(r)} = \begin{bmatrix} \widetilde{u}_1^{(r)} \\ \widetilde{u}_2^{(r)} \\ \widetilde{u}_3^{(r)} \end{bmatrix}, \quad \boldsymbol{\sigma}_n^{(r)} = \begin{bmatrix} \sigma_{11}^{(r)} \\ \sigma_{21}^{(r)} \\ \sigma_{31}^{(r)} \end{bmatrix}. \qquad [5.76]$$

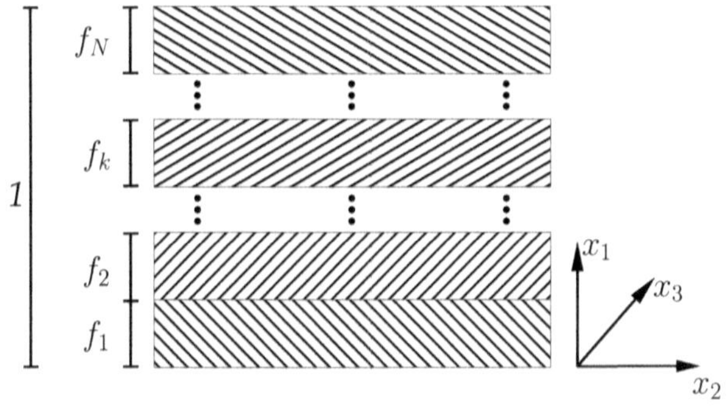

Figure 5.5. *Unit cell of a multilayered composite with N distinct layers*

Due to the continuity of tractions and normal heat fluxes, the system [5.75] after integration is written as

$$\frac{\mathrm{d}\delta\widetilde{\boldsymbol{u}}^{(r)}}{\mathrm{d}x_1} = \left[\boldsymbol{D}_{nn}^{\varepsilon(r)}\right]^{-1} \cdot \left[\boldsymbol{m}^{\varepsilon} - \boldsymbol{\sigma}_n^{(r)}\right],$$

$$\frac{\mathrm{d}\delta\widetilde{\theta}^{(r)}}{\mathrm{d}x_1} = \left[\kappa_{11}^{(r)}\right]^{-1} \left[m^{\theta} + q_1^{(r)}\right], \qquad [5.77]$$

where $\boldsymbol{m}^{\varepsilon}$ and m^{θ} are constant vectors and scalars, respectively, with the same values for all layers. Integrating once more, the solution of the system is written in the form

$$\delta\widetilde{\boldsymbol{u}}^{(r)} = \left[\boldsymbol{D}_{nn}^{\varepsilon(r)}\right]^{-1} \cdot \left[\boldsymbol{m}^{\varepsilon} - \boldsymbol{\sigma}_{n}^{(r)}\right] x_1 + \boldsymbol{e}^{\varepsilon(r)} ,$$

$$\delta\widetilde{\theta}^{(r)} = \left[\kappa_{11}^{\varepsilon(r)}\right]^{-1} \left[m^{\theta} - q_1^{(r)}\right] x_1 + e^{\theta(r)} , \qquad [5.78]$$

where $\boldsymbol{e}^{\varepsilon(r)}$ and $e^{\theta(r)}$ are constant vectors and scalars, respectively, different from layer to layer. The final expressions indicate that $\delta\widetilde{\boldsymbol{u}}^{(r)}$ and $\delta\widetilde{\theta}^{(r)}$ are linear functions of x_1. Due to the periodicity conditions, we can assume, without loss of generality, that $\delta\widetilde{\boldsymbol{u}}^{(1)}(0) = \delta\widetilde{\boldsymbol{u}}^{(N)}(1) = \boldsymbol{0}$ and $\delta\widetilde{\theta}^{(1)}(0) = \delta\widetilde{\theta}^{(N)}(1) = \boldsymbol{0}$, which lead to

$$\boldsymbol{e}^{\varepsilon(1)} = \boldsymbol{0} \,, \boldsymbol{e}^{\varepsilon(N)} = - \left[\boldsymbol{D}_{nn}^{\varepsilon(N)}\right]^{-1} \cdot \left[\boldsymbol{m}^{\varepsilon} - \boldsymbol{\sigma}_{n}^{(N)}\right] ,$$

$$e^{\theta(1)} = \boldsymbol{0} \,, e^{\theta(N)} = - \left[\kappa_{11}^{(N)}\right]^{-1} \left[m^{\theta} + q_1^{(N)}\right] . \qquad [5.79]$$

The continuity of $\delta\widetilde{\boldsymbol{u}}$ and $\delta\widetilde{\theta}$ at all the interfaces is expressed by the relations

$$\delta\widetilde{\boldsymbol{u}}^{(r)}(f_r) = \delta\widetilde{\boldsymbol{u}}^{(r+1)}(f_r) , \quad \delta\widetilde{\theta}^{(r)}(f_r) = \delta\widetilde{\theta}^{(r+1)}(f_r) ,$$
$$r = 1, 2, ..., N-1 . \qquad [5.80]$$

Using [5.78], equations [5.80] are expressed for each interface as

$$\left[\boldsymbol{D}_{nn}^{\varepsilon(1)}\right]^{-1} \cdot \left[\boldsymbol{m}^{\varepsilon} - \boldsymbol{\sigma}_{n}^{(1)}\right] f_1 = \left[\boldsymbol{D}_{nn}^{\varepsilon(2)}\right]^{-1} \cdot \left[\boldsymbol{m}^{\varepsilon} - \boldsymbol{\sigma}_{n}^{(2)}\right] f_1 + \boldsymbol{e}^{\varepsilon(2)} ,$$

$$\vdots$$

$$\left[\boldsymbol{D}_{nn}^{\varepsilon(k)}\right]^{-1} \cdot \left[\boldsymbol{m}^{\varepsilon} - \boldsymbol{\sigma}_{n}^{(k)}\right] \sum_{r=1}^{k} f_r + \boldsymbol{e}^{\varepsilon(k)}$$

$$= \left[\boldsymbol{D}_{nn}^{\varepsilon(k+1)}\right]^{-1} \cdot \left[\boldsymbol{m}^{\varepsilon} - \boldsymbol{\sigma}_{n}^{(k+1)}\right] \sum_{r=1}^{k} f_r + \boldsymbol{e}^{\varepsilon(k+1)} ,$$

$$\vdots$$

$$\left[\boldsymbol{D}_{nn}^{\varepsilon(N-1)}\right]^{-1} \cdot \left[\boldsymbol{m}^{\varepsilon} - \boldsymbol{\sigma}_{n}^{(N-1)}\right] \sum_{r=1}^{N-1} f_r + \boldsymbol{e}^{\varepsilon(N-1)}$$

$$= \left[\boldsymbol{D}_{nn}^{\varepsilon(N)}\right]^{-1} \cdot \left[\boldsymbol{m}^{\varepsilon} - \boldsymbol{\sigma}_{n}^{(N)}\right] \sum_{r=1}^{N-1} f_r - \left[\boldsymbol{D}_{nn}^{\varepsilon(N)}\right]^{-1} \cdot \left[\boldsymbol{m}^{\varepsilon} - \boldsymbol{\sigma}_{n}^{(N)}\right],$$

and

$$\left[\kappa_{11}^{(1)}\right]^{-1} \left[m^{\theta} + q_1^{(1)}\right] f_1 = \left[\kappa_{11}^{(2)}\right]^{-1} \left[m^{\theta} + q_1^{(2)}\right] f_1 + e^{\theta(2)},$$

$$\vdots$$

$$\left[\kappa_{11}^{(k)}\right]^{-1} \left[m^{\theta} + q_1^{(k)}\right] \sum_{r=1}^{k} f_r + e^{\theta(k)}$$

$$= \left[\kappa_{11}^{(k+1)}\right]^{-1} \left[m^{\theta} + q_1^{(k+1)}\right] \sum_{r=1}^{k} f_r + e^{\theta(k+1)},$$

$$\vdots$$

$$\left[\kappa_{11}^{(N-1)}\right]^{-1} \left[m^{\theta} + q_1^{(N-1)}\right] \sum_{r=1}^{N-1} f_r + e^{\theta(N-1)}$$

$$= \left[\kappa_{11}^{(N)}\right]^{-1} \left[m^{\theta} + q_1^{(N)}\right] \sum_{r=1}^{N-1} f_r - \left[\kappa_{11}^{(N)}\right]^{-1} \left[m^{\theta} + q_1^{(N)}\right].$$

Through summation, the previous expressions lead to

$$\boldsymbol{m}^{\varepsilon} = \left[\sum_{r=1}^{N} f_r \left[\boldsymbol{D}_{nn}^{\varepsilon(r)}\right]^{-1}\right]^{-1} \cdot \left[\sum_{r=1}^{N} f_r \left[\boldsymbol{D}_{nn}^{\varepsilon(r)}\right]^{-1} \cdot \boldsymbol{\sigma}_{n}^{(r)}\right],$$

$$m^{\theta} = \left[\sum_{r=1}^{N} f_r \left[\kappa_{11}^{\varepsilon(r)}\right]^{-1}\right]^{-1} \left[\sum_{r=1}^{N} f_r \left[\kappa_{11}^{\varepsilon(r)}\right]^{-1} q_1^{(r)}\right]. \qquad [5.81]$$

As is observed in equations [5.77], the knowledge of the vector $\boldsymbol{m}^{\varepsilon}$ and scalar m^{θ} is sufficient to compute the increments of the micro-strains and micro-temperature gradients at each layer, since

$$\delta\varepsilon_{11}^{(r)} = \frac{\mathrm{d}\delta\widetilde{u}_1^{(r)}}{\mathrm{d}x_1}, \quad 2\delta\varepsilon_{21}^{(r)} = \frac{\mathrm{d}\delta\widetilde{u}_2^{(r)}}{\mathrm{d}x_1}, \quad 2\delta\varepsilon_{31}^{(r)} = \frac{\mathrm{d}\delta\widetilde{u}_3^{(r)}}{\mathrm{d}x_1},$$

$$\delta\varepsilon_{kl}^{(r)} = 0 \text{ for the rest of the cases},$$

$$\delta\nabla\theta_1^{(r)} = \frac{\mathrm{d}\delta\widetilde{\theta}_1^{(r)}}{\mathrm{d}x_1}, \quad \delta\nabla\theta_2^{(r)} = \delta\nabla\theta_3^{(r)} = 0. \qquad [5.82]$$

With these increments, the actual values of the microscopic strains and temperature gradients at each point of the unit cell are obtained. Then, with this information and the constitutive law for each material, the microscopic stresses, internal variables and heat fluxes are calculated.

5.4.2. *Unit cell problem: concentration tensors for obtaining the effective tangent moduli*

For the simplicity of the subsequent expressions, the following matrices for each layer are defined:

$$\boldsymbol{D}_{nn}^{\varepsilon(r)} = \begin{bmatrix} D_{1111}^{\varepsilon(r)} & D_{1121}^{\varepsilon(r)} & D_{1131}^{\varepsilon(r)} \\ D_{2111}^{\varepsilon(r)} & D_{2121}^{\varepsilon(r)} & D_{2131}^{\varepsilon(r)} \\ D_{3111}^{\varepsilon(r)} & D_{3121}^{\varepsilon(r)} & D_{3131}^{\varepsilon(r)} \end{bmatrix},$$

$$\boldsymbol{D}_{nt}^{\varepsilon(r)} = \begin{bmatrix} D_{1122}^{\varepsilon(r)} & D_{1133}^{\varepsilon(r)} & D_{1123}^{\varepsilon(r)} \\ D_{2122}^{\varepsilon(r)} & D_{2133}^{\varepsilon(r)} & D_{2123}^{\varepsilon(r)} \\ D_{3122}^{\varepsilon(r)} & D_{3133}^{\varepsilon(r)} & D_{3123}^{\varepsilon(r)} \end{bmatrix}, \quad \boldsymbol{D}_{n}^{\theta(r)} = \begin{bmatrix} D_{11}^{\theta(r)} \\ D_{21}^{\theta(r)} \\ D_{31}^{\theta(r)} \end{bmatrix}. \qquad [5.83]$$

Following similar steps with those for the derivation of the microscopic fields, equations $[5.64]_{1,2}$ for each material r take the forms,

$$\frac{\mathrm{d}}{\mathrm{d}x_1}\left(\boldsymbol{D}_{nn}^{\varepsilon(r)} \cdot \frac{\mathrm{d}\boldsymbol{\chi}_n^{\varepsilon(r)}}{\mathrm{d}x_1} + \boldsymbol{D}_{nn}^{\varepsilon(r)}\right) = 0,$$

$$\frac{\mathrm{d}}{\mathrm{d}x_1}\left(\boldsymbol{D}_{nn}^{\varepsilon(r)}\frac{\mathrm{d}\boldsymbol{\chi}_t^{\varepsilon(r)}}{\mathrm{d}x_1}+\boldsymbol{D}_{nt}^{\varepsilon(r)}\right)=0\,,$$

$$\frac{\mathrm{d}}{\mathrm{d}x_1}\left(\boldsymbol{D}_{nn}^{\varepsilon(r)}\frac{\mathrm{d}\boldsymbol{\chi}^{\theta(r)}}{\mathrm{d}x_1}+\boldsymbol{D}_{n}^{\theta(r)}\right)=0\,, \qquad [5.84]$$

with

$$\boldsymbol{\chi}_n^{\varepsilon(r)}=\begin{bmatrix}\chi_{111}^{\varepsilon(r)} & \chi_{211}^{\varepsilon(r)} & \chi_{311}^{\varepsilon(r)}\\ \chi_{112}^{\varepsilon(r)} & \chi_{212}^{\varepsilon(r)} & \chi_{312}^{\varepsilon(r)}\\ \chi_{113}^{\varepsilon(r)} & \chi_{213}^{\varepsilon(r)} & \chi_{313}^{\varepsilon(r)}\end{bmatrix},$$

$$\boldsymbol{\chi}_t^{\varepsilon(r)}=\begin{bmatrix}\chi_{221}^{\varepsilon(r)} & \chi_{331}^{\varepsilon(r)} & \chi_{231}^{\varepsilon(r)}\\ \chi_{222}^{\varepsilon(r)} & \chi_{332}^{\varepsilon(r)} & \chi_{232}^{\varepsilon(r)}\\ \chi_{223}^{\varepsilon(r)} & \chi_{333}^{\varepsilon(r)} & \chi_{233}^{\varepsilon(r)}\end{bmatrix}, \quad \boldsymbol{\chi}^{\theta(r)}=\begin{bmatrix}\chi_1^{\theta(r)}\\ \chi_2^{\theta(r)}\\ \chi_3^{\theta(r)}\end{bmatrix}. \qquad [5.85]$$

It is noted that the micro-tangent moduli used here are the same with micro-tangent moduli at the last iteration step of the unit cell problem systems [5.84] have exactly the same structure with [5.75], thus their s has similar form with [5.78], i.e.

$$\frac{\mathrm{d}\boldsymbol{\chi}_n^{\varepsilon(r)}}{\mathrm{d}x_1}=\left[\boldsymbol{D}_{nn}^{\varepsilon(r)}\right]^{-1}\cdot\left[\boldsymbol{m}_n^{\varepsilon}-\boldsymbol{D}_{nn}^{\varepsilon(r)}\right],$$

$$\frac{\mathrm{d}\boldsymbol{\chi}_t^{\varepsilon(r)}}{\mathrm{d}x_1}=\left[\boldsymbol{D}_{nn}^{\varepsilon(r)}\right]^{-1}\cdot\left[\boldsymbol{m}_t^{\varepsilon}-\boldsymbol{D}_{nt}^{\varepsilon(r)}\right],$$

$$\frac{\mathrm{d}\boldsymbol{\chi}^{\theta(r)}}{\mathrm{d}x_1}=\left[\boldsymbol{D}_{nn}^{\varepsilon(r)}\right]^{-1}\cdot\left[\boldsymbol{m}_\theta^{\varepsilon}-\boldsymbol{D}_{n}^{\theta(r)}\right],$$

where the second-order tensors $\boldsymbol{m}_n^{\varepsilon}$, $\boldsymbol{m}_t^{\varepsilon}$ and the vector
expressions

$$\boldsymbol{m}_n^{\varepsilon}=\left[\sum_{r=1}^{N}f_r\left[\boldsymbol{D}_{nn}^{\varepsilon(r)}\right]^{-1}\right]^{-1},$$

$$\boldsymbol{m}_t^\varepsilon = \boldsymbol{m}_n^\varepsilon \cdot \left[\sum_{r=1}^{N} f_r \left[\boldsymbol{D}_{nn}^{\varepsilon(r)}\right]^{-1} \cdot \boldsymbol{D}_{nt}^{\varepsilon(r)}\right],$$

$$\boldsymbol{m}_\theta^\varepsilon = \boldsymbol{m}_n^\varepsilon \cdot \left[\sum_{r=1}^{N} f_r \left[\boldsymbol{D}_{nn}^{\varepsilon(r)}\right]^{-1} \cdot \boldsymbol{D}_{n}^{\theta(r)}\right]. \qquad [5.87]$$

With regard to the thermal conductivity, the appropriate system of equations [5.64]$_3$ is reduced to

$$\frac{\mathrm{d}}{\mathrm{d}x_1}\left(\boldsymbol{\kappa}_n^{(r)} + \kappa_{11}^{(r)} \frac{\mathrm{d}\boldsymbol{\psi}^{\theta(r)}}{\mathrm{d}x_1}\right) = 0, \quad \text{with}$$

$$\boldsymbol{\kappa}_n^{(r)} = \begin{bmatrix} \kappa_{11}^{(r)} \\ \kappa_{12}^{(r)} \\ \kappa_{13}^{(r)} \end{bmatrix}^T, \quad \boldsymbol{\psi}^{\theta(r)} = \begin{bmatrix} \psi_1^{\theta(r)} \\ \psi_2^{\theta(r)} \\ \psi_3^{\theta(r)} \end{bmatrix}^T. \qquad [5.88]$$

Following similar steps with those for the derivation of the microscopic fields, the spatial derivatives of $\boldsymbol{\psi}^{\theta(r)}$ can be expressed as

$$\frac{\mathrm{d}\boldsymbol{\psi}^{\theta(r)}}{\mathrm{d}x_1} = \left[\kappa_{11}^{(r)}\right]^{-1}\left[\boldsymbol{m}^\theta - \boldsymbol{\kappa}_n^{(r)}\right], \qquad [5.89]$$

where

$$\boldsymbol{m}^\theta = \left[\sum_{r=1}^{N} f_r \left[\kappa_{11}^{(r)}\right]^{-1}\right]^{-1} \sum_{r=1}^{N} f_r \left[\kappa_{11}^{(r)}\right]^{-1} \boldsymbol{\kappa}_n^{(r)}. \qquad [5.90]$$

From these results and considering the minor symmetries of all fourth-order [tens]ors, the mechanical-related concentration tensors are written on the Voigt [nota]tion as

$$\widetilde{\boldsymbol{A}}^{\varepsilon(r)} = \widetilde{\boldsymbol{I}} + \widetilde{\boldsymbol{\chi}}^{*(r)},$$

$$\widetilde{\boldsymbol{\chi}}^{*(r)} = \begin{bmatrix} \frac{\mathrm{d}\chi_{111}^{\varepsilon(r)}}{\mathrm{d}x_1} & \frac{\mathrm{d}\chi_{221}^{\varepsilon(r)}}{\mathrm{d}x_1} & \frac{\mathrm{d}\chi_{331}^{\varepsilon(r)}}{\mathrm{d}x_1} & \frac{\mathrm{d}\chi_{211}^{\varepsilon(r)}}{\mathrm{d}x_1} & \frac{\mathrm{d}\chi_{311}^{\varepsilon(r)}}{\mathrm{d}x_1} & \frac{\mathrm{d}\chi_{231}^{\varepsilon(r)}}{\mathrm{d}x_1} \\ 0 & 0 & 0 & 0 & 0 & 0 \\ 0 & 0 & 0 & 0 & 0 & 0 \\ \frac{\mathrm{d}\chi_{112}^{\varepsilon(r)}}{\mathrm{d}x_1} & \frac{\mathrm{d}\chi_{222}^{\varepsilon(r)}}{\mathrm{d}x_1} & \frac{\mathrm{d}\chi_{332}^{\varepsilon(r)}}{\mathrm{d}x_1} & \frac{\mathrm{d}\chi_{212}^{\varepsilon(r)}}{\mathrm{d}x_1} & \frac{\mathrm{d}\chi_{312}^{\varepsilon(r)}}{\mathrm{d}x_1} & \frac{\mathrm{d}\chi_{232}^{\varepsilon(r)}}{\mathrm{d}x_1} \\ \frac{\mathrm{d}\chi_{113}^{\varepsilon(r)}}{\mathrm{d}x_1} & \frac{\mathrm{d}\chi_{223}^{\varepsilon(r)}}{\mathrm{d}x_1} & \frac{\mathrm{d}\chi_{333}^{\varepsilon(r)}}{\mathrm{d}x_1} & \frac{\mathrm{d}\chi_{213}^{\varepsilon(r)}}{\mathrm{d}x_1} & \frac{\mathrm{d}\chi_{313}^{\varepsilon(r)}}{\mathrm{d}x_1} & \frac{\mathrm{d}\chi_{233}^{\varepsilon(r)}}{\mathrm{d}x_1} \\ 0 & 0 & 0 & 0 & 0 & 0 \end{bmatrix},$$

$$\widetilde{\boldsymbol{A}}^{\theta(r)} = \begin{bmatrix} \frac{\mathrm{d}\chi_1^{\theta(r)}}{\mathrm{d}x_1} & 0 & 0 & \frac{\mathrm{d}\chi_2^{\theta(r)}}{\mathrm{d}x_1} & \frac{\mathrm{d}\chi_3^{\theta(r)}}{\mathrm{d}x_1} & 0 \end{bmatrix}^T. \quad [5.91]$$

Additionally, the following concentration tensor can be constructed per each layer,

$$\boldsymbol{A}^{\kappa(r)} = \begin{bmatrix} 1 + \frac{\mathrm{d}\psi_1^{\theta(r)}}{\mathrm{d}x_1} & \frac{\mathrm{d}\psi_2^{\theta(r)}}{\mathrm{d}x_1} & \frac{\mathrm{d}\psi_3^{\theta(r)}}{\mathrm{d}x_1} \\ 0 & 1 & 0 \\ 0 & 0 & 1 \end{bmatrix}. \quad [5.92]$$

It is noted that, in Voigt notation, the strain-related concentration tensors $\boldsymbol{A}^{\varepsilon(r)}$ are expressed in the $\widetilde{\boldsymbol{A}}^{\varepsilon(r)}$ form as described in section 1.1.4. Additionally, $\boldsymbol{A}^{\theta(r)}$ are written as strain type second-order tensors $\widetilde{\boldsymbol{A}}^{\theta(r)}$.

The concentration tensors in expressions [5.91] and [5.92] allow us to compute analytically all the necessary macroscopic moduli through equations [5.67].

5.5. Numerical applications

5.5.1. *1-D mechanical problem of a multilayered composite*

Consider the multilayered material of Figure 5.6, consisting of a repeated unit cell with two elastic isotropic layers, one made of steel 316 (Young's modulus $E_1 = 185$ GPa) and one made of SiC (Young's modulus $E_2 = 410$ GPa). The two layers have the same volume fraction ($f_1 = f_2 = 0.5$). The global structure is fixed at one edge and it is subjected to uniform tension stress $\sigma = 10$ MPa at the other edge. Moreover, the steel 316 and the SiC are subjected to body forces per unit volume $\rho_1 b_1 = 78$ MPa/m and $\rho_2 b_2 = 31$ MPa/m, respectively. The size of the total structure is equal to 1 m, while the dimensionless characteristic length of the unit cell is ϵ (i.e. the actual size of the cell is $\epsilon \cdot 1$). Neglecting inertia and thermal effects, the 1-D differential form of the equilibrium equation $\mathrm{div}^{\epsilon}\boldsymbol{\sigma}^{\epsilon} + \rho^{\epsilon}\boldsymbol{b}^{\epsilon} = \mathbf{0}$ is written as

$$E_i \frac{\mathrm{d}^2 u_i}{\mathrm{d}\overline{x}^2} + \rho_i b_i = 0,$$

where i takes the values 1 and 2, according to the layers. The stress is continuous throughout the multilayered structure due to the traction continuity hypothesis. The differential equation is accompanied by the displacement continuity between two layers and the boundary condition $u = 0$ at $\overline{x} = 0$. This problem can be solved analytically, since the above differential equation leads to express the displacement at each layer in the form

$$u_i = -\frac{\rho_i b_i}{2E_i}\overline{x}^2 + \frac{c_1}{E_i}\overline{x} + c_2.$$

The unkonwn constants c_1 and c_2 vary between the different layers of the overall composite. They can be computed by using the boundary conditions and the continuity of tractions and displacements at the interfaces.

In Figure 5.7, the displacement and the stress as functions of the position are presented for various unit cell sizes. As is observed, a decrease in the unit cell size causes lower variations in the displacement and stress fields between the layers. When ϵ is much smaller compared to the size of the structure, the solution becomes independent of the unit cell size and obtains only a macroscopic variation with the position. For small values of ϵ, the composite can be substituted by an equivalent homogeneous medium that obeys the 1-D differential equation [HAS 98]

$$\overline{E}\frac{\mathrm{d}^2 \overline{u}}{\mathrm{d}\overline{x}^2} + \overline{\rho b} = 0,$$

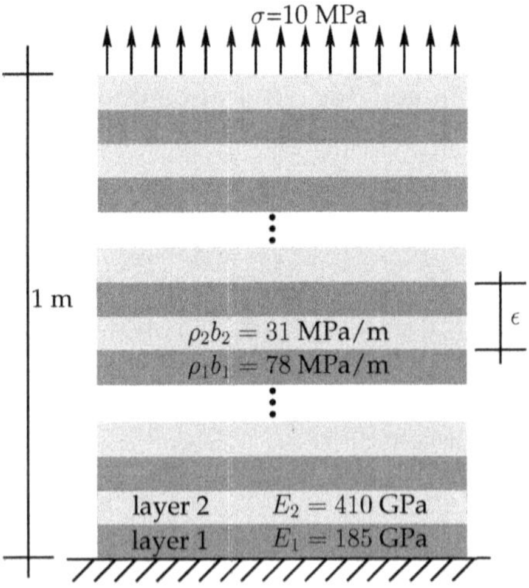

Figure 5.6. *Multilayered material subjected to constant stress. For a color version of this figure, see www.iste.co.uk/chatzigeorgiou/thermomechanical.zip*

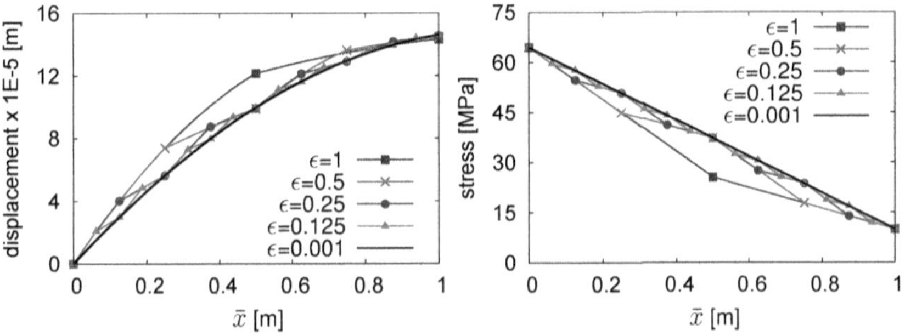

a) b)

Figure 5.7. *Variation of a) the displacement and b) the stress with respect to the position of the multilayered material for various unit cell sizes. For a color version of this figure, see www.iste.co.uk/chatzigeorgiou/thermomechanical.zip*

where $\overline{E} = \left\langle E_i^{-1} \right\rangle^{-1} = 254.96$ GPa and $\overline{\rho b} = \langle \rho_i b_i \rangle = 54.5$ MPa/m. The expression for the $\overline{E}$ is obtained by using the methodology of section 5.4 and reducing the problem to 1-D. The homogenization results follow accurately the displacement-position and the stress-position curves for very small ϵ (Figure 5.8).

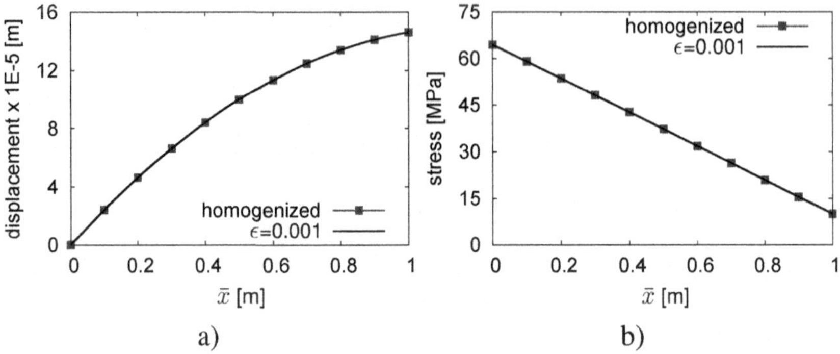

Figure 5.8. *Variation of a) the displacement and b) the stress with respect to the position for the actual multilayered material and the homogenized material. For a color version of this figure, see www.iste.co.uk/chatzigeorgiou/thermomechanical.zip*

The homogenization framework also allows us to identify the microscopic strains at each macroscopic position, which, according to the concentration tensor formulas derived in section 5.4, can be computed from the expression $\varepsilon_i(\overline{x}) = \overline{\varepsilon}(\overline{x})\ \overline{E}/E_i = \overline{\sigma}(\overline{x})/E_i$. Figure 5.9 shows the variation of the microscopic strains for the two layers with respect to the macro-position, as well as the analytical solution for $\epsilon = 0.02$. As was expected, the microscopic strains are highly oscillating. These results indicate that the homogenization framework can describe very accurately both scales.

5.5.2. *1-D thermal problem of a multilayered composite*

Consider the multilayered material of Figure 5.10, consisting of a repeated unit cell with two thermally isotropic layers, one made of steel 316 with thermal properties

$$\kappa_1 = 1.5\text{E-5 MN/[s K]}, \quad \tilde{c}_1 = \rho_1 c_{p1} = 3.9 \text{ MPa/K}$$

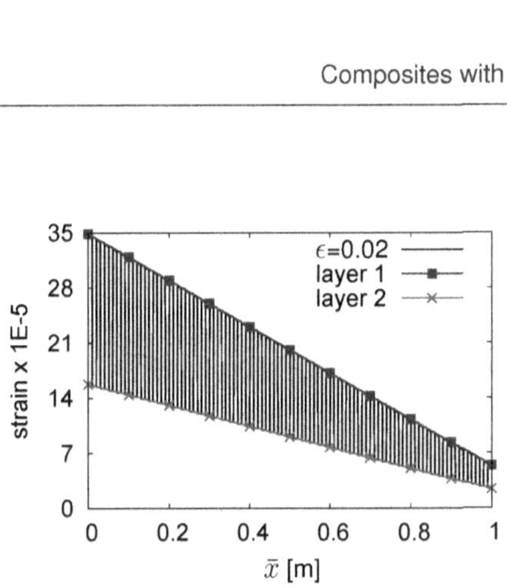

Figure 5.9. *Variation of strain for the actual multilayered material ($\epsilon = 0.02$) and microscopic strains for the homogenized material (for layers 1 and 2). For a color version of this figure, see www.iste.co.uk/chatzigeorgiou/thermomechanical.zip*

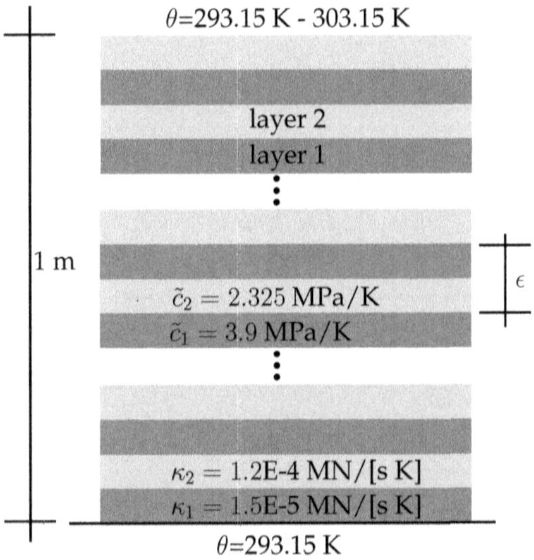

Figure 5.10. *Multilayered material subjected to temperature increase at one end. For a color version of this figure, see www.iste.co.uk/chatzigeorgiou/thermomechanical.zip*

and the other made of SiC ceramic with thermal properties

$$\kappa_2 = 1.2\text{E-4 MN/[s K]}, \quad \tilde{c}_2 = \rho_2 c_{p2} = 2.325 \text{ MPa/K}.$$

The two different materials have equal volume fractions ($f_1 = f_2 = 0.5$). Initially, the global structure is under uniform temperature equal to 293.15 K. The temperature remains fixed to 293.15 K at one edge, while it increases linearly at the other edge from 293.15 to 303.15 K in 1000 s. The size of the total structure is equal to 1 m, while the dimensionless characteristic length of the unit cell is ϵ (i.e. the actual size of the cell is $\epsilon = 1$). Neglecting heat sources and mechanical effects, the 1-D expression of the energy equation can be written as the classical parabolic partial differential equation

$$\tilde{c}_i \frac{\partial \theta_i}{\partial t} = \kappa_i \frac{\partial^2 \theta_i}{\partial \overline{x}^2},$$

where i takes the values 1 and 2, according to the layers. Considering continuity of temperature and heat flux at the interface between two neighbor layers, the energy equation can be solved numerically using the finite differences method.

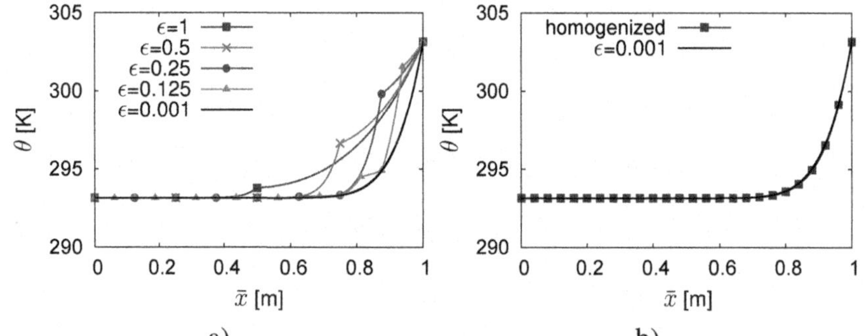

Figure 5.11. *a) Variation of temperature with the position for various ϵ and b) temperature profile comparison between the actual composite and the homogenized material at t = 1000 s. For a color version of this figure, see www.iste.co.uk/chatzigeorgiou/thermomechanical.zip*

In Figure 5.11(a), the temperature as a function of the position at the end of the analysis is presented for various unit cell sizes. As can be observed, for large values of ϵ, the temperature profile changes significantly during the change from one layer to the other. For very small values of ϵ, the fluctuations vanish and a smooth, macroscopic temperature, is observed. For small values of ϵ, the composite can be substituted by an equivalent homogeneous medium that obeys the 1-D partial differential equation

$$\tilde{\bar{c}}\frac{\partial\bar{\theta}}{\partial t} = \bar{\kappa}\frac{\partial^2\bar{\theta}}{\partial\bar{x}^2},$$

where $\tilde{\bar{c}} = \langle\tilde{c}_i\rangle = 3.1125$ MPa/K and $\bar{\kappa} = \left\langle\kappa_i^{-1}\right\rangle^{-1} = 2.667$E-5 MN/[s K]. The expressions for $\tilde{\bar{c}}$ and $\bar{\kappa}$ are obtained by using the methodology of section 5.4 and reducing the problem to 1-D. The homogenization results follow closely the temperature-position curve for very small ϵ (Figure 5.11(b)).

5.5.3. *3-D mechanical analysis: elastoplastic multilayered composite*

Consider a multilayered material whose unit cell consists of two layers with elastoplastic materials, having equal volume fraction ($f_1 = f_2 = 0.5$). The elastoplastic constitutive laws of these materials include only isotropic hardening and the von Mises stress σ^{vM} is expressed in terms of the accumulated plastic strain p according to the power law relation

$$\sigma^{\mathrm{vM}} = Y + kp^m,$$

where Y is the elastic limit and k, m are plastic hardening parameters. The parameters chosen for this example are summarized in Table 5.8.

The composite is assumed to be under isothermal conditions, while mechanically a tension stress is applied in the $\bar{x}_1$ direction (Figure 5.12). This stress reaches the value of 40 MPa and then the material is unloaded. In the other two directions, the composite is traction-free. The microscopic and macroscopic mechanical responses are shown in Figure 5.13. As is observed in Figure 5.13(b), microscopically, the two constituents develop stress in the $\bar{x}_2$ (and $\bar{x}_3$) direction, even though the overall composite has zero stress $\bar{\sigma}_{22}$ during the whole loading. Due to equal volume fractions, the two layers microscopic stresses σ_{22} are opposite. These off-axis stresses cause a complex stress condition in the constituents during unloading, leading to plastification of the first material even before the $\bar{\sigma}_{11}$ reaches zero (Figure 5.13(a)) and creating a kinematic-like hardening effect in the composite.

Property	Material 1	Material 2
Young's modulus [MPa]	31034	3034
Poisson's ratio	0.3	0.3
elastic limit [MPa]	10	20
k [MPa]	20	150
m	0.3	0.3

Table 5.8. *Elastoplastic properties of multilayered composite constituents*

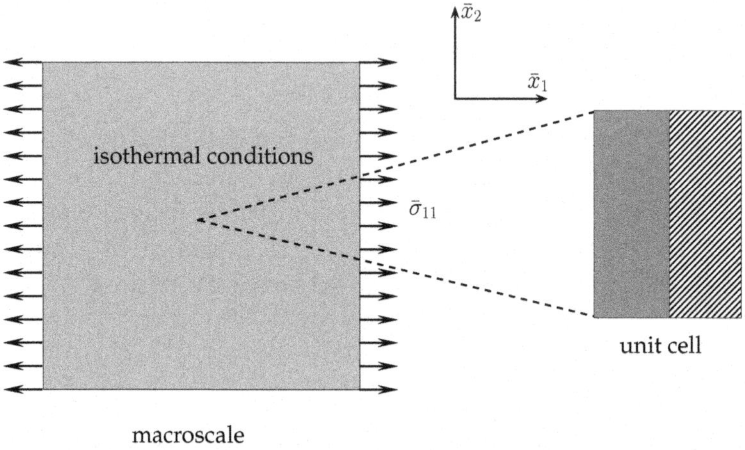

Figure 5.12. *Multilayered elastoplastic composite: uniaxial loading under isothermal conditions. For a color version of this figure, see www.iste.co.uk/chatzigeorgiou/thermomechanical.zip*

Such effects demonstrate the difficulties involved in developing constitutive models at the macroscale from the knowledge of the mechanical response of the constituents. Indeed, as has been demonstrated for power law viscoplasticity [CHA 10c], the macroscopic response exhibits a rheology that is not present at the local scale, but which is strongly dependent on the microstructure.

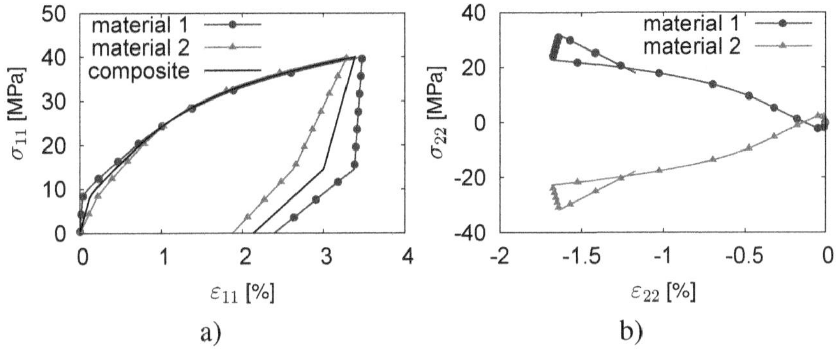

Figure 5.13. *a) Microscopic and overall mechanical response for the layers and composite, respectively, in the direction of loading. b) Microscopic stress versus microscopic strain in the direction 2 for the layers. For a color version of this figure, see www.iste.co.uk/chatzigeorgiou/thermomechanical.zip*

5.5.4. *Fully coupled 3-D thermomechanical analysis: viscoplastic multilayered composite*

Consider a multilayered material whose unit cell consists of two layers: the first layer (80% volume fraction) is made by ceramic SiC, behaving elastically, while the second (20% volume fraction) is made by steel 316, behaving as viscoplastic material. When steel 316 is plastically deformed, the von Mises stress σ^{vM} is expressed in terms of the accumulated plastic strain p and its rate according to the relation

$$\sigma^{\text{vM}} = Y + Q_1 p + Q_2[1 - \exp(-bp)] + K_a \dot{p}^{1/N_a},$$

where Y is the elastic limit, Q_1, Q_2, b are plastic hardening parameters and $K_{\dot{a}}$, N_a are viscous parameters. The material parameters of both the metal and ceramic are shown in Table 5.9.

The multilayered material is initially under a uniform temperature equal to 293.15 K. In this example, it is assumed that the composite operates under adiabatic conditions (div$\boldsymbol{q}$ = 0 everywhere) and a uniaxial compression in the direction $\overline{x}_1$, normal to the layers, is applied (Figure 5.14). The other two directions are considered as traction-free. The uniaxial stress loading is conducted under a stress rate 0.1 MPa/s and includes the following steps:

1) compressive loading up to 300 MPa and subsequent unloading;

Property	Steel 316	SiC
Young's modulus [GPa]	185	410
Poisson's ratio	0.27	0.14
thermal expansion [1/K]	1.5E-5	4.0E-6
density [t/m^3]	7.8	3.1
specific heat capacity [MJ/[t K]]	0.5	0.75
thermal conductivity [MN/[s K]]	1.5E-5	1.2E-4
elastic limit [MPa]	84	
K_a [MPa]	151	
N_a	24	
Q_1 [MPa]	6400	
Q_2 [MPa]	270	
b	25	

Table 5.9. *Thermomechanical properties of multilayered composite constituents. The ceramic properties have been obtained by Accuratus [ACC 13], while the steel properties by Lemaitre and Chaboche [LEM 02]*

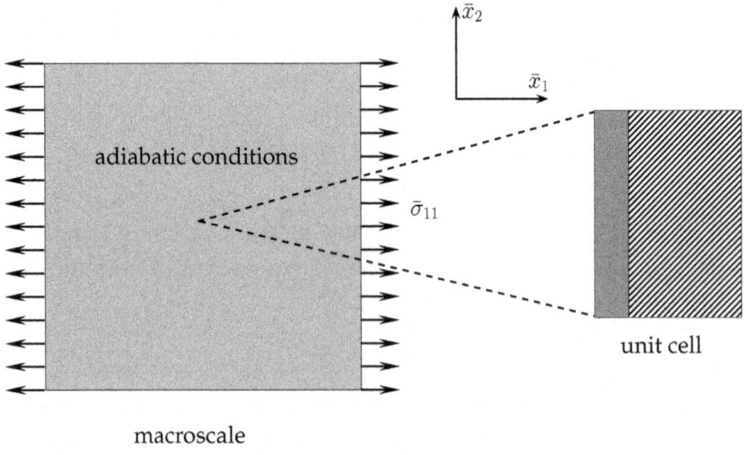

Figure 5.14. *Multilayered viscoplastic composite: uniaxial loading under adiabatic conditions. For a color version of this figure, see www.iste.co.uk/chatzigeorgiou/thermomechanical.zip*

2) compressive loading up to 320 MPa and subsequent unloading;

3) compressive loading up to 340 MPa and subsequent unloading.

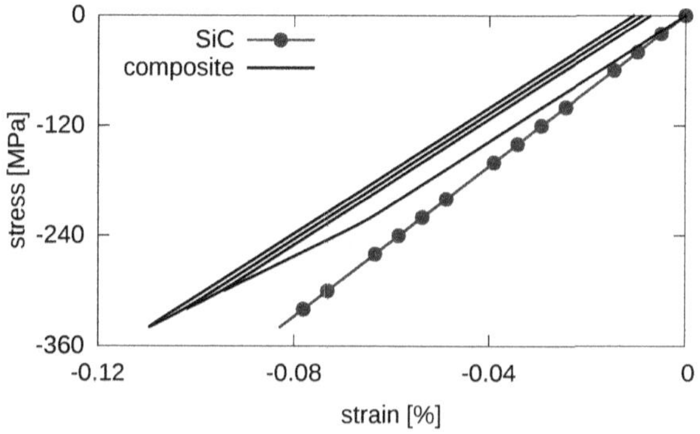

a)

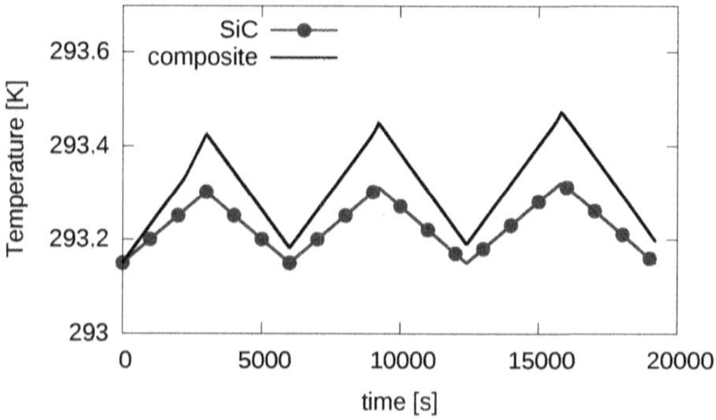

b)

Figure 5.15. *Comparison between the composite response and the response of the SiC when subjected to the same thermomechanical conditions. Variation of a) stress with respect to strain and b) temperature with respect to time. For a color version of this figure, see www.iste.co.uk/chatzigeorgiou/thermomechanical.zip*

Figure 5.15 shows the mechanical and thermal response of the composite, compared with the corresponding response of the pure ceramic when subjected to the same thermomechanical conditions. As expected, the composite presents a behavior similar to the ceramic, but at the same time it shows viscoplastic effects due to the influence of the steel. In Figure 5.15(b), the change of the curve's slope for the composite during the initial loading is due to the presence of dissipation.

6

Composites with Random Structure

Nowadays, it is well established that the overall behavior of a composite strongly depends on the properties of the material constituents and the microscopic geometry, i.e. the volume fraction, shape and orientation of constituents. Homogenization methods, as pioneered by Hill [HIL 63], Hill and Rice [HIL 72b] and Hashin [HAS 83], allow us to study the overall mechanical behavior of polycrystalline metals and composites with a microstructure that contains particles or/and fibers. Many homogenization methods rely on the famous Eshelby's equivalent inclusion theory, which [ESH 57] gave rise to the micromechanics of media with random microstructure. Since then, more advanced techniques, most popular among them the self-consistent and the Mori-Tanaka techniques [MOR 73], have been developed and extended in order to study composites with many types of particles and various particle orientations. The differences between the classical mean-field theories (Eshelby dilute approach, Mori-Tanaka and self-consistent) have been extensively discussed in several books [MUR 87, NEM 99b, QU 06].

Several extensions of the classical mean-field theories provide solutions for composites with coated inclusions, example the methods proposed by Hori and Nemat-Nasher [HOR 93] and by Cherkaoui *et al.* [CHE 95]. Apart from the Mori-Tanaka and the self-consistent approach, there are other established methodologies in the micromechanics literature for composites with ellipsoidal type particles [PON 95]. Also, there are several techniques in the literature to obtain analytical formulas for the overall behavior of composites with long cylindrical fibers or spherical particles [HAS 64, CHR 79, BEN 89, FIS 01, QU 06].

The overall response of heterogeneous, nonlinear, dissipative composites with random microstructure has been studied extensively in the literature. Several methodologies, based on mean-field theories, have been developed in order to examine the macroscopic behavior of viscoelastic, elastoplastic, viscoplastic or damaged composite materials [PON 91, MER 95, PON 97, FIS 01, DES 01, MER 02, DOG 03, CHA 05, LOV 06, JEN 09, MER 09, YVO 09, TEK 10, LAH 13, DES 16]. Reviews and comparisons of different multiscale techniques in both periodic and random media can be found in the literature [KAN 09, PIN 09, CHA 10a, GEE 10, MER 12].

In principle, the homogenization theories for generalized standard materials with nonlinear dissipative mechanisms do not lead to a unique way of expressing thermodynamic potentials (internal energy, Helmholtz or Gibbs) of the macroscopic composite response. Many proposed frameworks in the literature overcome this difficulty by postulating the specific form of the macroscopic constitutive law and then obtaining a proper connection with the microscopic constituents behavior [HUT 76, SUN 04, SUQ 12]. These works often limit the discussion on the mechanical part of the response. The coupling appears frequently only in one-way, by assuming the microscopic stress as a function of the macroscopic temperature [TAK 85, BEN 91, MAG 03]. Most of the thermomechanical studies in the literature focus on thermoelastic materials ROS 70, BEN 91, WAK 91.

The scope of this chapter is to present a general micromechanics methodology for fully coupled thermomechanical processes in random media. The general concepts of the Eshelby problems are presented and various Eshelby-based approaches for thermoelastic composites are discussed. Additionally, a numerical scheme and a solution strategy are proposed for composites with material nonlinearities and dissipative mechanisms. The computational framework closely follows the analogous framework for periodic media, though taking into account that in mean-field theories only average quantities are known and not the local fields at every point of the microstructure. The discussion in the chapter considers exclusively small deformation processes. A proper extension of the theory for large deformations has been presented by Nemat-Nasser [NEM 99a].

6.1. Inclusion problems

The Eshelby's inclusion problems [ESH 57] were revolutionary for the development of the micromechanics of random media. Eshelby identified an analytical solution for the elasticity problem regarding the existence of an arbitrary inclusion of ellipsoidal shape in an infinite medium. This problem provides the necessary background for the study of composites in which ellipsoidal particles are distributed randomly in a matrix phase.

Here, the Eshelby inclusion problem is briefly discussed. More details can be found in the book by Mura [MUR 87] or in the book by Qu and Cherkaoui [QU 06]. The extension of the Eshelby results to large deformation processes which is presented by Nemat-Nasser [NEM 99a].

First, the difference between inclusion and inhomogeneity needs to be clarified:

– An inclusion Ω of a uniform (homogeneous) body $\mathcal{B}$ is a region where uniform eigenstrains $\boldsymbol{\varepsilon}^*$ appear. These eigenstrains are considered generally inelastic and they disturb the total strains $\boldsymbol{\varepsilon}$ in the body;

– An inhomogeneity Ω of a body $\mathcal{B}$ is a region where the material properties are different with regard to the rest of the uniform body.

6.1.1. *Eshelby's inclusion problem*

The Eshelby inclusion problem considers an inclusion Ω in an infinite uniform elastic body. From a practical point of view, this is equivalent to the case of a very small inclusion Ω inside a finite elastic body $\mathcal{B}$ with stiffness tensor $\boldsymbol{D}_0^{\varepsilon}$ (Figure 6.1). At the boundary of the body, the tractions and the displacement field are assumed to be zero. Moreover, no body forces are considered. In mathematical formalism, the Eshelby problem solves the set of equations

$$\begin{aligned}
\operatorname{div}\boldsymbol{\sigma} &= \mathbf{0} \quad \text{in } \mathcal{B}\,,\\
\boldsymbol{\sigma} &= \boldsymbol{D}_0^{\varepsilon}\!:\![\boldsymbol{\varepsilon} - \boldsymbol{\varepsilon}^*] \quad \text{in } \mathcal{B}\,,\\
\boldsymbol{\sigma}\cdot\boldsymbol{n} &= \mathbf{0} \quad \text{in } \partial\mathcal{B}\,,\\
\boldsymbol{\varepsilon}^* &= \begin{cases} \text{constant} \neq \mathbf{0} & \text{in } \Omega\,,\\ \mathbf{0} & \text{in } \mathcal{B}-\Omega\,. \end{cases}
\end{aligned} \qquad [6.1]$$

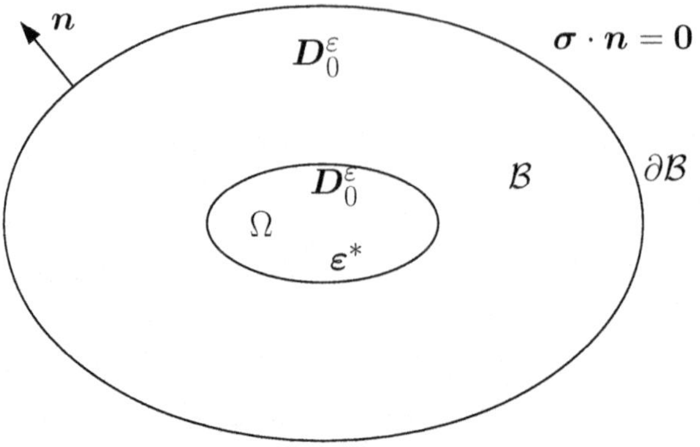

Figure 6.1. *Eshelby inclusion problem*

Of course, the total strains have different distributions inside and outside the inclusion. Eshelby [ESH 57] has proven that inside an inclusion with ellipsoidal shape, the total strains are uniform and they are given by the expression

$$\boldsymbol{\varepsilon} = \text{constant} = \boldsymbol{S}:\boldsymbol{\varepsilon}^* \quad \text{in } \Omega\,, \qquad [6.2]$$

where $\boldsymbol{S}$ is the fourth-order Eshelby tensor that depends on the geometry of the inclusion and the material properties $\boldsymbol{D}_0^\varepsilon$. For an ellipsoidal inclusion with semi-axes a_1, a_2, a_3 (Figure 6.2), the Eshelby tensor is given by the expression [MUR 87]

$$\boldsymbol{S} = \boldsymbol{P}:\boldsymbol{D}_0^\varepsilon,$$
$$\boldsymbol{P} = \frac{1}{8\pi}\int_{-1}^{1}\int_{0}^{2\pi}\left[\boldsymbol{Z}^{-1}\,\overline{\underline{\otimes}}\,\left[\widehat{\zeta}\otimes\widehat{\zeta}\right] + \left[\widehat{\zeta}\otimes\widehat{\zeta}\right]\underline{\otimes}\,\boldsymbol{Z}^{-1}\right]\mathrm{d}\omega\mathrm{d}\zeta_3,$$
$$\boldsymbol{Z} = \boldsymbol{D}_0^\varepsilon\,\overline{:}\,\left[\widehat{\zeta}\otimes\widehat{\zeta}\right]. \qquad [6.3]$$

In the above expression $\widehat{\zeta}$ is a vector with components

$$\widehat{\zeta}_1 = \sqrt{1-\zeta_3^2}\,\frac{\cos\omega}{a_1}, \quad \widehat{\zeta}_2 = \sqrt{1-\zeta_3^2}\,\frac{\sin\omega}{a_2}, \quad \widehat{\zeta}_3 = \frac{\zeta_3}{a_3}. \qquad [6.4]$$

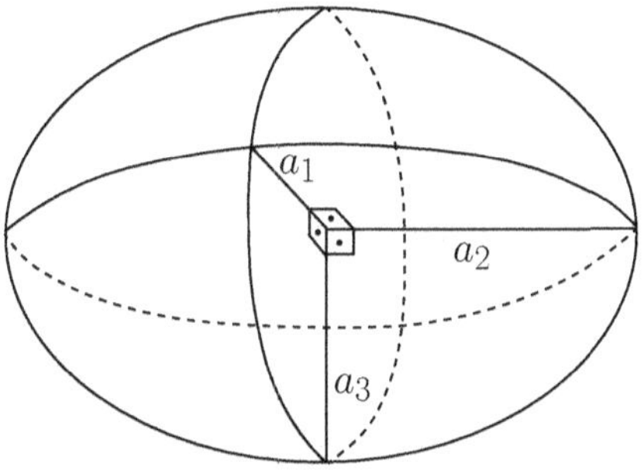

Figure 6.2. *Ellipsoidal inclusion*

For special inclusion geometries and isotropic materials, this tensor has analytical form [MUR 87]. For anisotropic materials, the Eshelby tensor can be evaluated numerically [GAV 90].

6.1.2. *Inhomogeneity problem and Eshelby's equivalent inclusion principle*

Consider a small inhomogeneity Ω with stiffness $\boldsymbol{D}_r^\varepsilon$ in an infinite elastic body $\mathcal{B}$ of stiffness $\boldsymbol{D}_0^\varepsilon$ which has a stress $\boldsymbol{\sigma}_0$ and a strain ε_0 in the far field (Figure 6.3). In mathematical formalism, the following set of equations needs to be solved:

$$\begin{aligned} &\mathrm{div}\boldsymbol{\sigma} = \mathbf{0} \quad \text{in}\, \mathcal{B}\,, \\ &\boldsymbol{\sigma} = \begin{cases} \boldsymbol{D}_0^\varepsilon{:}\boldsymbol{\varepsilon} \ \text{in}\, \mathcal{B} - \Omega\,, \\ \boldsymbol{D}_r^\varepsilon{:}\boldsymbol{\varepsilon} \quad \text{in}\, \Omega\,, \end{cases} \\ &\boldsymbol{\sigma} = \boldsymbol{\sigma}_0 \quad \text{and} \quad \boldsymbol{\varepsilon} = \boldsymbol{\varepsilon}_0 \quad \text{in far field.} \end{aligned} \qquad [6.5]$$

Eshelby [ESH 57] expresses this problem as the sum of two simpler problems:

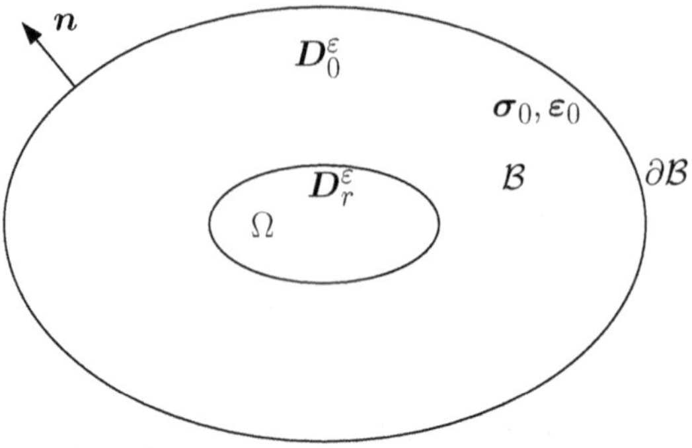

Figure 6.3. *Inhomogeneity problem*

1) a problem with constant stresses $\boldsymbol{\sigma}_0$ and strains $\boldsymbol{\varepsilon}_0$ at the far field, where the inhomogeneity has the same properties as the rest of the material (Figure 6.4a). In mathematical formalism, this is expressed as

$$\operatorname{div}\boldsymbol{\sigma}_0 = \mathbf{0} \quad \text{in } \mathcal{B}, \qquad \boldsymbol{\sigma}_0 = \boldsymbol{D}_0^{\varepsilon}:\boldsymbol{\varepsilon}_0 \quad \text{in } \mathcal{B}. \tag{6.6}$$

The solution of this problem is trivial and is expressed as uniform stress $\boldsymbol{\sigma}_0$ and uniform strain $\boldsymbol{\varepsilon}_0$ inside the entire body $\mathcal{B}$.

2) a boundary value problem with zero surface tractions, where the inhomogeneity has been substituted by an equivalent inclusion problem. Under these conditions, it is assumed that the region Ω has the same properties as the rest of the body and the disturbance due to the inhomogeneity is recovered by an imaginary eigenstrain (Figure 6.4b). In mathematical formalism, the second problem can be expressed as

$$\begin{aligned}
\operatorname{div}\tilde{\boldsymbol{\sigma}} &= \mathbf{0} \quad \text{in } \mathcal{B}, \\
\tilde{\boldsymbol{\sigma}} &= \boldsymbol{D}_0^{\varepsilon}:[\tilde{\boldsymbol{\varepsilon}} - \boldsymbol{\varepsilon}^*] \quad \text{in } \mathcal{B}, \\
\tilde{\boldsymbol{\sigma}}\cdot\boldsymbol{n} &= \mathbf{0} \quad \text{in } \partial\mathcal{B}, \\
\boldsymbol{\varepsilon}^* &= \begin{cases} \text{constant} \neq \mathbf{0} & \text{in } \Omega, \\ \mathbf{0} & \text{in } \mathcal{B} - \Omega. \end{cases}
\end{aligned} \tag{6.7}$$

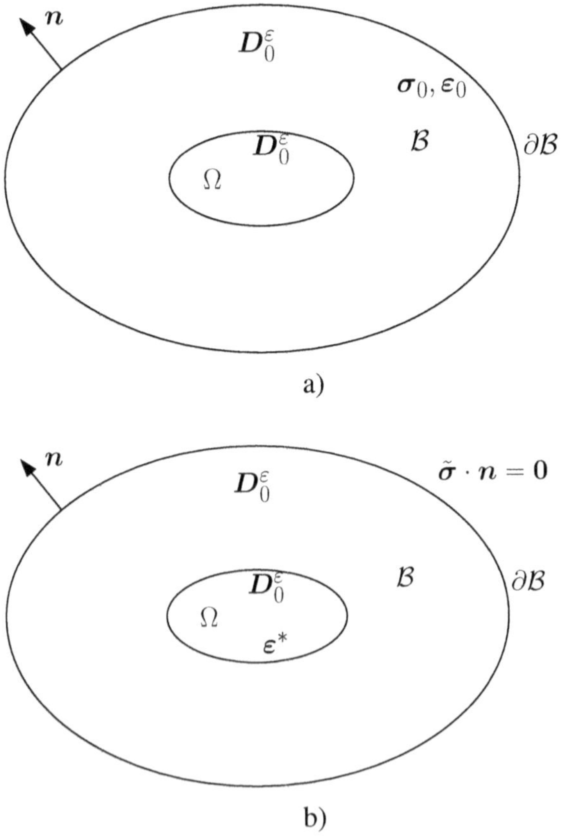

Figure 6.4. *Inhomogeneity problem split into two problems*

This is the classical Eshelby problem, which means that inside the inclusion holds

$$\tilde{\varepsilon} = \text{constant} = \boldsymbol{S}:\boldsymbol{\varepsilon}^* \quad \text{in}\ \Omega\,. \qquad [6.8]$$

All the problems are elastic and the principle of superposition holds. Thus, the solution of the initial problem is the sum of the solutions of the two simpler problems. Focusing only on the region inside the inhomogeneity, we can write

$$\boldsymbol{\sigma} = \boldsymbol{\sigma}_0 + \tilde{\boldsymbol{\sigma}} \quad \text{in}\ \Omega\,,$$
$$\boldsymbol{\varepsilon} = \boldsymbol{\varepsilon}_0 + \tilde{\boldsymbol{\varepsilon}} \quad \text{in}\ \Omega\,. \qquad [6.9]$$

Equation $[6.9]_1$, with the help of $[6.5]_2$, $[6.6]_2$, $[6.7]_2$ and $[6.9]_2$, is written as

$$\boldsymbol{D}_r^{\varepsilon}\!:\![\boldsymbol{\varepsilon}_0 + \tilde{\boldsymbol{\varepsilon}}] = \boldsymbol{D}_0^{\varepsilon}\!:\!\boldsymbol{\varepsilon}_0 + \boldsymbol{D}_0^{\varepsilon}\!:\![\tilde{\boldsymbol{\varepsilon}} - \boldsymbol{\varepsilon}^*] \quad \text{in}\, \Omega\,. \qquad [6.10]$$

Substituting $\tilde{\boldsymbol{\varepsilon}}$ with $\boldsymbol{\varepsilon}^*$ from equation [6.8], the following result is obtained

$$[[\boldsymbol{D}_r^{\varepsilon} - \boldsymbol{D}_0^{\varepsilon}]\!:\!\boldsymbol{S} + \boldsymbol{D}_0^{\varepsilon}]\,:\!\boldsymbol{\varepsilon}^* = -[\boldsymbol{D}_r^{\varepsilon} - \boldsymbol{D}_0^{\varepsilon}]\!:\!\boldsymbol{\varepsilon}_0 \quad \text{in}\, \Omega\,. \qquad [6.11]$$

The last relation allows us to connect the eigenstrain $\boldsymbol{\varepsilon}^*$ with the far field constant strain $\boldsymbol{\varepsilon}_0$ [QU 06],

$$\boldsymbol{\varepsilon}^* = -\left[\boldsymbol{S} + [\boldsymbol{D}_r^{\varepsilon} - \boldsymbol{D}_0^{\varepsilon}]^{-1}\!:\!\boldsymbol{D}_0^{\varepsilon}\right]^{-1}\!:\!\boldsymbol{\varepsilon}_0 \quad \text{in}\, \Omega\,. \qquad [6.12]$$

Thus, from equation $[6.9]_2$ the total strain inside the inhomogeneity is expressed as

$$\boldsymbol{\varepsilon} = \text{constant} = \boldsymbol{T}\!:\!\boldsymbol{\varepsilon}_0 \quad \text{in}\, \Omega\,, \qquad [6.13]$$

where [QU 06]

$$\begin{aligned} \boldsymbol{T} &= \boldsymbol{\mathcal{I}} - \boldsymbol{S}\!:\,[[\boldsymbol{D}_r^{\varepsilon} - \boldsymbol{D}_0^{\varepsilon}]\!:\!\boldsymbol{S} + \boldsymbol{D}_0^{\varepsilon}]^{-1}\,:\![\boldsymbol{D}_r^{\varepsilon} - \boldsymbol{D}_0^{\varepsilon}] \\ &= \boldsymbol{\mathcal{I}} - \boldsymbol{S}\!:\,\left[\boldsymbol{S} + [\boldsymbol{D}_r^{\varepsilon} - \boldsymbol{D}_0^{\varepsilon}]^{-1}\!:\!\boldsymbol{D}_0^{\varepsilon}\right]^{-1} \\ &= \left[\boldsymbol{\mathcal{I}} + \boldsymbol{S}\!:\!\boldsymbol{D}_0^{\varepsilon-1}\!:\![\boldsymbol{D}_r^{\varepsilon} - \boldsymbol{D}_0^{\varepsilon}]\right]^{-1}. \end{aligned} \qquad [6.14]$$

It is recalled that $\boldsymbol{\mathcal{I}}$ is the fourth-order symmetric identity tensor. The Eshelby equivalence principle, which constitutes finding a prescribed eigenstrain in an inclusion to represent the effect of an inhomogeneity, allows us to utilize its important result (i.e. that the total strain inside the inclusion is uniform and is connected with the far field strain through the Eshelby tensor) for heterogeneous materials. $\boldsymbol{T}$, the interaction tensor, is the practical outcome of this principle and plays a central role in the Eshelby-based approaches for obtaining the overall mechanical properties of random composites, as it is shown later in this chapter.

6.1.3. *Inhomogeneous inhomogeneity problem: thermoelastic case*

Here, an approach analogous to the one described by Lester *et al.* [LES 11] for inelastic media is followed. Takao and Taya [TAK 85] used a similar (but not the same) concept to identify the thermal expansion coefficient of composites with random microstructure.

Consider an infinite medium (index 0) under uniform temperature θ. Assuming a reference temperature θ_{ref}, the difference $\Delta\theta = \theta - \theta_{\text{ref}}$ is defined. In the medium, there is an inhomogeneity r that has its own eigenstrain

$$\boldsymbol{a}_r^* = [\boldsymbol{\alpha}_r - \boldsymbol{\alpha}_0]\,\Delta\theta, \qquad [6.15]$$

due to thermal expansion misfit between the two materials. The total strain in the far field is split into a thermal and an elastic part, i.e.

$$\boldsymbol{\varepsilon}_0 = \boldsymbol{\varepsilon}_0^{\text{el}} + \boldsymbol{\alpha}_0\Delta\theta. \qquad [6.16]$$

The total strain $\boldsymbol{\varepsilon}_r$ in the inhomogeneity is considered to have two contributions: (1) the strain of the medium in the far field and (2) a disturbance strain $\tilde{\boldsymbol{\varepsilon}}_r$, i.e.

$$\boldsymbol{\varepsilon}_r = \boldsymbol{\varepsilon}_0 + \tilde{\boldsymbol{\varepsilon}}_r. \qquad [6.17]$$

Using the above decompositions and the constitutive law equations for thermoelastic materials (see section 2.5.3 and section 3.3),

$$\boldsymbol{\sigma} = \boldsymbol{D}^{\varepsilon} : \boldsymbol{\varepsilon} + \boldsymbol{D}^{\theta}\Delta\theta, \quad \boldsymbol{D}^{\theta} = -\boldsymbol{D}^{\varepsilon} : \boldsymbol{\alpha}, \qquad [6.18]$$

the Eshelby equivalence principle states the following: In an elastic medium under a far field strain $\boldsymbol{\varepsilon}_0^{\text{el}}$ that has an inhomogeneity subjected to an eigenstrain $\boldsymbol{a}_r^*$, the stress in the inhomogeneity can be written as

$$\boldsymbol{\sigma}_r = \boldsymbol{D}_r^{\varepsilon} : \left[\boldsymbol{\varepsilon}_0^{\text{el}} + \tilde{\boldsymbol{\varepsilon}}_r - \boldsymbol{a}_r^*\right] = \boldsymbol{D}_0^{\varepsilon} : \left[\boldsymbol{\varepsilon}_0^{\text{el}} + \tilde{\boldsymbol{\varepsilon}}_r - \boldsymbol{a}_r^* - \boldsymbol{\varepsilon}_r^*\right], \qquad [6.19]$$

where ε_r^* is a perturbed strain. The Eshelby solution provides the relation

$$\tilde{\varepsilon}_r = S_r\colon [a_r^* + \varepsilon_r^*] = S_r{:}\varepsilon_r^{**}, \qquad [6.20]$$

with S_r being the Eshelby tensor. Substituting [6.20] in [6.19] yields

$$\varepsilon_r^{**} = -\left[[D_r^\varepsilon - D_0^\varepsilon]:S_r + D_0^\varepsilon\right]^{-1} : \left[[D_r^\varepsilon - D_0^\varepsilon]:\varepsilon_0^{\text{el}} - D_r^\varepsilon{:}a_r^*\right]. \qquad [6.21]$$

Substituting [6.21] and [6.16] in [6.20] gives

$$\tilde{\varepsilon}_r = [T_r^\varepsilon - \mathcal{I}] : \varepsilon_0 + T_r^\theta \Delta\theta, \qquad [6.22]$$

with

$$T_r^\varepsilon = \left[\mathcal{I} + S_r{:}D_0^{\varepsilon-1}{:}[D_r^\varepsilon - D_0^\varepsilon]\right]^{-1},$$
$$T_r^\theta = [\mathcal{I} - T_r^\varepsilon] : [D_0^\varepsilon - D_r^\varepsilon]^{-1} : \left[D_r^\theta - D_0^\theta\right]. \qquad [6.23]$$

The tensors T_r^ε and T_r^θ are called interaction tensors. Finally, substitution of [6.22] in [6.17] provides the strain in the inhomogeneity as a function of the far field total strain and the temperature difference,

$$\varepsilon_r = T_r^\varepsilon{:}\varepsilon_0 + T_r^\theta \Delta\theta. \qquad [6.24]$$

This result is consistent with the one obtained by Benveniste and coworkers [BEN 90, BEN 91].

6.1.4. *Inhomogeneity problem for thermal conductivity*

Consider a small inhomogeneity Ω with thermal conductivity κ_r in an infinite body $\mathcal{B}$ of thermal conductivity κ_0 which has a heat flux q_0 and a temperature gradient $\nabla\theta_0$ in the far field (Figure 6.5). Ignoring heat sources and heat capacity, the following set of equations needs to be solved:

$$\text{div}q = \mathbf{0} \quad \text{in } \mathcal{B},$$

$$q = \begin{cases} -\boldsymbol{\kappa}_0 \cdot \boldsymbol{\nabla\theta} & \text{in } \mathcal{B} - \Omega\,, \\ -\boldsymbol{\kappa}_r \cdot \boldsymbol{\nabla\theta} & \text{in } \Omega\,, \end{cases}$$

$$q = \boldsymbol{q}_0 \quad \text{and} \quad \boldsymbol{\nabla\theta} = \boldsymbol{\nabla\theta}_0 \quad \text{in far field}\,. \qquad [6.25]$$

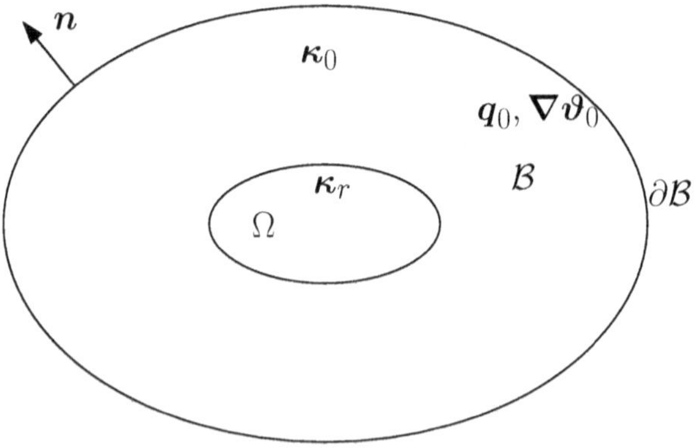

Figure 6.5. *Inhomogeneity problem for thermal conductivity*

It has been shown [HAT 86] that there is a complete analogy between the mechanical equilibrium and the heat conduction inhomogeneity problems. Thus, following the same steps as in section 6.1.2 it is found that the temperature gradient in the inhomogeneity can be expressed in terms of the temperature gradient in the far field as

$$\boldsymbol{\nabla\theta} = \boldsymbol{T}^{\kappa} \cdot \boldsymbol{\nabla\theta}_0 \quad \text{in } \Omega\,, \qquad [6.26]$$

where

$$\boldsymbol{T}^{\kappa} = \left[\boldsymbol{I} + \boldsymbol{S}^{\kappa} \cdot \boldsymbol{\kappa}_0^{-1} \cdot [\boldsymbol{\kappa}_r - \boldsymbol{\kappa}_0]\right]^{-1}\,, \qquad [6.27]$$

and $\boldsymbol{S}^{\kappa}$ is the second-order Eshelby's conduction tensor that depends on the shape of the inhomogeneity and the material properties $\boldsymbol{\kappa}_0$ [HAT 86, LEQ 08, STR 11]. Utilizing the Fourier transform and following the same procedure discussed in Mura's book [MUR 87] for the mechanical problem, the Eshelby's conduction tensor for an ellipsoidal inclusion with semi-axes a_1, a_2, a_3 (Figure 6.2) embedded in the body $\mathcal{B}$ is given by the expression

$$\boldsymbol{S}^{\kappa} = \boldsymbol{P}^{\kappa}\cdot\boldsymbol{\kappa}_0,$$

$$\boldsymbol{P}^{\kappa} = \frac{1}{4\pi}\int_{-1}^{1}\int_{0}^{2\pi} Z^{-1}\left[\widehat{\zeta}\otimes\widehat{\zeta}\right]\mathrm{d}\omega\mathrm{d}\zeta_3, \quad Z = \boldsymbol{\kappa}_0 : \left[\widehat{\zeta}\otimes\widehat{\zeta}\right], \qquad [6.28]$$

where $\widehat{\zeta}$ is a vector with components

$$\widehat{\zeta}_1 = \sqrt{1-\zeta_3^2}\,\frac{\cos\omega}{a_1}, \quad \widehat{\zeta}_2 = \sqrt{1-\zeta_3^2}\,\frac{\sin\omega}{a_2}, \quad \widehat{\zeta}_3 = \frac{\zeta_3}{a_3}. \qquad [6.29]$$

For special inclusion geometries and isotropic materials, this tensor has analytical form [HAT 86]. For anisotropic materials, $\boldsymbol{S}^{\kappa}$ can be computed numerically with a procedure similar to the one described by [GAV 90].

6.2. Eshelby-based approaches: Linear thermoelastic composites

6.2.1. *Random media with distinct phases and concentration tensors*

For a fully disordered heterogeneous material, we can assume that its RVE is under uniform macroscopic temperature $\overline{\theta}$. Based on the discussion in section 5.3 for periodic media, such an assumption is very reasonable. Moreover, in the previous chapter, it has been proven that the microscale problem for periodic media is decoupled and the body forces and heat sources appear only on the macroscale. Thus, it can be reasonably assumed here that this conclusion holds also for media with random microstructure.

In the case of linear thermoelastic heterogeneous materials, the micro-stresses, micro-strains and macro-temperature are connected through the relations [BEN 91] (see section 2.5.3),

$$\boldsymbol{\sigma}(\boldsymbol{x}) = \boldsymbol{D}^{\varepsilon}(\boldsymbol{x}):\boldsymbol{\varepsilon}(\boldsymbol{x}) + \boldsymbol{D}^{\theta}(\boldsymbol{x})[\overline{\theta} - \overline{\theta}_{\text{ref}}],$$

where the tensors $\boldsymbol{D}^{\varepsilon}$ and $\boldsymbol{D}^{\theta}$ depend on the position. However, the micro-heat flux and the micro-temperature gradient are connected through the relations [STR 11]

$$\boldsymbol{q}(\boldsymbol{x}) = -\boldsymbol{\kappa}(\boldsymbol{x})\cdot\boldsymbol{\nabla\theta}(\boldsymbol{x}),$$

where the tensor $\boldsymbol{\kappa}$ also depends on the position.

The passage from the microscale to the macroscale can be obtained by defining:

– A relation between the microscopic strains, the macroscopic strains and the macroscopic temperature of the form [BEN 91]

$$\boldsymbol{\varepsilon}(\boldsymbol{x}) = \boldsymbol{A}^{\varepsilon}(\boldsymbol{x}):\overline{\boldsymbol{\varepsilon}} + \boldsymbol{A}^{\theta}(\boldsymbol{x})[\overline{\theta} - \overline{\theta}_{\text{ref}}],$$

where $\overline{\theta}_{\text{ref}}$ is a reference temperature, $\boldsymbol{A}^{\varepsilon}$ is the fourth-order strain – strain concentration tensor and $\boldsymbol{A}^{\theta}$ is the second order strain – temperature concentration tensor. From this definition, it can easily be shown that

$$\langle \boldsymbol{A}^{\varepsilon} \rangle = \boldsymbol{\mathcal{I}}, \quad \left\langle \boldsymbol{A}^{\theta} \right\rangle = \boldsymbol{0},$$

where $\boldsymbol{\mathcal{I}}$ denotes the fourth-order symmetric identity tensor.

– A relation between the microscopic and macroscopic temperature gradient of the form [STR 11]

$$\boldsymbol{\nabla\theta}(\boldsymbol{x}) = \boldsymbol{A}^{\kappa}(\boldsymbol{x})\cdot\overline{\boldsymbol{\nabla\theta}},$$

where $\boldsymbol{A}^{\kappa}$ is the second-order temperature gradient concentration tensor. Obviously, it holds

$$\langle \boldsymbol{A}^{\kappa} \rangle = \boldsymbol{I}.$$

The above formulation is suitable for a disordered heterogeneous material, knowing that the fields are heterogeneous from one point to another in the medium. The majority of the composites with random microstructure can, however, be seen as media whose RVE consists of several distinct phases. In

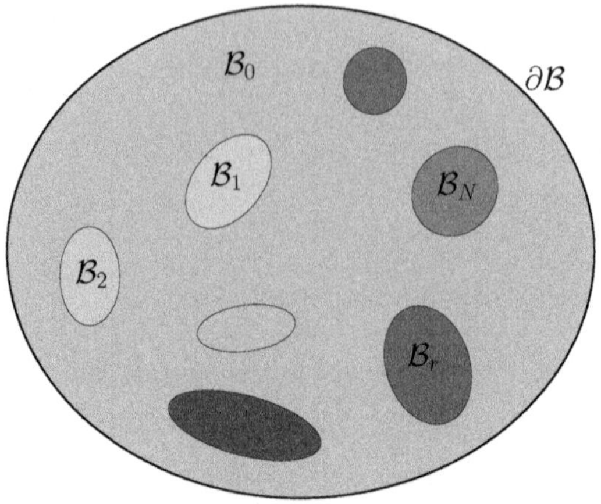

Figure 6.6. *Composite with multiple inhomogeneities (random medium). For a color version of this figure, see www.iste.co.uk/chatzigeorgiou/thermomechanical.zip*

such cases, certain simplification of the global homogenization problem can be obtained.

Consider a composite material, composed of a thermoelastic matrix material (index 0) and N thermoelastic inhomogeneities of ellipsoidal shape. In the RVE of this composite, the matrix is occupying the space $\mathcal{B}_0$ and has volume V_0, while the r_{th} inhomogeneity occupies the space $\mathcal{B}_r$ and has volume V_r (Figure 6.6). Obviously,

$$\mathcal{B} = \mathcal{B}_0 \bigcup \mathcal{B}_1 \bigcup ... \bigcup \mathcal{B}_r \bigcup ... \bigcup \mathcal{B}_N, \quad \mathcal{B}_i \bigcap \mathcal{B}_j = \emptyset \text{ for } i \neq j,$$
$$\text{and} \quad V = V_0 + V_1 + ... + V_r + ... + V_N.$$

The RVE is assumed to be under uniform temperature $\overline{\theta}$.

When the medium is constructed by $N+1$ separate phases, it is possible to define the average quantities per phase,

$$\boldsymbol{\varepsilon}_r = \frac{1}{V_r} \int_{\mathcal{B}_r} \boldsymbol{\varepsilon} \mathrm{d}v, \boldsymbol{\sigma}_r = \frac{1}{V_r} \int_{\mathcal{B}_r} \boldsymbol{\sigma} \mathrm{d}v,$$

$$\nabla\boldsymbol{\theta}_r = \frac{1}{V_r}\int_{\mathcal{B}_r} \nabla\boldsymbol{\theta}\mathrm{d}v, \boldsymbol{q}_r = \frac{1}{V_r}\int_{\mathcal{B}_r} \boldsymbol{q}\mathrm{d}v, \tag{6.30}$$

for $r = 0, 1, 2, ..., N$. In this formalism, the constitutive laws per phase that connect the stress with strain and temperature and the heat flux with the temperature gradient are written as

$$\boldsymbol{\sigma}_r = \boldsymbol{D}_r^{\varepsilon}:\boldsymbol{\varepsilon}_r + \boldsymbol{D}_r^{\theta}[\bar{\theta} - \bar{\theta}_{\text{ref}}], \quad \boldsymbol{q}_r = -\boldsymbol{\kappa}_r\cdot\nabla\boldsymbol{\theta}_r. \tag{6.31}$$

Moreover, we can identify strain and temperature gradient concentration tensors per phase that are independent of the position vector,

$$\boldsymbol{\varepsilon}_r = \boldsymbol{A}_r^{\varepsilon}:\bar{\boldsymbol{\varepsilon}} + \boldsymbol{A}_r^{\theta}[\bar{\theta} - \bar{\theta}_{\text{ref}}], \quad \nabla\boldsymbol{\theta}_r = \boldsymbol{A}_r^{\kappa}\cdot\overline{\nabla\boldsymbol{\theta}}. \tag{6.32}$$

From these definitions and equations [4.1], it can easily be shown that

$$\bar{\boldsymbol{\varepsilon}} = \sum_{r=0}^{N} f_r\boldsymbol{\varepsilon}_r, \quad \bar{\boldsymbol{\sigma}} = \sum_{r=0}^{N} f_r\boldsymbol{\sigma}_r, \quad \overline{\nabla\boldsymbol{\theta}} = \sum_{r=0}^{N} f_r\nabla\boldsymbol{\theta}_r, \quad \bar{\boldsymbol{q}} = \sum_{r=0}^{N} f_r\boldsymbol{q}_r,$$

$$\sum_{r=0}^{N} f_r\boldsymbol{A}_r^{\varepsilon} = \boldsymbol{\mathcal{I}}, \quad \sum_{r=0}^{N} f_r\boldsymbol{A}_r^{\theta} = \boldsymbol{0}, \quad \sum_{r=0}^{N} f_r\boldsymbol{A}_r^{\kappa} = \boldsymbol{I}, \quad f_r = \frac{V_r}{V}. \tag{6.33}$$

When considering the composite in total, we could expect that the macroscopic constitutive laws should have a similar form, i.e.

$$\bar{\boldsymbol{\sigma}} = \overline{\boldsymbol{D}}^{\varepsilon}:\bar{\boldsymbol{\varepsilon}} + \overline{\boldsymbol{D}}^{\theta}[\bar{\theta} - \bar{\theta}_{\text{ref}}], \quad \bar{\boldsymbol{q}} = -\bar{\boldsymbol{\kappa}}\cdot\overline{\nabla\boldsymbol{\theta}}. \tag{6.34}$$

With the help of the concentration tensors, the macroscopic tensors $\overline{\boldsymbol{D}}^{\varepsilon}$, $\overline{\boldsymbol{D}}^{\theta}$ and $\bar{\boldsymbol{\kappa}}$ can be easily shown that are given by the expressions

$$\overline{\boldsymbol{D}}^{\varepsilon} = \sum_{r=0}^{N} f_r\boldsymbol{D}_r^{\varepsilon}:\boldsymbol{A}_r^{\varepsilon}, \quad \overline{\boldsymbol{D}}^{\theta} = \sum_{r=0}^{N} f_r\left[\boldsymbol{D}_r^{\theta} + \boldsymbol{D}_r^{\varepsilon}:\boldsymbol{A}_r^{\theta}\right],$$

$$\bar{\boldsymbol{\kappa}} = \sum_{r=0}^{N} f_r\boldsymbol{\kappa}_r\cdot\boldsymbol{A}_r^{\kappa}. \tag{6.35}$$

The last relations are in agreement with equations $[5.67]_{1,2,3}$ that refer to periodic media. An alternative way to obtain the effective properties of thermomechanical composites is provided by Levin [LEV 67, ROS 70] and leads to equivalent expressions, as has been illustrated by Benveniste and coworkers [BEN 91].

The identification of the concentration tensors is the main task of the various Eshelby-based methods that are going to be discussed in the following sections.

NOTE.– The consistency between displacement and traction boundary conditions has been discussed extensively in the literature [NEM 99b, QU 06]. It is very common to introduce stress concentration tensors in complete analogy with the strain concentration tensors. A formalism based on stress and heat flux concentration tensors follows analogous steps and is not going to be presented here. Both Mori-Tanaka and self-consistent approaches, discussed later in this section, are proven to respect the consistency.

6.2.2. *Mori-Tanaka method*

The presentation of the Mori-Tanaka approach is based on the discussion in the literature [QU 06, BEN 91].

Consider the composite of Figure 6.6. For a typical inhomogeneity with mechanical stiffness tensor $\boldsymbol{D}_r^{\varepsilon}$, thermal stiffness tensor $\boldsymbol{D}_r^{\theta}$ and thermal conductivity $\boldsymbol{\kappa}_r$, the effects of the other inhomogeneities are communicated to it through the strain, stress, temperature gradient and heat flux fields in its surrounding matrix material with mechanical stiffness tensor $\boldsymbol{D}_0^{\varepsilon}$, thermal stiffness tensor $\boldsymbol{D}_0^{\theta}$ and thermal conductivity $\boldsymbol{\kappa}_0$. Although the various fields vary from one location to another inside the matrix, the average strain $\boldsymbol{\varepsilon}_0$, stress $\boldsymbol{\sigma}_0$, temperature gradient $\boldsymbol{\nabla\theta}_0$ and heat flux $\boldsymbol{q}_0$ in the matrix represent good approximations, at a sufficient distance, of the actual matrix fields surrounding each inhomogeneity, when a large number of inhomogeneities exist and are randomly distributed in the matrix. Also, it would be reasonable to assume that the absence of only one inhomogeneity will not affect the overall thermoelastic behavior of the composite. In other words, when the r_{th} inhomogeneity is removed and replaced by the matrix material, the average fields will not change. Therefore, as far as the r_{th} inhomogeneity is concerned, it can be viewed as an ellipsoidal inhomogeneity, placed within the matrix which is subjected to the uniform fields ε_0 and $\nabla\boldsymbol{\theta}_0$. Under this assumption, and using the relations presented in section 6.1 for the thermoelastic case and the thermal conductivity problem, the strain ε_r and the temperature gradient $\nabla\boldsymbol{\theta}_r$ in the r_{th} inhomogeneity are expressed as

$$\varepsilon_r = \boldsymbol{T}_r^{\varepsilon}:\varepsilon_0 + \boldsymbol{T}_r^{\theta}[\bar{\theta} - \bar{\theta}_{\text{ref}}], \quad \nabla\boldsymbol{\theta}_r = \boldsymbol{T}_r^{\kappa}\cdot\nabla\boldsymbol{\theta}_0, \tag{6.36}$$

where

$$\begin{aligned}
\boldsymbol{T}_r^{\varepsilon} &= \left[\boldsymbol{\mathcal{I}} + \boldsymbol{S}_r:\boldsymbol{D}_0^{\varepsilon-1}:[\boldsymbol{D}_r^{\varepsilon} - \boldsymbol{D}_0^{\varepsilon}]\right]^{-1},\\
\boldsymbol{T}_r^{\theta} &= [\boldsymbol{\mathcal{I}} - \boldsymbol{T}_r^{\varepsilon}] : [\boldsymbol{D}_0^{\varepsilon} - \boldsymbol{D}_r^{\varepsilon}]^{-1} : \left[\boldsymbol{D}_r^{\theta} - \boldsymbol{D}_0^{\theta}\right],\\
\boldsymbol{T}_r^{\kappa} &= \left[\boldsymbol{I} + \boldsymbol{S}_r^{\kappa}\cdot\boldsymbol{\kappa}_0^{-1}\cdot[\boldsymbol{\kappa}_r - \boldsymbol{\kappa}_0]\right]^{-1},
\end{aligned} \tag{6.37}$$

and $\boldsymbol{S}_r$, $\boldsymbol{S}_r^{\kappa}$ are the Eshelby tensors, which depend on the matrix properties $\boldsymbol{D}_0^{\varepsilon}$, $\boldsymbol{\kappa}_0$ and the shape of the r_{th} inhomogeneity. Obviously,

$$\boldsymbol{T}_0^{\varepsilon} = \boldsymbol{\mathcal{I}}, \quad \boldsymbol{T}_0^{\theta} = \boldsymbol{0}, \quad \boldsymbol{T}_0^{\kappa} = \boldsymbol{I}. \tag{6.38}$$

Averaging equations [6.36] over the whole RVE and considering equation $[6.33]_{1,3}$ yields

$$\bar{\varepsilon} = \sum_{r=0}^{N} f_r\boldsymbol{T}_r^{\varepsilon}:\varepsilon_0 + \sum_{r=0}^{N} f_r\boldsymbol{T}_r^{\theta}[\bar{\theta} - \bar{\theta}_{\text{ref}}], \quad \overline{\nabla\boldsymbol{\theta}} = \sum_{r=0}^{N} f_r\boldsymbol{T}_r^{\kappa}\cdot\nabla\boldsymbol{\theta}_0,$$

which allows us to express the average strain and temperature gradient in the matrix as

$$\begin{aligned}
\varepsilon_0 &= \left[\sum_{n=0}^{N} f_n\boldsymbol{T}_n^{\varepsilon}\right]^{-1}:\bar{\varepsilon} - \left[\sum_{n=0}^{N} f_n\boldsymbol{T}_n^{\varepsilon}\right]^{-1}:\sum_{n=0}^{N} f_n\boldsymbol{T}_n^{\theta}[\bar{\theta} - \bar{\theta}_{\text{ref}}],\\
\nabla\boldsymbol{\theta}_0 &= \left[\sum_{n=0}^{N} f_n\boldsymbol{T}_n^{\varepsilon}\right]^{-1}\cdot\overline{\nabla\boldsymbol{\theta}}.
\end{aligned} \tag{6.39}$$

Combining equations [6.36], [6.39] and considering [6.32], we eventually obtain

$$\boldsymbol{A}_r^{\varepsilon} = \boldsymbol{T}_r^{\varepsilon} : \left[\sum_{n=0}^{N} f_n \boldsymbol{T}_n^{\varepsilon}\right]^{-1},$$

$$\boldsymbol{A}_r^{\theta} = -\boldsymbol{A}_r^{\varepsilon} : \left[\sum_{n=0}^{N} f_n \boldsymbol{T}_n^{\theta}\right] + \boldsymbol{T}_r^{\theta},$$

$$\boldsymbol{A}_r^{\kappa} = \boldsymbol{T}_r^{\kappa} \cdot \left[\sum_{n=0}^{N} f_n \boldsymbol{T}_n^{\kappa}\right]^{-1}. \qquad [6.40]$$

All the necessary concentration tensors for the matrix phase are calculated with the help of the general relations $[6.33]_{5,6,7}$.

The advantage of Mori-Tanaka is that it provides analytical formulas for the response of the composite. It is a very efficient and accurate method in the case of matrix-particle type composites, where the particles can have arbitrary orientation, with a relatively low-volume fraction (typically less than 30%).

6.2.3. *Self-consistent method*

The presentation of the self-consistent approach is based on the discussion in the literature [QU 06, BEN 91].

Consider the composite of Figure 6.6. In the self-consistent method, it is assumed that the macroscopic tensors $\overline{\boldsymbol{D}}^{\varepsilon}$, $\overline{\boldsymbol{D}}^{\theta}$ and $\overline{\boldsymbol{\kappa}}$ are already known. If there are numerous inhomogeneities in the composite, the macroscopic properties are not affected by the absence of one inhomogeneity. When the composite is subjected to displacement and temperature gradient boundary conditions, we may envision that the effects of the applied loads and the interaction between the inhomogeneities can be accounted for by assuming that the r_{th} inhomogeneity is placed within a homogeneous matrix with properties $\overline{\boldsymbol{D}}^{\varepsilon}$, $\overline{\boldsymbol{D}}^{\theta}$ and $\overline{\boldsymbol{\kappa}}$ that is subjected to the strain tensor $\overline{\boldsymbol{\varepsilon}}$ and the temperature gradient vector $\overline{\boldsymbol{\nabla\theta}}$. In a similar manner to the Mori-Tanaka method, a reasonable assumption is that the absence of only one inhomogeneity does not affect the overall thermoelastic behavior of the composite. Under these hypotheses, and using the relations presented in section 6.1 for the thermoelastic case and the thermal conductivity problem, the strain $\boldsymbol{\varepsilon}_r$ and the temperature gradient $\boldsymbol{\nabla\theta}_r$ in the r_{th} inhomogeneity are expressed as

$$\boldsymbol{\varepsilon}_r = \boldsymbol{A}_r^{\varepsilon}:\overline{\boldsymbol{\varepsilon}} + \boldsymbol{A}_r^{\theta}[\overline{\theta} - \overline{\theta}_{\text{ref}}], \quad \boldsymbol{\nabla\theta}_r = \boldsymbol{A}_r^{\kappa}\cdot\overline{\boldsymbol{\nabla\theta}}, \tag{6.41}$$

where

$$\begin{aligned}
\boldsymbol{A}_r^{\varepsilon} &= \left[\boldsymbol{\mathcal{I}} + \overline{\boldsymbol{S}}_r:\overline{\boldsymbol{D}}^{\varepsilon-1}:[\boldsymbol{D}_r^{\varepsilon} - \overline{\boldsymbol{D}}^{\varepsilon}]\right]^{-1}, \\
\boldsymbol{A}_r^{\theta} &= [\boldsymbol{\mathcal{I}} - \boldsymbol{A}_r^{\varepsilon}] : \left[\overline{\boldsymbol{D}}^{\varepsilon} - \boldsymbol{D}_r^{\varepsilon}\right]^{-1} : \left[\boldsymbol{D}_r^{\theta} - \overline{\boldsymbol{D}}^{\theta}\right], \\
\boldsymbol{A}_r^{\kappa} &= \left[\boldsymbol{I} + \overline{\boldsymbol{S}}_r^{\kappa}\cdot\overline{\boldsymbol{\kappa}}^{-1}\cdot[\boldsymbol{\kappa}_r - \overline{\boldsymbol{\kappa}}]\right]^{-1},
\end{aligned} \tag{6.42}$$

and $\overline{\boldsymbol{S}}_r$, $\overline{\boldsymbol{S}}_r^{\kappa}$ are the Eshelby tensors, which depend on the homogenized medium properties $\overline{\boldsymbol{D}}^{\varepsilon}$, $\overline{\boldsymbol{\kappa}}$ and the shape of the r_{th} inhomogeneity. All the necessary concentration tensors for the matrix phase are calculated with the help of the general relations [6.33]$_{5,6,7}$.

The self-consistent approach yields implicit equations for the macroscopic properties, requiring iterative computational schemes even in the case of thermoelasticity. Its main advantage is that it does not require the presence of a matrix phase (the role of "matrix" is played by the effective medium), thus it is an efficient method for studying polycrystalline material structures.

6.2.4. *Ponte-Castañeda and Willis method*

The Ponte-Castañeda and Willis approach (PCW) is based on the Eshelby problem. The main difference with the Mori-Tanaka method is illustrated in Figure 6.7 [PON 95, HU 00b].

In a composite, each inhomogeneity is considered to be distributed in a specific manner with regard to other inhomogeneities. This distribution forms each individual "shape". The Mori-Tanaka method always considers that an inhomogeneity is distributed inside the matrix with a distribution of the same shape as the inhomogeneity. Thus, in a multi-inhomogeneity composite with different shapes and orientations, each inhomogeneity is embedded in a a larger space of similar shape (Figure 6.7a). Ponte-Castañeda and Willis separated the shapes of the inhomogeneities and their distributions. Their hypothesis is that while the inhomogeneities can have arbitrary ellipsoidal shape and orientation, they all have a distribution of the same shape (Figure 6.7b). In a spherical particle composite or a composite with aligned particles of the same shape, the two methods render exactly the same solution as long as the PCW method chooses the shapes of the inhomogeneity and the distribution to be the same (which is a logical assumption).

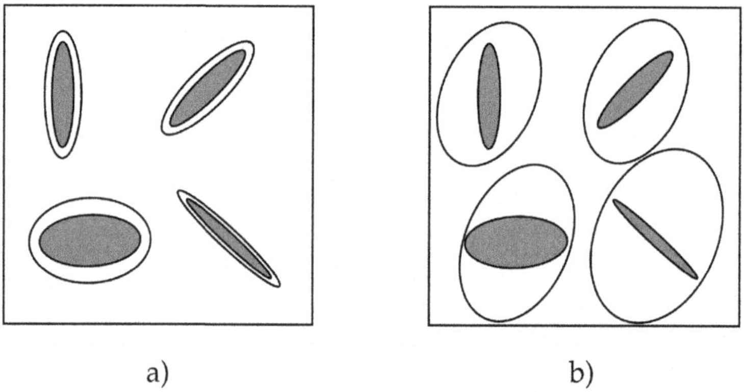

Figure 6.7. *a) Mori-Tanaka and b) Ponte-Castañeda and Willis method*

The PCW method is Hashin–Shtrikman estimate. The RVE of the composite is subjected to linear displacement (i.e. uniform strain) at the far field. The Ponte-Castañeda and Willis approach can be described as follows [HU 00a]:

A thermoelastic material is characterized by a constitutive law of the form

$$\boldsymbol{\sigma} = \boldsymbol{D}^{\varepsilon}:\boldsymbol{\varepsilon} + \boldsymbol{D}^{\theta}[\theta - \theta_{\text{ref}}]. \quad [6.43]$$

Consider a composite with N+1 phases (the matrix has index 0). Assume also that the composite is under uniform (macroscopic) temperature. Using the methodology developed by Willis, it can be shown that the constitutive law can be written alternatively as

$$\boldsymbol{\sigma} = \boldsymbol{D}^{\varepsilon}:\boldsymbol{\varepsilon} + \boldsymbol{D}^{\theta}[\bar{\theta} - \bar{\theta}_{\text{ref}}] = \boldsymbol{D}_0^{\varepsilon}:\boldsymbol{\varepsilon} + \boldsymbol{D}_0^{\theta}[\bar{\theta} - \bar{\theta}_{\text{ref}}] + \boldsymbol{\tau}, \qquad [6.44]$$

where $\boldsymbol{\tau}$ is the (spatially dependent) polarization stress, given by

$$\boldsymbol{\tau} = [\boldsymbol{D}^{\varepsilon} - \boldsymbol{D}_0^{\varepsilon}]:\boldsymbol{\varepsilon} + [\boldsymbol{D}^{\theta} - \boldsymbol{D}_0^{\theta}][\bar{\theta} - \bar{\theta}_{\text{ref}}]. \qquad [6.45]$$

However, the strain at each position is expressed as

$$\boldsymbol{\varepsilon} = \bar{\boldsymbol{\varepsilon}} - \boldsymbol{\Gamma}^0:\boldsymbol{\tau} = \bar{\boldsymbol{\varepsilon}} - \int_{\text{RVE}} \boldsymbol{\Gamma}^0(\boldsymbol{x}, \boldsymbol{x}'):[\boldsymbol{\tau}(\boldsymbol{x}') - \bar{\boldsymbol{\tau}}]\mathrm{d}\boldsymbol{x}', \qquad [6.46]$$

where $\boldsymbol{\Gamma}^0$ is related to the Green's function and $\bar{\boldsymbol{\tau}} = \langle \boldsymbol{\tau} \rangle$. Combining the last two expressions, we obtain

$$\begin{aligned} &[\boldsymbol{D}^{\varepsilon} - \boldsymbol{D}_0^{\varepsilon}]^{-1}:\boldsymbol{\tau} + \int_{\text{RVE}} \boldsymbol{\Gamma}^0(\boldsymbol{x}, \boldsymbol{x}'):[\boldsymbol{\tau}(\boldsymbol{x}') - \bar{\boldsymbol{\tau}}]\mathrm{d}\boldsymbol{x}' \\ &= \bar{\boldsymbol{\varepsilon}} + [\boldsymbol{D}^{\varepsilon} - \boldsymbol{D}_0^{\varepsilon}]^{-1}:[\boldsymbol{D}^{\theta} - \boldsymbol{D}_0^{\theta}][\bar{\theta} - \bar{\theta}_{\text{ref}}]. \end{aligned} \qquad [6.47]$$

Averaging equation [6.44] over the volume of the r_{th} inhomogeneity yields

$$\boldsymbol{\sigma}_r = \boldsymbol{D}_r^{\varepsilon}:\boldsymbol{\varepsilon}_r + \boldsymbol{D}_r^{\theta}[\bar{\theta} - \bar{\theta}_{\text{ref}}] = \boldsymbol{D}_0^{\varepsilon}:\boldsymbol{\varepsilon}_r + \boldsymbol{D}_0^{\theta}[\bar{\theta} - \bar{\theta}_{\text{ref}}] + \boldsymbol{\tau}_r, \qquad [6.48]$$

Obviously, $\boldsymbol{\tau}_0 = \boldsymbol{0}$. When the distribution functions of all inhomogeneities have the same shape, averaging equation [6.47] over the volume of the r_{th} inhomogeneity yields

$$\begin{aligned} &\left[[\boldsymbol{D}_r^{\varepsilon} - \boldsymbol{D}_0^{\varepsilon}]^{-1} + \boldsymbol{P}_r\right]:\boldsymbol{\tau}_r - \boldsymbol{P}_d: \sum_{q=1}^{N} c_q \boldsymbol{\tau}_q \\ &= \bar{\boldsymbol{\varepsilon}} + [\boldsymbol{D}_r^{\varepsilon} - \boldsymbol{D}_0^{\varepsilon}]^{-1}:[\boldsymbol{D}_r^{\theta} - \boldsymbol{D}_0^{\theta}][\bar{\theta} - \bar{\theta}_{\text{ref}}], \end{aligned} \qquad [6.49]$$

where $\boldsymbol{P}_r = \boldsymbol{S}_r:[\boldsymbol{D}_0^{\varepsilon}]^{-1}$ and $\boldsymbol{P}_d = \boldsymbol{S}_d:[\boldsymbol{D}_0^{\varepsilon}]^{-1}$. Note that $\boldsymbol{S}_r$ is the Eshelby tensor of the r_{th} inhomogeneity, depending on the matrix properties and the shape of the inhomogeneity, while $\boldsymbol{S}_d$ is the Eshelby tensor of the distribution, depending on the matrix properties and the shape of the distribution. Averaging equation [6.49] over the volume of all inhomogeneities yields

$$\sum_{r=1}^{N} f_r \boldsymbol{\tau}_r = \overline{\mathcal{T}}^{\varepsilon}:\bar{\boldsymbol{\varepsilon}} + \overline{\mathcal{T}}^{\theta}[\bar{\theta} - \bar{\theta}_{\text{ref}}], \qquad [6.50]$$

where f_r is the volume fraction of the r_{th} inhomogeneity and

$$\overline{\mathcal{T}}^{\varepsilon} = \overline{\mathcal{T}}:\left[\sum_{r=1}^{N} f_r \boldsymbol{T}_r\right],$$

$$\overline{\mathcal{T}}^{\theta} = \overline{\mathcal{T}}:\left[\sum_{r=1}^{N} f_r \boldsymbol{T}_r:[\boldsymbol{D}_r^{\varepsilon} - \boldsymbol{D}_0^{\varepsilon}]^{-1}:[\boldsymbol{D}_r^{\theta} - \boldsymbol{D}_0^{\theta}]\right],$$

$$\overline{\mathcal{T}} = \left[\boldsymbol{\mathcal{I}} - \sum_{r=1}^{N} f_r \boldsymbol{T}_r:\boldsymbol{P}_d\right]^{-1}. \qquad [6.51]$$

Moreover,

$$\boldsymbol{T}_r = \left[[\boldsymbol{D}_r^{\varepsilon} - \boldsymbol{D}_0^{\varepsilon}]^{-1} + \boldsymbol{P}_r\right]^{-1}. \qquad [6.52]$$

6.2.4.1. *Macroscopic properties*

Averaging [6.48] over the whole RVE yields

$$\bar{\boldsymbol{\sigma}} = \boldsymbol{D}_0^{\varepsilon}:\bar{\boldsymbol{\varepsilon}} + \boldsymbol{D}_0^{\theta}[\bar{\theta} - \bar{\theta}_{\text{ref}}] + \sum_{r=1}^{N} f_r \boldsymbol{\tau}_r, \qquad [6.53]$$

which, combined with [6.50], leads to the expression

$$\bar{\boldsymbol{\sigma}} = \overline{\boldsymbol{D}}^{\varepsilon}:\bar{\boldsymbol{\varepsilon}} + \overline{\boldsymbol{D}}^{\theta}[\bar{\theta} - \bar{\theta}_{\text{ref}}], \qquad [6.54]$$

with

$$\overline{\boldsymbol{D}}^{\varepsilon} = \boldsymbol{D}_0^{\varepsilon} + \overline{\mathcal{T}}{:}\left[\sum_{r=1}^{N} f_r \boldsymbol{T}_r\right],$$

$$\overline{\boldsymbol{D}}^{\theta} = \boldsymbol{D}_0^{\theta} + \overline{\mathcal{T}}{:}\left[\sum_{r=1}^{N} f_r \boldsymbol{T}_r{:}[\boldsymbol{D}_r^{\varepsilon} - \boldsymbol{D}_0^{\varepsilon}]^{-1}{:}[\boldsymbol{D}_r^{\theta} - \boldsymbol{D}_0^{\theta}]\right]. \qquad [6.55]$$

6.2.4.2. *Concentration tensors*

With the help of [6.50], equation [6.49] is written as

$$\begin{aligned}\boldsymbol{\tau}_r &= \left[\boldsymbol{T}_r{:}\boldsymbol{P}_d{:}\overline{\mathcal{T}}^{\varepsilon} + \boldsymbol{T}_r\right]{:}\bar{\boldsymbol{\varepsilon}} \\ &\quad + \left[\boldsymbol{T}_r{:}\boldsymbol{P}_d{:}\overline{\mathcal{T}}^{\theta} + \boldsymbol{T}_r{:}[\boldsymbol{D}_r^{\varepsilon} - \boldsymbol{D}_0^{\varepsilon}]^{-1}{:}[\boldsymbol{D}_r^{\theta} - \boldsymbol{D}_0^{\theta}]\right][\bar{\theta} - \bar{\theta}_{\text{ref}}] \\ &= \mathcal{T}_r^{\varepsilon}{:}\bar{\boldsymbol{\varepsilon}} + \mathcal{T}_r^{\theta}[\bar{\theta} - \bar{\theta}_{\text{ref}}]. \end{aligned} \qquad [6.56]$$

Finally, combining the last expression with equation [6.48] yields

$$[\boldsymbol{D}_r^{\varepsilon} - \boldsymbol{D}_0^{\varepsilon}]{:}\boldsymbol{\varepsilon}_r = [\boldsymbol{D}_0^{\theta} - \boldsymbol{D}_r^{\theta}][\bar{\theta} - \bar{\theta}_{\text{ref}}] + \mathcal{T}_r^{\varepsilon}{:}\bar{\boldsymbol{\varepsilon}} + \mathcal{T}_r^{\theta}[\bar{\theta} - \bar{\theta}_{\text{ref}}],$$

or

$$\begin{aligned}\boldsymbol{\varepsilon}_r &= [\boldsymbol{D}_r^{\varepsilon} - \boldsymbol{D}_0^{\varepsilon}]^{-1}{:}\mathcal{T}_r^{\varepsilon}{:}\bar{\boldsymbol{\varepsilon}} + [\boldsymbol{D}_r^{\varepsilon} - \boldsymbol{D}_0^{\varepsilon}]^{-1}{:}[\mathcal{T}_r^{\theta} + \boldsymbol{D}_0^{\theta} - \boldsymbol{D}_r^{\theta}][\bar{\theta} - \bar{\theta}_{\text{ref}}] \\ &= \boldsymbol{A}_r^{\varepsilon}{:}\bar{\boldsymbol{\varepsilon}} + \boldsymbol{A}_r^{\theta}[\bar{\theta} - \bar{\theta}_{\text{ref}}].\end{aligned}$$

NOTE.– The effective thermal conductivity $\bar{\boldsymbol{\kappa}}$ and the concentration tensors $\boldsymbol{A}_r^{\kappa}$ of each inhomogeneity are obtained by following analogous steps and they have similar forms with the $\overline{\boldsymbol{D}}^{\varepsilon}$ and $\boldsymbol{A}_r^{\varepsilon}$, respectively.

6.3. Nonlinear thermomechanical processes

When nonlinear processes are considered, the framework described in the previous section needs to be extended in order to account for: (1) an incremental formulation that allows the computation at each step of appropriate tangent moduli and (2) the conservation of energy, since the heat exchange can increase drastically due to mechanical dissipation.

The proposed methodology follows similar steps to the linearized incremental method introduced for composites with periodic microstructure (section 5.2.1). The main difference is that the RVE problem is solved "analytically" with the help of the concentration tensors.

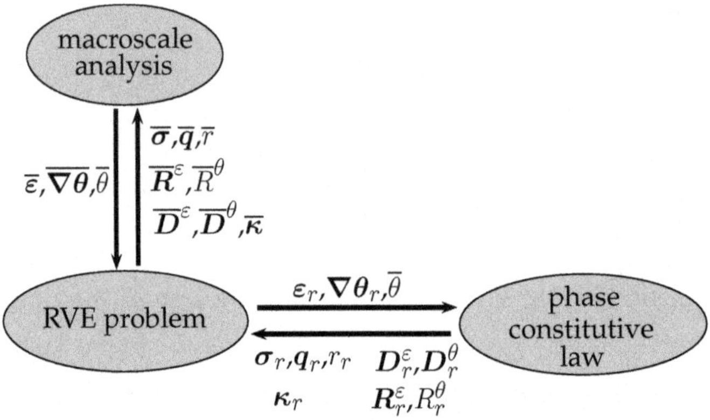

Figure 6.8. *Computational homogenization scheme*

As Figure 6.8 depicts, the macroscale analysis provides the macroscopic strain, temperature and temperature gradient. The microscale (RVE) problem utilizes this information and the appropriate concentration tensors to obtain the macro-stresses and the macro-heat fluxes through an iterative process, as well as the necessary tangent moduli.

6.3.1. *Microscale problem*

Once the macroscopic strain, temperature and temperature gradient are provided by the macroscale analysis, the microscale problem is split in the following parts:

1) For a thermoelastic prediction, the microscale strain and temperature gradient increments at each phase of the composite are calculated by the linearized form of the relations [6.32],

$$\Delta\varepsilon_r = \boldsymbol{A}_r^{\varepsilon}:\Delta\bar{\varepsilon} + \boldsymbol{A}_r^{\theta}\Delta\bar{\theta}, \quad \Delta\boldsymbol{\nabla}\boldsymbol{\theta}_r = \boldsymbol{A}_r^{\kappa}\cdot\Delta\overline{\boldsymbol{\nabla}\boldsymbol{\theta}},$$

where the tensors $\boldsymbol{A}_r^{\varepsilon}$, $\boldsymbol{A}_r^{\theta}$ and $\boldsymbol{A}_r^{\kappa}$ are provided by certain explicit formulas, according to the followed approach (Mori-Tanaka, self-consistent or PCW). The microscopic tangent moduli $\boldsymbol{D}_r^{\varepsilon}$, $\boldsymbol{D}_r^{\theta}$ at the first iteration are considered those of thermoelastic materials (elastic prediction). During the iteration process, their value changes according to the evolution of the nonlinear mechanisms.

2) Using the microscopic strain $\boldsymbol{\varepsilon}_r$, the microscopic temperature gradient $\boldsymbol{\nabla}\boldsymbol{\theta}_r$ and the macro-temperature $\bar{\theta}$, the constitutive law of each material phase is utilized in order to calculate the micro-stresses $\boldsymbol{\sigma}_r$, the micro-heat fluxes $\boldsymbol{q}_r$ and the quantities $r_r = -\bar{\theta}\dot{\eta}_r + \gamma_{\text{loc}_r}$. For nonlinear materials, the stress is usually computed iteratively (see, for instance, the general scheme of section 3.2.3 and the various specific schemes for elastoplastic media in section 3.4). The thermomechanical concentration tensors are also evaluated and if they differ signifficantly from those of the previous iteration increment, then the first two parts are repeated.

3) Once the iteration process is completed, the updated tangent moduli $\boldsymbol{D}_r^{\varepsilon}$, $\boldsymbol{D}_r^{\theta}$, $\boldsymbol{R}_r^{\varepsilon}$, R_r^{θ}, and $\boldsymbol{\kappa}_r$, as well as the updated concentration tensors $\boldsymbol{A}_r^{\varepsilon}$, $\boldsymbol{A}_r^{\theta}$, $\boldsymbol{A}_r^{\kappa}$, for each phase are utilized in order to obtain the macroscopic thermomechanical tangent moduli. The latter are computed through analogous relations with the periodic media, i.e.

$$\overline{\boldsymbol{D}}^{\varepsilon} = \sum_{r=0}^{N} f_r \boldsymbol{D}_r^{\varepsilon}:\boldsymbol{A}_r^{\varepsilon}, \quad \overline{\boldsymbol{D}}^{\theta} = \sum_{r=0}^{N} f_r \left[\boldsymbol{D}_r^{\theta} + \boldsymbol{D}_r^{\varepsilon}:\boldsymbol{A}_r^{\theta}\right],$$

$$\bar{\boldsymbol{\kappa}} = \sum_{r=0}^{N} f_r \boldsymbol{\kappa}_r\cdot\boldsymbol{A}_r^{\kappa},$$

$$\overline{\boldsymbol{R}}^{\varepsilon} = \sum_{r=0}^{N} f_r \boldsymbol{R}_r^{\varepsilon}:\boldsymbol{A}_r^{\varepsilon}, \quad \overline{R}^{\theta} = \sum_{r=0}^{N} f_r \left[R_r^{\theta} + \boldsymbol{R}_r^{\varepsilon}:\boldsymbol{A}_r^{\theta}\right]. \qquad [6.57]$$

Table 6.1 presents the algorithmic scheme for solving the microscale (RVE) problem. Table 6.2 provides the general scheme that describes the constitutive law of a material phase in a representative volume element.

6.3.2. *Macroscale problem*

Ignoring inertia forces, body forces and heat sources, the macroscopic problem can be written in linearized incremental form as (Table 5.1)

$$\begin{aligned} \eth\bar{\varepsilon} &= \overline{\text{grad}}_{\text{sym}}\eth\bar{\boldsymbol{u}}, \quad \eth\overline{\boldsymbol{\nabla\theta}} = \overline{\text{grad}}\eth\bar{\theta}, \\ \text{div}\,(\bar{\boldsymbol{\sigma}} + \eth\bar{\boldsymbol{\sigma}}) &= \mathbf{0}, \quad \text{div}\,(\bar{\boldsymbol{q}} + \eth\bar{\boldsymbol{q}}) = \bar{r} + \eth\bar{r}, \end{aligned} \qquad [6.58]$$

where

$$\bar{\boldsymbol{\sigma}} = \sum_{r=0}^{N} f_r \boldsymbol{\sigma}_r, \quad \bar{\boldsymbol{q}} = \sum_{r=0}^{N} f_r \boldsymbol{q}_r, \quad \bar{r} = \sum_{r=0}^{N} f_r r_r, \qquad [6.59]$$

and

$$\eth\bar{\boldsymbol{\sigma}} = \overline{\boldsymbol{D}}^{\varepsilon}:\eth\bar{\varepsilon} + \overline{\boldsymbol{D}}^{\theta}\eth\bar{\theta}, \quad \eth\bar{\boldsymbol{q}} = -\bar{\boldsymbol{\kappa}}\cdot\eth\overline{\boldsymbol{\nabla\theta}}, \quad \eth\bar{r} = \overline{\boldsymbol{R}}^{\varepsilon}:\eth\bar{\varepsilon} + \overline{R}^{\theta}\eth\bar{\theta}. \qquad [6.60]$$

The numerical procedure for the macroscale problem, which is shown in Table 6.3, follows similar steps to that in periodic media.

6.4. Discussion

Before proceeding to the numerical examples, the following points are worthy of discussion:

– The Eshelby-based homogenization methods are very efficient in providing the macroscopic properties, since they involve analytical (in the case of thermoelasticity) or semi-analytical (in the case of nonlinear mechanisms) solutions to the RVE problem. The main criticism though for these techniques is based on two issues:

1) A crucial assumption in mean field approaches when dealing with nonlinear mechanisms is that the average concentration tensors are approximately equal to the concentration tensors based on the phase averages

1. At time step n everything is known in both scales.

2. At time step $n+1$ and at a specific macro-iteration, $\Delta\bar{\theta}$, $\Delta\bar{\varepsilon}$ and $\Delta\overline{\boldsymbol{\nabla}\boldsymbol{\theta}}$ are provided by the **macroscale analysis**. At the beginning of the micro-iterations set at every phase r all the average microscopic variables, the microscopic tangent moduli and the concentration tensors equal to their corresponding values at time step n.

3. Compute the time increments of the strain and the temperature gradient at each phase r from

$$\Delta\boldsymbol{\varepsilon}_r = \boldsymbol{A}^{\varepsilon}_r : \Delta\bar{\boldsymbol{\varepsilon}} + \boldsymbol{A}^{\theta}_r \Delta\bar{\theta}, \quad \Delta\boldsymbol{\nabla}\boldsymbol{\theta}_r = \boldsymbol{A}^{\kappa}_r \cdot \Delta\overline{\boldsymbol{\nabla}\boldsymbol{\theta}}.$$

4. At every phase r evaluate $\boldsymbol{\sigma}_r$, r_r, $\boldsymbol{q}_r$, as well as the microscopic thermomechanical tangent moduli $\boldsymbol{D}^{\varepsilon}_r$, $\boldsymbol{D}^{\theta}_r$, $\boldsymbol{R}^{\varepsilon}_r$, R^{θ}_r, $\boldsymbol{\kappa}_r$ using the **phase constitutive law**.

5. Evaluate the concentration tensors $\boldsymbol{A}^{\varepsilon}_r$, $\boldsymbol{A}^{\theta}_r$ and $\boldsymbol{A}^{\kappa}_r$ at every phase r according to the chosen micromechanics method.

6. If the convergence criterion is satisfied then continue with step 7, else return to step 3.

7. Compute the macroscopic variables

$$\bar{\boldsymbol{\sigma}} = \sum_{r=0}^{N} f_r \boldsymbol{\sigma}_r, \quad \bar{\boldsymbol{q}} = \sum_{r=0}^{N} f_r \boldsymbol{q}_r, \quad \bar{r} = \sum_{r=0}^{N} f_r r_r.$$

8. Compute the macroscopic moduli

$$\begin{aligned}
\overline{\boldsymbol{D}}^{\varepsilon} &= \sum_{r=0}^{N} f_r \boldsymbol{D}^{\varepsilon}_r : \boldsymbol{A}^{\varepsilon}_r, \quad \overline{\boldsymbol{D}}^{\theta} = \sum_{r=0}^{N} f_r \left[\boldsymbol{D}^{\theta}_r + \boldsymbol{D}^{\varepsilon}_r : \boldsymbol{A}^{\theta}_r\right], \\
\overline{\boldsymbol{R}}^{\varepsilon} &= \sum_{r=0}^{N} f_r \boldsymbol{R}^{\varepsilon}_r : \boldsymbol{A}^{\varepsilon}_r, \quad \overline{R}^{\theta} = \sum_{r=0}^{N} f_r \left[R^{\theta}_r + \boldsymbol{R}^{\varepsilon}_r : \boldsymbol{A}^{\theta}_r\right], \\
\bar{\boldsymbol{\kappa}} &= \sum_{r=0}^{N} f_r \boldsymbol{\kappa}_r \cdot \boldsymbol{A}^{\kappa}_r.
\end{aligned}$$

Table 6.1. *RVE problem*

of the state variables [LAG 91]. When nonlinearities are present, the tangent moduli $\boldsymbol{D}^{\varepsilon}$ and $\boldsymbol{D}^{\theta}$ can vary spatially inside a single material phase (especially the matrix phase), which could lead to significant errors, since the Eshelby-based approaches only calculate average quantities. We need to be cautious about the validity of these methods for inelastic composites [SEG 02, GON 04, LES 11].

1. At time step n everything is known for the material phase.
2. At time step $n+1$ the time increments $\Delta\boldsymbol{\varepsilon}_r$ and $\Delta\boldsymbol{\nabla}\boldsymbol{\theta}_r$ are provided by the **RVE problem**. Moreover the time increment $\Delta\bar{\theta}$ is obtained from the **macroscale analysis**.
3. The stress tensor $\boldsymbol{\sigma}_r$, the internal variables $\boldsymbol{\xi}_r$, the scalar r_r and the heat flux $\boldsymbol{q}_r$ are computed through a **constitutive law algorithm**, described in subsection 3.2.3.
4. The thermomechanical moduli $\boldsymbol{D}^{\varepsilon}_r$, $\boldsymbol{D}^{\theta}_r$, $\boldsymbol{R}^{\varepsilon}_r$, R^{θ}_r and $\boldsymbol{\kappa}_r$ are computed from a **tangent moduli algorithm**, described in subsection 3.2.4.

Table 6.2. *Algorithm for the phase constitutive law*

2) The Mori-Tanaka and self-consistent methods ensure consistency in the constitutive law for both displacement and traction boundary conditions [QU 06], but the satisfaction of the Hill–Mandel theorem cannot be ensured. While it is true that $\bar{\boldsymbol{\varepsilon}} = \sum_{r=0}^{N} f_r \boldsymbol{\varepsilon}_r$ and $\bar{\boldsymbol{\sigma}} = \sum_{r=0}^{N} f_r \boldsymbol{\sigma}_r$, there is no proof in the Eshelby-based techniques that the sum of the products $\sum_{r=0}^{N} f_r \boldsymbol{\sigma}_r : \boldsymbol{\varepsilon}_r$ is equal to the macroscopic product $\bar{\boldsymbol{\sigma}} : \bar{\boldsymbol{\varepsilon}}$. Actually, the non-satisfaction of the Hill–Mandel theorem is illustrated by the lack of major symmetry on the macroscopic elasticity tensor in several composites with multiple inhomogeneities of different shapes [BEN 91]. Indeed, in the case of linear elasticity, the Hill–Mandel theorem is essential to verify the existence of a macro-potential that connects the stresses with the strains [NEM 99b]. Such a macro-potential is sufficient to guarantee the major symmetry of the macroscopic elasticity tensor [SUQ 87]. This issue may cause problems in extreme cases for the macroscopic energy balance.

1. At time step n everything is known. At time step $n+1$ begin the macro-iterations by setting the values of all quantities and tangent moduli in both scales equal to the previous time step values. Also set $\Delta\overline{\boldsymbol{u}} = 0$, $\Delta\overline{\theta} = 0$.

2. Compute the virtual increments of $\overline{\boldsymbol{u}}$ and $\overline{\theta}$ from the macroscopic equilibrium and macroscopic energy balance

$$\begin{aligned} \operatorname{div}\left(\overline{\boldsymbol{\sigma}} + \overline{\boldsymbol{D}}^{\varepsilon} : \overline{\operatorname{grad}}\eth\overline{\boldsymbol{u}} + \overline{\boldsymbol{D}}^{\theta}\eth\overline{\theta}\right) &= \boldsymbol{0}, \\ \overline{\operatorname{div}}\left(\overline{\boldsymbol{q}} - \overline{\boldsymbol{\kappa}}\cdot\overline{\operatorname{grad}}\eth\overline{\theta}\right) &= \overline{r} + \overline{\boldsymbol{R}}^{\varepsilon} : \overline{\operatorname{grad}}\eth\overline{\boldsymbol{u}} + \overline{R}^{\theta}\eth\overline{\theta}, \end{aligned}$$

with appropriate macroscopic boundary conditions.

3. Update the macro-quantities

$$\begin{aligned} \Delta\overline{\boldsymbol{u}} &= \Delta\overline{\boldsymbol{u}} + \eth\overline{\boldsymbol{u}}, \quad \eth\overline{\boldsymbol{\varepsilon}} = \overline{\operatorname{grad}}_{\text{sym}}\eth\overline{\boldsymbol{u}}, \quad \Delta\overline{\boldsymbol{\varepsilon}} = \Delta\overline{\boldsymbol{\varepsilon}} + \eth\overline{\boldsymbol{\varepsilon}}, \\ \Delta\overline{\theta} &= \Delta\overline{\theta} + \eth\overline{\theta}, \quad \eth\overline{\boldsymbol{\nabla\theta}} = \overline{\operatorname{grad}}\eth\overline{\theta}, \quad \Delta\overline{\boldsymbol{\nabla\theta}} = \Delta\overline{\boldsymbol{\nabla\theta}} + \eth\overline{\boldsymbol{\nabla\theta}}. \end{aligned}$$

4. At each macroscopic point compute $\overline{\boldsymbol{\sigma}}$, $\overline{\boldsymbol{q}}$, $\overline{r}$, $\overline{\boldsymbol{D}}^{\varepsilon}$, $\overline{\boldsymbol{D}}^{\theta}$, $\overline{\boldsymbol{R}}^{\varepsilon}$, $\overline{R}^{\theta}$ and $\overline{\boldsymbol{\kappa}}$ through its corresponding representative volume element, using the **RVE problem**.

5. If the convergence criterion is satisfied
   then update the values of all the microscopic variables in the phases
   of the RVEs linked with the macroscopic points,
   set $n = n+1$ and return to step 1 for the next time increment,
   else return to step 2.

Table 6.3. *Algorithm for the macroscale analysis*

– With regard to calculation issues, in the case of anisotropic and non-symmetric (in terms of major symmetries) fourth-order tensors, the Voigt notation significantly simplifies the tensor operations in the homogenization, as long as the known tensors are written from the beginning in the correct form:

1) For all material phases: (i) The fourth-order tangent modulus tensors connecting stress with strain should be written in the net form $\boldsymbol{D}^{\varepsilon}_{r}$. (ii) The second-order tangent modulus tensors connecting stress with temperature should be written in the stress-type form $\boldsymbol{D}^{\theta}_{r}$, or, equivalently, the second-order thermal expansion coefficient tensors should be written in the strain-type form $\widetilde{\boldsymbol{\alpha}}_{r}$. (iii) The second-order tangent modulus tensors connecting r with strain should be written in stress type form $\boldsymbol{R}^{\varepsilon}_{r}$.

2) For all inhomogeneities, the fourth-order Eshelby tensors should be written in the $\widetilde{\boldsymbol{S}}_r$ form.

3) The symmetric fourth-order identity tensor should be written in the $\widetilde{\mathcal{I}} = \overline{\mathcal{I}}$ form.

Using the above representations and the properties of section 1.1.4, we can easily show that, for all the mean field methods discussed in the chapter (Mori-Tanaka, self-consistent, Ponte-Castañeda and Willis), the matrix operations involved lead to fourth-order strain – strain concentration tensors of the $\widetilde{\boldsymbol{A}}_r^{\varepsilon}$ form and second-order strain – temperature concentration tensors $\widetilde{\boldsymbol{A}}_r^{\theta}$ of strain type. These, in turn, implemented directly in $[6.57]_{1,2,4,5}$, provide through matrix multiplications the corresponding tangent moduli in their correct form, i.e. the fourth-order tensor $\overline{\boldsymbol{D}}^{\varepsilon}$ in net form, the second-order tensors $\overline{\boldsymbol{D}}^{\theta}$ and $\overline{\boldsymbol{R}}^{\varepsilon}$ as stress types and $\overline{R}^{\theta}$ as scalar.

6.5. Example: composite with spherical particles

The following example refers to isotropic thermoelastic constituents. The obtained results could be extended to the case of nonlinear composites, as long as the inelastic mechanisms do not alter the constituents' isotropy. The purely mechanical part of the problem has been presented in detail, using various micromechanics techniques, by Qu and Cherkaoui [QU 06].

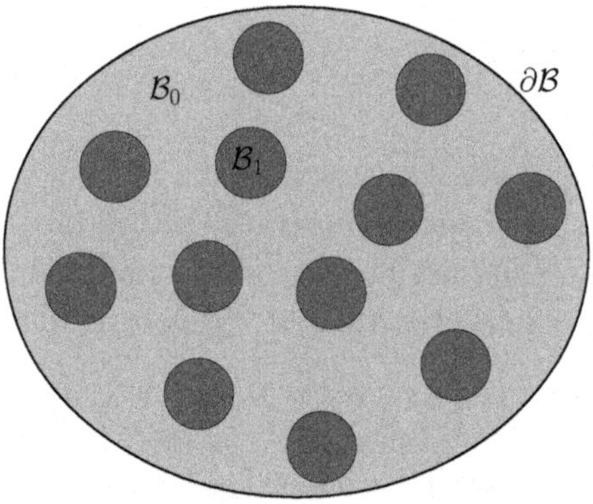

Figure 6.9. *Particulate composite with spherical particles*

Consider the composite structure of Figure 6.9. The matrix (index 0) is a thermoelastic isotropic material with bulk modulus K_0, shear modulus μ_0, thermal expansion coefficient α_0 and thermal conductivity coefficient κ_0. Similarly, the spherical particles (index 1) are also from a thermoelastic isotropic material with bulk modulus K_1, shear modulus μ_1, thermal expansion coefficient α_1 and thermal conductivity coefficient κ_1. Moreover, the scalar f_1 denotes the volume fraction of the particles in the matrix phase.

6.5.1. *Mori-Tanaka*

Assuming that the macroscopic strain, temperature and temperature gradient fields are known, the aim of Mori-Tanaka is to identify the average fields per phase and the various macroscopic tangent moduli. In order to achieve this goal, it is sufficient to calculate the thermomechanical concentration tensors. Indeed, the strain and the temperature gradient per material phase can be identified through equations [6.32], while the stress and the heat flux per phase are computed by the constitutive relations [6.31]. However, the macroscopic tangent moduli are calculated with the help of the expressions [6.35].

In Voigt notation, the fourth-order elasticity tensor and the second-order thermal expansion coefficients tensor for an isotropic elastic material (phase r in the composite) can be written in the forms [LAI 10]

$$\boldsymbol{D}_r^{\varepsilon} = \begin{bmatrix} K_r + \frac{4}{3}\mu_r & K_r - \frac{2}{3}\mu_r & K_r - \frac{2}{3}\mu_r & 0 & 0 & 0 \\ K_r - \frac{2}{3}\mu_r & K_r + \frac{4}{3}\mu_r & K_r - \frac{2}{3}\mu_r & 0 & 0 & 0 \\ K_r - \frac{2}{3}\mu_r & K_r - \frac{2}{3}\mu_r & K_r + \frac{4}{3}\mu_r & 0 & 0 & 0 \\ 0 & 0 & 0 & \mu_r & 0 & 0 \\ 0 & 0 & 0 & 0 & \mu_r & 0 \\ 0 & 0 & 0 & 0 & 0 & \mu_r \end{bmatrix}, \quad [6.61]$$

$$\boldsymbol{D}_r^{\theta} = -\boldsymbol{D}_r^{\varepsilon}{:}\boldsymbol{\alpha}_r = \begin{bmatrix} -3K_r\alpha_r & -3K_r\alpha_r & -3K_r\alpha_r & 0 & 0 & 0 \end{bmatrix}^T. \quad [6.62]$$

As it is already discussed in the theoretical part of this chapter, the Eshelby-based homogenization methods require the Eshelby tensor, which depends on the matrix properties and the particles shape. The fourth-order mechanical Eshelby tensor for isotropic matrix with spherical-type particles is expressed in Voigt notation as [QU 06]

$$\widetilde{\boldsymbol{S}}_1 = \begin{bmatrix} \gamma_0 + \frac{4}{3}\delta_0 & \gamma_0 - \frac{2}{3}\delta_0 & \gamma_0 - \frac{2}{3}\delta_0 & 0 & 0 & 0 \\ \gamma_0 - \frac{2}{3}\delta_0 & \gamma_0 + \frac{4}{3}\delta_0 & \gamma_0 - \frac{2}{3}\delta_0 & 0 & 0 & 0 \\ \gamma_0 - \frac{2}{3}\delta_0 & \gamma_0 - \frac{2}{3}\delta_0 & \gamma_0 + \frac{4}{3}\delta_0 & 0 & 0 & 0 \\ 0 & 0 & 0 & 2\delta_0 & 0 & 0 \\ 0 & 0 & 0 & 0 & 2\delta_0 & 0 \\ 0 & 0 & 0 & 0 & 0 & 2\delta_0 \end{bmatrix}, \quad [6.63]$$

with

$$\gamma_0 = \frac{K_0}{3K_0 + 4\mu_0}, \quad \delta_0 = \frac{3K_0 + 6\mu_0}{15K_0 + 20\mu_0}. \quad [6.64]$$

According to the Mori-Tanaka method, the interaction tensors for the particles can be calculated from equations $[6.37]_{1,2}$. Using the Hill's notation [HIL 65, QU 06], which is suitable for isotropic materials, the calculations involving fourth-order tensors can be drastically reduced. Using the special notation and its properties, as discussed in section 1.1.5, the final expressions for the interaction tensors are obtained in the form

$$\widetilde{\boldsymbol{T}}_1^{\varepsilon} = \begin{bmatrix} \tau_a + \frac{4}{3}\tau_b & \tau_a - \frac{2}{3}\tau_b & \tau_a - \frac{2}{3}\tau_b & 0 & 0 & 0 \\ \tau_a - \frac{2}{3}\tau_b & \tau_a + \frac{4}{3}\tau_b & \tau_a - \frac{2}{3}\tau_b & 0 & 0 & 0 \\ \tau_a - \frac{2}{3}\tau_b & \tau_a - \frac{2}{3}\tau_b & \tau_a + \frac{4}{3}\tau_b & 0 & 0 & 0 \\ 0 & 0 & 0 & 2\tau_b & 0 & 0 \\ 0 & 0 & 0 & 0 & 2\tau_b & 0 \\ 0 & 0 & 0 & 0 & 0 & 2\tau_b \end{bmatrix}, \quad [6.65]$$

$$\widetilde{\boldsymbol{T}}_1^{\theta} = \begin{bmatrix} \tau_c & \tau_c & \tau_c & 0 & 0 & 0 \end{bmatrix}^T, \qquad [6.66]$$

where

$$\begin{aligned} \tau_a &= \frac{K_0}{3K_0 + 9\gamma_0[K_1 - K_0]}, \\ \tau_b &= \frac{\mu_0}{2\mu_0 + 4\delta_0[\mu_1 - \mu_0]}, \\ \tau_c &= 3\gamma_0\frac{K_1\alpha_1 - K_0\alpha_0}{K_0 + 3\gamma_0[K_1 - K_0]}. \end{aligned} \qquad [6.67]$$

With the help of relations $[6.40]_{1,2}$ and $[6.38]_{1,2}$ the mechanical concentration tensors for the particles (i.e. the concentration tensors that connect the average strain in the particles with the macroscopic strain and temperature) are written as

$$\widetilde{\boldsymbol{A}}_1^{\varepsilon} = \begin{bmatrix} a_a + \frac{4}{3}a_b & a_a - \frac{2}{3}a_b & a_a - \frac{2}{3}a_b & 0 & 0 & 0 \\ a_a - \frac{2}{3}a_b & a_a + \frac{4}{3}a_b & a_a - \frac{2}{3}a_b & 0 & 0 & 0 \\ a_a - \frac{2}{3}a_b & a_a - \frac{2}{3}a_b & a_a + \frac{4}{3}a_b & 0 & 0 & 0 \\ 0 & 0 & 0 & 2a_b & 0 & 0 \\ 0 & 0 & 0 & 0 & 2a_b & 0 \\ 0 & 0 & 0 & 0 & 0 & 2a_b \end{bmatrix}, \qquad [6.68]$$

$$\widetilde{\boldsymbol{A}}_1^{\theta} = \begin{bmatrix} a_c & a_c & a_c & 0 & 0 & 0 \end{bmatrix}^T, \qquad [6.69]$$

where

$$a_a = \frac{\tau_a}{1 - f_1 + 3f_1\tau_a}, \quad a_b = \frac{\tau_b}{1 - f_1 + 2f_1\tau_b}, \quad a_c = \tau_c[1 - 3f_1a_a]. \qquad [6.70]$$

With regard to the thermal conductivity, the corresponding tensor for an isotropic material phase r takes the form

$$\boldsymbol{\kappa}_r = \kappa_r \boldsymbol{I}. \qquad [6.71]$$

Additionally, the second-order thermal Eshelby tensor for isotropic matrix with spherical inhomogeneities is given by [HAT 86]

$$\boldsymbol{S}_1^{\kappa} = \frac{1}{3}\boldsymbol{I}. \qquad [6.72]$$

According to the Mori-Tanaka method, the temperature gradient interaction tensor for the particles is calculated from equation [6.37]$_3$,

$$\boldsymbol{T}_1^{\kappa} = \frac{3\kappa_0}{2\kappa_0 + \kappa_1}\boldsymbol{I}, \qquad [6.73]$$

while the corresponding concentration tensor is computed from [6.40]$_3$ (with the help of [6.38]$_3$),

$$\boldsymbol{A}_1^{\kappa} = \frac{3\kappa_0}{3\kappa_0 + [1 - f_1][\kappa_1 - \kappa_0]}\boldsymbol{I}. \qquad [6.74]$$

All the necessary concentration tensors for the matrix phase are calculated with the help of the general relations [6.33]$_{5,6,7}$. It is also worth noticing that the Ponte-Castañeda and Willis approach provides the same results as long as we choose a distribution of spherical shape.

6.5.2. *Self-consistent*

When considering the self-consistent approach, the Eshelby tensor needs to be calculated for a "matrix" phase equivalent to the overall composite, i.e. a phase that corresponds to the macroscopic response. A composite consisting of spherical isotropic particles and an isotropic matrix presents isotropic behavior, leading to write the Eshelby tensor as

$$\widetilde{\overline{\boldsymbol{S}}}_1 = \begin{bmatrix} \bar{\gamma} + \frac{4}{3}\bar{\delta} & \bar{\gamma} - \frac{2}{3}\bar{\delta} & \bar{\gamma} - \frac{2}{3}\delta_0 & 0 & 0 & 0 \\ \bar{\gamma} - \frac{2}{3}\delta_0 & \bar{\gamma} + \frac{4}{3}\bar{\delta} & \bar{\gamma} - \frac{2}{3}\bar{\delta} & 0 & 0 & 0 \\ \bar{\gamma} - \frac{2}{3}\bar{\delta} & \bar{\gamma} - \frac{2}{3}\bar{\delta} & \bar{\gamma} + \frac{4}{3}\bar{\delta} & 0 & 0 & 0 \\ 0 & 0 & 0 & 2\bar{\delta} & 0 & 0 \\ 0 & 0 & 0 & 0 & 2\bar{\delta} & 0 \\ 0 & 0 & 0 & 0 & 0 & 2\bar{\delta} \end{bmatrix}, \qquad [6.75]$$

with

$$\bar{\gamma} = \frac{\overline{K}}{3\overline{K} + 4\bar{\mu}}, \quad \bar{\delta} = \frac{3\overline{K} + 6\bar{\mu}}{15\overline{K} + 20\bar{\mu}}. \qquad [6.76]$$

With the help of the Hill's notation, the mechanical concentration tensors for the particles are obtained as

$$\widetilde{\overline{\overline{\boldsymbol{A}}}}{}_1^{\varepsilon} = \widetilde{\overline{\overline{\boldsymbol{T}}}}{}_1^{\varepsilon}, \qquad [6.77]$$

with

$$\widetilde{\overline{\overline{\boldsymbol{T}}}}{}_1^{\varepsilon} = \begin{bmatrix} \bar{\tau}_a + \frac{4}{3}\bar{\tau}_b & \bar{\tau}_a - \frac{2}{3}\bar{\tau}_b & \bar{\tau}_a - \frac{2}{3}\bar{\tau}_b & 0 & 0 & 0 \\ \bar{\tau}_a - \frac{2}{3}\bar{\tau}_b & \bar{\tau}_a + \frac{4}{3}\bar{\tau}_b & \bar{\tau}_a - \frac{2}{3}\bar{\tau}_b & 0 & 0 & 0 \\ \bar{\tau}_a - \frac{2}{3}\bar{\tau}_b & \bar{\tau}_a - \frac{2}{3}\bar{\tau}_b & \bar{\tau}_a + \frac{4}{3}\bar{\tau}_b & 0 & 0 & 0 \\ 0 & 0 & 0 & 2\bar{\tau}_b & 0 & 0 \\ 0 & 0 & 0 & 0 & 2\bar{\tau}_b & 0 \\ 0 & 0 & 0 & 0 & 0 & 2\bar{\tau}_b \end{bmatrix}, \qquad [6.78]$$

and

$$\widetilde{\overline{\boldsymbol{A}}}{}_1^{\theta} = \widetilde{\overline{\boldsymbol{T}}}{}_1^{\theta} = \begin{bmatrix} \bar{\tau}_c & \bar{\tau}_c & \bar{\tau}_c & 0 & 0 & 0 \end{bmatrix}^T. \qquad [6.79]$$

In the above expressions,

$$\overline{\tau}_a = \frac{\overline{K}}{3\overline{K} + 9\overline{\gamma}[K_1 - \overline{K}]},$$

$$\overline{\tau}_b = \frac{\overline{\mu}}{2\overline{\mu}_0 + 4\overline{\delta}[\mu_1 - \overline{\mu}]},$$

$$\overline{\tau}_c = 3\overline{\gamma}\frac{K_1\alpha_1 - \overline{K}\overline{\alpha}}{\overline{K} + 3\overline{\gamma}[K_1 - \overline{K}]}. \quad [6.80]$$

With regard to the thermal conductivity, the second-order thermal Eshelby tensor for macroscopically isotropic composite with spherical inhomogeneities is given by [HAT 86]

$$\overline{\boldsymbol{S}}_1^{\kappa} = \frac{1}{3}\boldsymbol{I}. \quad [6.81]$$

According to the self-consistent method, the temperature gradient concentration tensor for the particles is calculated from equation $[6.42]_3$,

$$\overline{\boldsymbol{A}}_1^{\kappa} = \overline{\boldsymbol{T}}_1^{\kappa} = \frac{3\overline{\kappa}}{2\overline{\kappa} + \kappa_1}\boldsymbol{I}. \quad [6.82]$$

All the necessary concentration tensors for the matrix phase are calculated with the help of the general relations $[6.33]_{5,6,7}$.

6.6. Numerical applications

In the numerical applications that follow, the composite response is computed using two different methods, the Mori-Tanaka and the self-consistent. For composites with one type of particle or fiber, the Ponte-Castañeda and Willis method provides the same results with the Mori-Tanaka method if the shapes of the distribution and the inhomogeneities are chosen to be the same.

6.6.1. *Isotropic matrix with isotropic spherical particles: thermoelastic case*

Considering the composite of Figure 6.9, it is assumed that the matrix is made of an epoxy while the particles are made of silica. The thermoelastic properties of both materials are summarized in Table 6.4[1].

Property	Epoxy	Silica
Young's modulus [GPa]	2.25	73
Poisson's ratio	0.19	0.19
thermal expansion coefficient [ppm/C]	88	0.5
thermal conductivity [W/[m K]]	0.195	1.5

Table 6.4. *Thermomechanical properties of particulate composite constituents [WON 99]*

Using the Mori-Tanaka or the self-consistent method, we can obtain the overall strain-related tangent modulus tensor $\overline{\boldsymbol{D}}^{\varepsilon}$ (and consequently the Young's modulus $\overline{E}$), the overall temperature-related tangent modulus tensor $\overline{\boldsymbol{D}}^{\theta}$ (and consequently the thermal expansion coefficient $\overline{\alpha}$) and the overall thermal conductivity $\overline{\kappa}$ of the composite for various particles' volume fractions. The obtained results are illustrated in the three graphs of Figure 6.10. In the same graphs, the experimental values obtained by Wong and Bollambally [WON 99] are also presented. As it is observed, the Mori-Tanaka shows good agreement with the experimental measurements of the thermal properties (with the exception of thermal conductivity at 50% volume fraction), while for the Young's modulus there is an underestimation for volume fractions of 30% and higher. However, the self-consistent method overestimates the Young's modulus and thermal conductivity and underestimates the thermal expansion coefficient, but it appears to provide a better estimation for Young's modulus, compared to Mori-Tanaka, at high values of particles' volume fraction.

1 1 ppm/C = 10^{-6} 1/K.

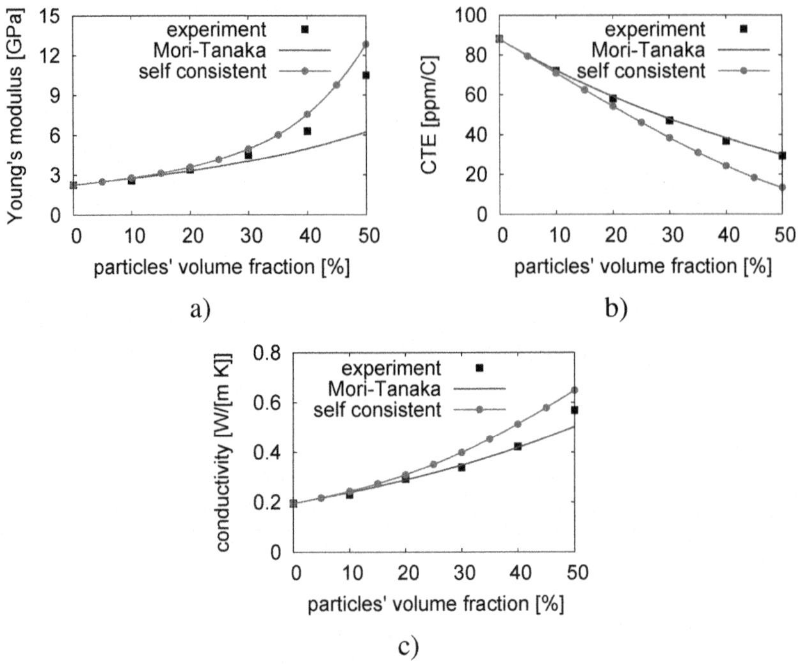

Figure 6.10. *Variation of a) Young's modulus, b) thermal expansion coefficient and c) thermal conductivity of the composite with respect to particles' volume fraction. The points represent experimentally measured values [WON 99]. For a color version of this figure, see www.iste.co.uk/chatzigeorgiou/thermomechanical.zip*

6.6.2. *Isotropic matrix with isotropic fibers: elastic case*

Unidirectional cylindrical fiber composites with isotropic constituents (matrix and fibers) behave as transversely isotropic materials. In the mean field theories discussed in this chapter, the loss of isotropy in the effective properties arises in the form of the Eshelby tensor, which is anisotropic and does not retain major symmetries [MUR 87].

Consider the unidirectional cylindrical fiber composite of Figure 6.11. The fibers are quite long with aspect ratio (length/diameter) equal to 100. The matrix is an isotropic material with Young's modulus $E_0 = 3.45$ GPa and Poisson's ratio $\nu_0 = 0.3$, while the fibers are also made of isotropic material with Young's modulus $E_1 = 73$ GPa and Poisson's ratio $\nu_1 = 0.22$. Figure 6.12 shows the composite's axial and transverse Young's moduli as a function of the fibers volume fraction, as predicted by the Mori-Tanaka and the self-consistent methods. As observed, the two methods provide almost the same axial Young's modulus, but self-consistent predicts generally higher transverse Young's modulus compared to Mori-Tanaka.

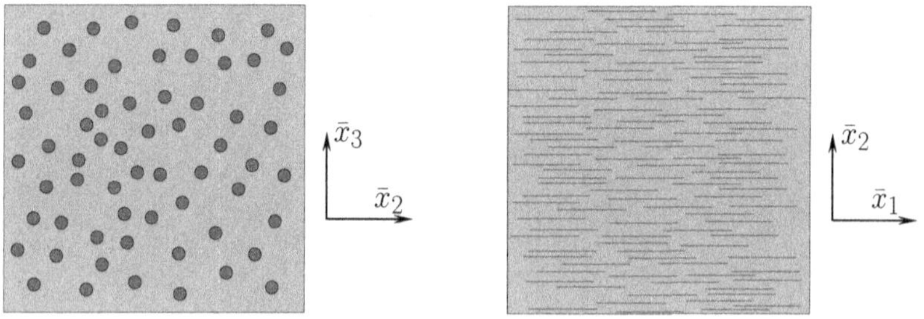

Figure 6.11. *Schematic illustration of two cross-sections in a unidirectional fiber composite with cylindrical fibers. For a color version of this figure, see www.iste.co.uk/chatzigeorgiou/thermomechanical.zip*

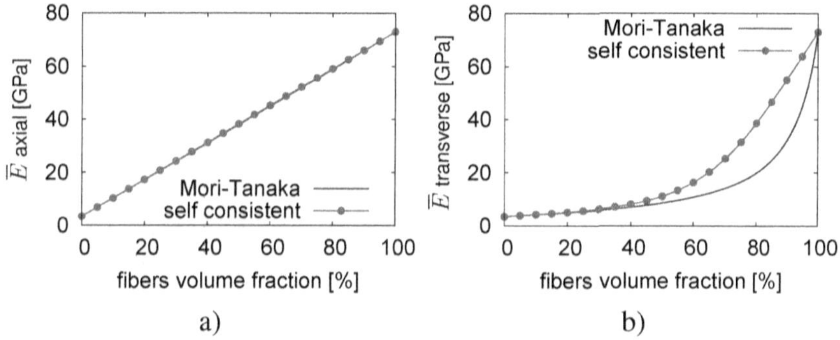

Figure 6.12. *Prediction of the Mori-Tanaka and the self-consistent methods of a) the axial Young's modulus and b) the transverse Young's modulus of the fiber composite as a function of the fibers' volume fraction. For a color version of this figure, see www.iste.co.uk/chatzigeorgiou/thermomechanical.zip*

It is also interesting to investigate the response of the fiber composite when the fibers axis is not aligned with the global coordinate axis (Figure 6.13). As expected, the Young's modulus in the global direction $\bar{x}_1$ varies with respect to the angle between $\bar{x}_1$ and the fibers axis, getting its higher value, equal to the axial $\overline{E}$, at zero degrees (fibers aligned parallel to $\bar{x}_1$) and its lower value, equal to the transverse $\overline{E}$, at 90 degrees (fibers aligned perpendicular to $\bar{x}_1$).

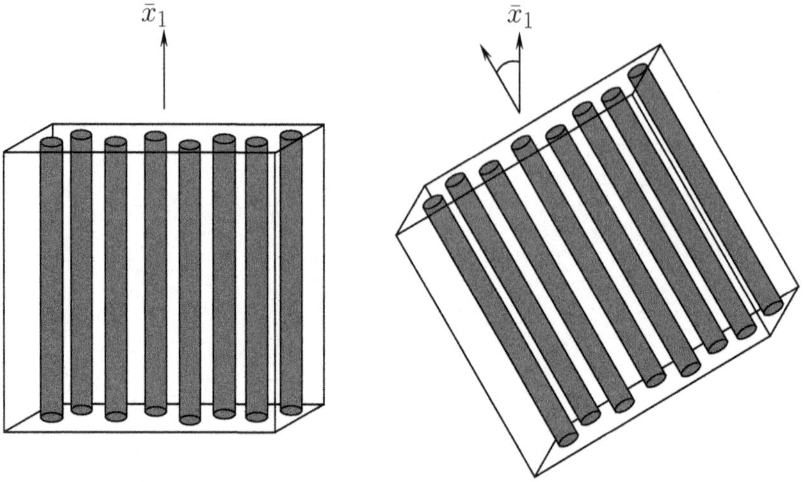

Figure 6.13. *Rotation of the local fiber axis with respect to the global coordinate $\bar{x}_1$. For a color version of this figure, see www.iste.co.uk/chatzigeorgiou/thermomechanical.zip*

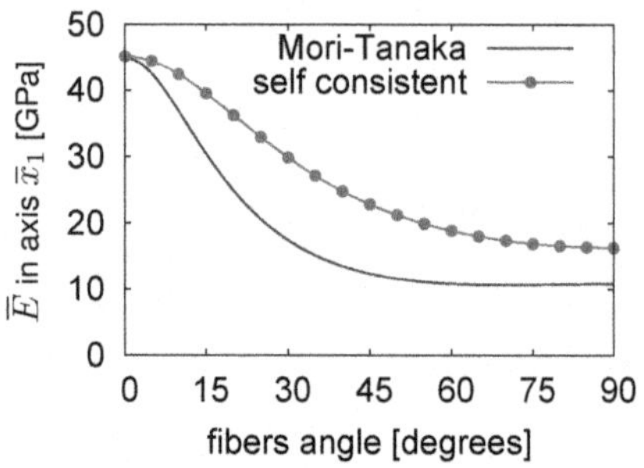

Figure 6.14. *Variation of the Young's modulus in the $\bar{x}_1$ direction with respect to the angle between $\bar{x}_1$ and the fibers axis. Predictions of Mori-Tanaka and self-consistent methods for fiber volume fraction equal to 60%. For a color version of this figure, see www.iste.co.uk/chatzigeorgiou/thermomechanical.zip*

6.6.3. *Isotropic matrix with isotropic elastoplastic spherical particles*

Considering the composite of Figure 6.15, it is assumed that the matrix is made of an elastic epoxy, while the spherical particles are made of an elastoplastic material with isotropic hardening. For the particles constitutive law, the von Mises stress σ^{vM} is expressed in terms of the accumulated plastic strain p according to the relation

$$\sigma^{\text{vM}} = Y + kp^m,$$

where Y is the elastic limit and k, m are plastic hardening parameters. The mechanical properties of both the matrix and particles are summarized in Table 6.5. The volume fraction of the particles is equal to 20%.

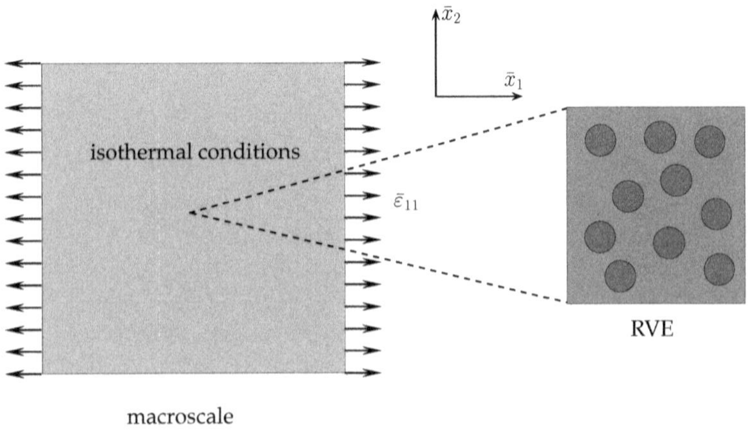

Figure 6.15. *Particulate composite under uniaxial stretching and isothermal conditions. For a color version of this figure, see www.iste.co.uk/chatzigeorgiou/thermomechanical.zip*

Property	Matrix	Particles
Young's modulus [GPa]	2.25	70
Poisson's ratio	0.19	0.3
elastic limit [MPa]		120
k [MPa]		400
m		0.3

Table 6.5. *Mechanical properties of particulate composite constituents*

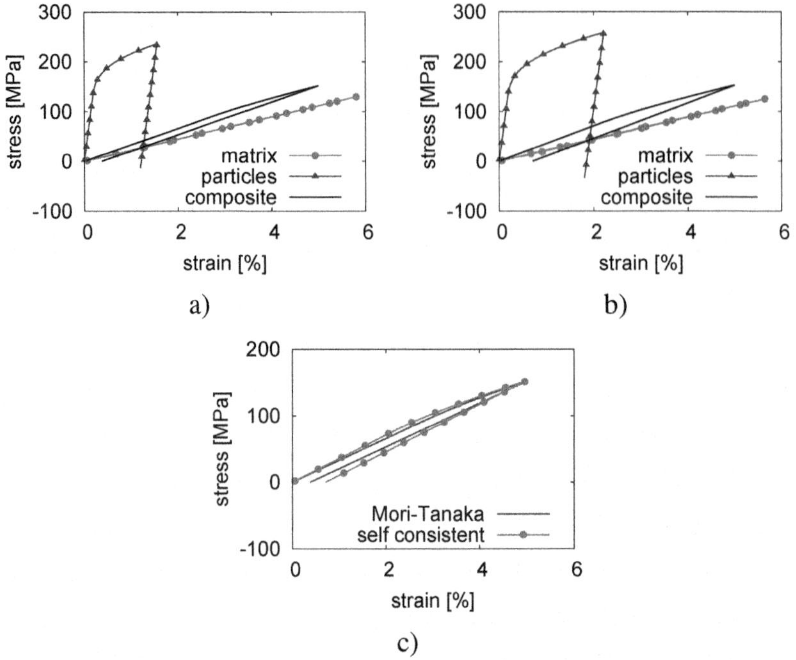

Figure 6.16. *Uniaxial stress–strain response of matrix, particles and composite according to a) Mori-Tanaka and b) self-consistent method. c) Comparison of macroscopic responses predicted by the two methods. For a color version of this figure, see www.iste.co.uk/chatzigeorgiou/thermomechanical.zip*

During the mechanical loading, the composite is under isothermal conditions. The material is subjected to uniaxial stretching up to 5% and then it is unloaded. The other two directions are assumed macroscopically stress free (Figure 6.15). The uniaxial stress–strain responses of the matrix, the particles and the composite according to the Mori-Tanaka and the self-consistent methods are illustrated in Figure 6.16. As shown in Figure 6.16c, the self-consistent method predicts the development of higher plastic strain at the end of the loading compared to the Mori-Tanaka approach.

Appendices

Appendix 1

Average Theorems in Large Deformations

In this appendix, the average theorems presented in Chapter 4 are extended to the case of large deformation processes. In the developed framework, all the operators are valid only in Cartesian coordinates and no curvilinear coordinates are considered.

A1.1. Preliminaries

A continuum body of volume V_0 at time t_0 is considered to be in the undeformed / reference configuration and occupies the space $\mathcal{D}_0$, bounded by the surface $\partial\mathcal{D}_0$ with unit vector $\boldsymbol{N}$ (see Figure A1.1). At time t, the body has been moved and deformed. In the deformed / current configuration, it occupies the space $\mathcal{D}$ (volume V_t), bounded by the surface $\partial\mathcal{D}$ with unit vector $\boldsymbol{n}$. Any point P on this body can be described with the help of a fixed point O in two ways: (1) with the position vector $\boldsymbol{X}$ of the reference configuration and (2) the position vector $\boldsymbol{x}$ of the current configuration (Figure A1.1). Both $\boldsymbol{X}$ and $\boldsymbol{x}$ refer to Cartesian coordinate systems. In the following, simplified notation is utilized:

$$\langle\{\bullet\}\rangle_0 = \frac{1}{V_0}\int_{\mathcal{D}_0}\{\bullet\}\,\mathrm{d}V\,, \quad \lceil\{\bullet\}\rfloor_0 = \frac{1}{V_0}\int_{\partial\mathcal{D}_0}\{\bullet\}\,\mathrm{d}S\,,$$

$$\langle\{\bullet\}\rangle_t = \frac{1}{V_t}\int_{\mathcal{D}}\{\bullet\}\,\mathrm{d}v\,, \quad \lceil\{\bullet\}\rfloor_t = \frac{1}{V_t}\int_{\partial\mathcal{D}}\{\bullet\}\,\mathrm{d}s\,.$$

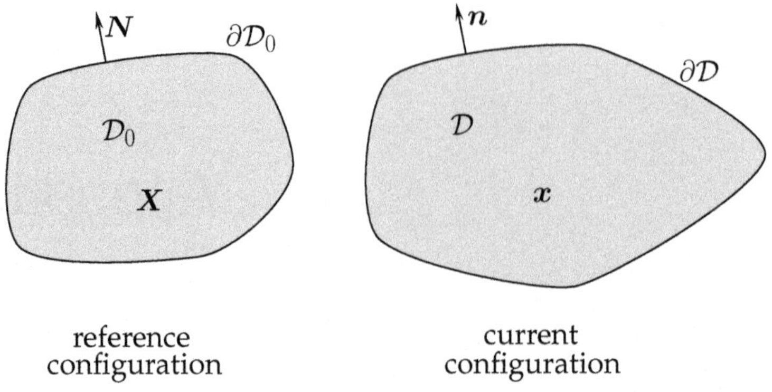

Figure A1.1. *Continuum body in reference and current configuration*

In this formalism, the divergence theorem [2.11] and [2.10] for a tensor $\boldsymbol{A}$ of arbitrary order can be written in compact form as

$$\langle \text{div}\boldsymbol{A} \rangle_0 = \lceil \boldsymbol{A} \cdot \boldsymbol{N} \rfloor_0 \quad \text{and} \quad \langle \text{div}\boldsymbol{A} \rangle_t = \lceil \boldsymbol{A} \cdot \boldsymbol{n} \rfloor_t .$$

For a second-order tensor $\boldsymbol{A}$, the following properties hold:

$$\langle \boldsymbol{A}^T \rangle_0 = \langle \boldsymbol{A} \rangle_0^T , \quad \lceil \boldsymbol{A}^T \rfloor_0 = \lceil \boldsymbol{A} \rfloor_0^T ,$$
$$\langle \boldsymbol{A}^T \rangle_t = \langle \boldsymbol{A} \rangle_t^T , \quad \lceil \boldsymbol{A}^T \rfloor_t = \lceil \boldsymbol{A} \rfloor_t^T .$$

Moreover, the periodic tensor properties described in Section 4.1 hold also in the large deformation framework (either in the reference space $\mathcal{D}_0$ or in the current space $\mathcal{D}$).

A1.2. Hill's Lemma and the Hill–Mandel theorem

In this section, there is frequent use of the tensor operations and the definitions presented in Chapter 1.

A1.2.1. *Mechanical problem*

The following notion for the reference configuration is adopted:

DEFINITION A1.1.– *Kinematically admissible deformation gradient* $\boldsymbol{F}$ *is considered to refer to every symmetric second-order tensor that is related with the displacement vector* $\boldsymbol{u}$ *through the relation*

$$\boldsymbol{F} = \boldsymbol{I} + \mathrm{Grad}\boldsymbol{u} = \mathrm{Grad}(\boldsymbol{u} + \boldsymbol{X})\,.$$

DEFINITION A1.2.– *Statically admissible Piola stress* $\boldsymbol{\Sigma}$ *refers to every symmetric second-order tensor that (ignoring body forces) satisfies the equilibrium equation*

$$\mathrm{Div}\boldsymbol{\Sigma} = \boldsymbol{0}\,.$$

Using these definitions and the divergence theorem [2.11], the volume average for a kinematically admissible deformation gradient is given by (see the properties in section 1.1):

$$\begin{aligned}\langle \boldsymbol{F}\rangle_0 &= \langle \mathrm{Grad}(\boldsymbol{u} + \boldsymbol{X})\rangle_0 = \langle \mathrm{Div}\left([\boldsymbol{u} + \boldsymbol{X}] \otimes \boldsymbol{I}\right)\rangle_0 \\ &= \lceil [\boldsymbol{u} + \boldsymbol{X}] \otimes \boldsymbol{N} \rfloor_0\,,\end{aligned} \qquad \text{[A1.1]}$$

and for a statically admissible Piola stress:

$$\begin{aligned}\langle \boldsymbol{\Sigma}\rangle_0 &= \langle \mathrm{Grad}\boldsymbol{X} \,\tilde{\cdot}\, \boldsymbol{\Sigma}\rangle_0^T = \langle \mathrm{Div}\left(\boldsymbol{X} \otimes \boldsymbol{\Sigma}\right)\rangle_0^T - \langle \boldsymbol{X} \otimes \mathrm{Div}\boldsymbol{\Sigma}\rangle_0^T \\ &= \lceil [\boldsymbol{\Sigma} \cdot \boldsymbol{N}] \otimes \boldsymbol{X} \rfloor_0\,.\end{aligned} \qquad \text{[A1.2]}$$

The average mechanical work on the body, produced by a kinematically admissible strain $\boldsymbol{F}$ and a statically admissible stress $\boldsymbol{\Sigma}$, is half of the quantity:

$$\begin{aligned}\langle \boldsymbol{\Sigma} : \boldsymbol{F}\rangle_0 &= \langle \boldsymbol{\Sigma} : \mathrm{Grad}(\boldsymbol{u} + \boldsymbol{X})\rangle_0 \\ &= \langle \mathrm{Div}\left([\boldsymbol{u} + \boldsymbol{X}] \cdot \boldsymbol{\Sigma}\right)\rangle_0 - \langle (\boldsymbol{u} + \boldsymbol{X}) \cdot \mathrm{Div}\boldsymbol{\Sigma}\rangle_0 \\ &= \lceil [\boldsymbol{u} + \boldsymbol{X}] \cdot [\boldsymbol{\Sigma} \cdot \boldsymbol{N}] \rfloor_0\,.\end{aligned} \qquad \text{[A1.3]}$$

Based on the above definitions and properties, the Hill's lemma is expressed in the following way.

LEMMA A1.1.– Let $\boldsymbol{F}$ be a kinematically admissible deformation gradient and $\boldsymbol{\Sigma}$ be a statically admissible Piola stress. Then, it holds that

$$\begin{aligned}&\langle\boldsymbol{\Sigma}:\boldsymbol{F}\rangle_0-\langle\boldsymbol{\Sigma}\rangle_0:\langle\boldsymbol{F}\rangle_0\\&=\lceil[\boldsymbol{u}+\boldsymbol{X}-\langle\boldsymbol{F}\rangle_0\cdot\boldsymbol{X}]\cdot[\boldsymbol{\Sigma}\cdot\boldsymbol{N}-\langle\boldsymbol{\Sigma}\rangle_0\cdot\boldsymbol{N}]\rfloor_0\,.\end{aligned} \qquad [A1.4]$$

Indeed, using the divergence theorem [2.11] and equations [A1.1], [A1.2] and [A1.3] we obtain

$$\begin{aligned}&\lceil[\boldsymbol{u}+\boldsymbol{X}-\langle\boldsymbol{F}\rangle_0\cdot\boldsymbol{X}]\cdot[\boldsymbol{\Sigma}\cdot\boldsymbol{N}-\langle\boldsymbol{\Sigma}\rangle_0\cdot\boldsymbol{N}]\rfloor_0\\&=\lceil[\boldsymbol{u}+\boldsymbol{X}]\cdot[\boldsymbol{\Sigma}\cdot\boldsymbol{N}]\rfloor_0-\langle\boldsymbol{F}\rangle_0:\lceil[\boldsymbol{\Sigma}\cdot\boldsymbol{N}]\otimes\boldsymbol{X}\rfloor_0\\&-\langle\boldsymbol{\Sigma}\rangle_0:\lceil[\boldsymbol{u}+\boldsymbol{X}]\otimes\boldsymbol{N}\rfloor_0+[\langle\boldsymbol{F}\rangle_0\mathbin{\underset{\sim}{:}}\langle\boldsymbol{\Sigma}\rangle_0]:\lceil\boldsymbol{X}\otimes\boldsymbol{N}\rfloor_0\\&=\langle\boldsymbol{\Sigma}:\boldsymbol{F}\rangle_0-\langle\boldsymbol{F}\rangle_0:\langle\boldsymbol{\Sigma}\rangle_0-\langle\boldsymbol{\Sigma}\rangle_0:\langle\boldsymbol{F}\rangle_0+\langle\boldsymbol{F}\rangle_0:\langle\boldsymbol{\Sigma}\rangle_0\\&=\langle\boldsymbol{\Sigma}:\boldsymbol{F}\rangle_0-\langle\boldsymbol{\Sigma}\rangle_0:\langle\boldsymbol{F}\rangle_0\,.\end{aligned}$$

Additionally, the Hill–Mandel theorem can be stated in the following way.

THEOREM A1.1.– Let $\boldsymbol{F}$ be a kinematically admissible deformation gradient and $\boldsymbol{\Sigma}$ be a statically admissible Piola stress. Then, the three types of conditions:

1) $\boldsymbol{u}=[\langle\boldsymbol{F}\rangle_0-\boldsymbol{I}]\cdot\boldsymbol{X}$ on the boundary $\partial\mathcal{D}_0$;

2) $\boldsymbol{\Sigma}\cdot\boldsymbol{N}=\langle\boldsymbol{\Sigma}\rangle_0\cdot\boldsymbol{N}$ on the boundary $\partial\mathcal{D}_0$;

3) $\boldsymbol{F}=\langle\boldsymbol{F}\rangle_0+\mathrm{Grad}\boldsymbol{z}$, with $\boldsymbol{z}$ periodic and $\boldsymbol{\Sigma}\cdot\boldsymbol{N}$ anti-periodic;

satisfy the energy equivalence

$$\langle\boldsymbol{\Sigma}:\boldsymbol{F}\rangle_0=\langle\boldsymbol{\Sigma}\rangle_0:\langle\boldsymbol{F}\rangle_0\,. \qquad [A1.5]$$

For the first two conditions, the proof is a direct consequence of the Hill's lemma [A1.4]. For the third condition, using the properties of tensor derivatives discussed in section 1.1, the divergence theorem [2.11], the definition A1.2 and the properties of the products between periodic and anti-periodic vectors described in Section 4.1, we obtain

$$\begin{aligned}\langle \boldsymbol{\Sigma}:\boldsymbol{F}\rangle_0 &= \langle \boldsymbol{\Sigma}\rangle_0 : \langle \boldsymbol{F}\rangle_0 + \langle \boldsymbol{\Sigma}:\mathrm{Grad}\boldsymbol{z}\rangle_0 \\ &= \langle \boldsymbol{\Sigma}\rangle_0 : \langle \boldsymbol{F}\rangle_0 + \langle \boldsymbol{\Sigma}:\mathrm{Grad}\boldsymbol{z} + \boldsymbol{z}\cdot\mathrm{Div}\boldsymbol{\Sigma}\rangle_0 \\ &= \langle \boldsymbol{\Sigma}\rangle_0 : \langle \boldsymbol{F}\rangle_0 + \langle \mathrm{Div}\,(\boldsymbol{z}\cdot\boldsymbol{\Sigma})\rangle_0 \\ &= \langle \boldsymbol{\Sigma}\rangle_0 : \langle \boldsymbol{F}\rangle_0 + \lceil \boldsymbol{z}\cdot [\boldsymbol{\Sigma}\cdot\boldsymbol{N}]\rfloor_0 = \langle \boldsymbol{\Sigma}\rangle_0 : \langle \boldsymbol{F}\rangle_0 .\end{aligned}$$

Similar results can be obtained for the current configuration by considering the inverse of deformation gradient $\boldsymbol{F}^{-1}$ as the kinematically admissible tensor and the Cauchy stress $\boldsymbol{\sigma}$ as the statically admissible tensor.

NOTE.– The Hill's lemma and the Hill–Mandel theorem are valid for the products $\boldsymbol{\Sigma}:\dot{\boldsymbol{F}}$ and $\dot{\boldsymbol{\Sigma}}:\boldsymbol{F}$. The proof is analogous.

A1.2.2. *Thermal and other problems*

The thermal problem is treated in exactly the same way as in the small deformations case. The only difference is that $\boldsymbol{\nabla}\boldsymbol{\theta}$ and $\boldsymbol{q}$ are substituted by $\boldsymbol{\nabla}\boldsymbol{\theta}_0$ and $\boldsymbol{q}_0$, respectively.

With regard to electric or magnetic phenomena, appropriate forms of the Hill's lemma and the Hill–Mandel theorem can be derived whether scalar or vector potentials are utilized [CHA 14].

A1.3. Useful identities

– For the boundary conditions of the theorem A1.1 and admissible fields, the following hold:

1) $\left\langle \boldsymbol{\Sigma}\cdot\boldsymbol{F}^T\right\rangle_0 = \langle \boldsymbol{\Sigma}\rangle_0 \cdot \langle \boldsymbol{F}\rangle_0^T$;

2) $\left\langle \boldsymbol{F}\cdot\boldsymbol{\Sigma}^T\right\rangle_0 = \langle \boldsymbol{F}\rangle_0 \cdot \langle \boldsymbol{\Sigma}\rangle_0^T$.

The proof of this statement is as follows. For any type of boundary condition, the properties of tensor derivatives discussed in section 1.1, the definitions A1.1, A1.2 and the divergence theorem [2.11] allow us to write

$$\begin{aligned}\langle \boldsymbol{\Sigma}\cdot \boldsymbol{F}^T\rangle_0 &= \left\langle \boldsymbol{\Sigma}\cdot [\mathrm{Grad}(\boldsymbol{u}+\boldsymbol{X})]^T\right\rangle_0\\
&= \langle \boldsymbol{\Sigma} \,\tilde{\because}\, \mathrm{Grad}(\boldsymbol{u}+\boldsymbol{X}) + \mathrm{Div}\boldsymbol{\Sigma}\otimes[\boldsymbol{u}+\boldsymbol{X}]\rangle_0\\
&= \left\langle [\mathrm{Div}\,([\boldsymbol{u}+\boldsymbol{X}]\otimes\boldsymbol{\Sigma})]^T\right\rangle_0 = \langle \mathrm{Div}\,([\boldsymbol{u}+\boldsymbol{X}]\otimes\boldsymbol{\Sigma})\rangle_0^T\\
&= \lceil[\boldsymbol{u}+\boldsymbol{X}]\otimes[\boldsymbol{\Sigma}\cdot\boldsymbol{N}]\rfloor_0^T = \lceil[\boldsymbol{\Sigma}\cdot\boldsymbol{N}]\otimes[\boldsymbol{u}+\boldsymbol{X}]\rfloor_0,\end{aligned}$$

$$\begin{aligned}\langle \boldsymbol{F}\cdot \boldsymbol{\Sigma}^T\rangle_0 &= \langle \mathrm{Grad}(\boldsymbol{u}+\boldsymbol{X}) \,\tilde{\because}\, \boldsymbol{\Sigma}\rangle_0\\
&= \langle \mathrm{Grad}(\boldsymbol{u}+\boldsymbol{X}) \,\tilde{\because}\, \boldsymbol{\Sigma} + [\boldsymbol{u}+\boldsymbol{X}]\otimes\mathrm{Div}\boldsymbol{\Sigma}\rangle_0\\
&= \langle \mathrm{Div}\,([\boldsymbol{u}+\boldsymbol{X}]\otimes\boldsymbol{\Sigma})\rangle_0 = \lceil[\boldsymbol{u}+\boldsymbol{X}]\otimes[\boldsymbol{\Sigma}\cdot\boldsymbol{N}]\rfloor_0.\end{aligned}$$

At this point, the proof is split in three parts:

1) For $\boldsymbol{u} = [\langle \boldsymbol{F}\rangle_0 - \boldsymbol{I}]\cdot\boldsymbol{X}$ on $\partial\mathcal{D}_0$, the use of equation [A1.2] yields

$$\begin{aligned}\langle \boldsymbol{\Sigma}\cdot \boldsymbol{F}^T\rangle_0 &= \lceil[\boldsymbol{\Sigma}\cdot\boldsymbol{N}]\otimes[\langle\boldsymbol{F}\rangle_0\cdot\boldsymbol{X}]\rfloor_0 = \lceil[\boldsymbol{\Sigma}\cdot\boldsymbol{N}]\otimes\boldsymbol{X}\rfloor_0\cdot\langle\boldsymbol{F}\rangle_0^T\\
&= \langle\boldsymbol{\Sigma}\rangle_0\cdot\langle\boldsymbol{F}\rangle_0^T,\end{aligned}$$

$$\begin{aligned}\langle \boldsymbol{F}\cdot \boldsymbol{\Sigma}^T\rangle_0 &= \lceil[\langle\boldsymbol{F}\rangle_0\cdot\boldsymbol{X}]\otimes[\boldsymbol{\Sigma}\cdot\boldsymbol{N}]\rfloor_0 = \langle\boldsymbol{F}\rangle_0\cdot\lceil\boldsymbol{X}\otimes[\boldsymbol{\Sigma}\cdot\boldsymbol{N}]\rfloor_0\\
&= \langle\boldsymbol{F}\rangle_0\cdot\lceil[\boldsymbol{\Sigma}\cdot\boldsymbol{N}]\otimes\boldsymbol{X}\rfloor_0^T = \langle\boldsymbol{F}\rangle_0\cdot\langle\boldsymbol{\Sigma}\rangle_0^T.\end{aligned}$$

2) For $\boldsymbol{\Sigma}\cdot\boldsymbol{N} = \langle\boldsymbol{\Sigma}\rangle_0\cdot\boldsymbol{N}$ on $\partial\mathcal{D}_0$, the use of equation [A1.1] yields

$$\begin{aligned}\langle \boldsymbol{\Sigma}\cdot \boldsymbol{F}^T\rangle_0 &= \lceil[\langle\boldsymbol{\Sigma}\rangle_0\cdot\boldsymbol{N}]\otimes[\boldsymbol{u}+\boldsymbol{X}]\rfloor_0 = \langle\boldsymbol{\Sigma}\rangle_0\cdot\lceil\boldsymbol{N}\otimes[\boldsymbol{u}+\boldsymbol{X}]\rfloor_0\\
&= \langle\boldsymbol{\Sigma}\rangle_0\cdot\lceil[\boldsymbol{u}+\boldsymbol{X}]\otimes\boldsymbol{N}\rfloor_0^T = \langle\boldsymbol{\Sigma}\rangle_0\cdot\langle\boldsymbol{F}\rangle_0^T,\end{aligned}$$

$$\begin{aligned}\langle \boldsymbol{F}\cdot \boldsymbol{\Sigma}^T\rangle_0 &= \lceil[\boldsymbol{u}+\boldsymbol{X}]\otimes[\langle\boldsymbol{\Sigma}\rangle_0\cdot\boldsymbol{N}]\rfloor_0 = \lceil[\boldsymbol{u}+\boldsymbol{X}]\otimes\boldsymbol{N}\rfloor_0\cdot\langle\boldsymbol{\Sigma}\rangle_0^T\\
&= \langle\boldsymbol{F}\rangle_0\cdot\langle\boldsymbol{\Sigma}\rangle_0^T.\end{aligned}$$

3) For $\boldsymbol{F} = \langle \boldsymbol{F} \rangle_0 + \text{Grad}\boldsymbol{z}$ with $\boldsymbol{z}$ periodic and $\boldsymbol{\Sigma} \cdot \boldsymbol{N}$ anti-periodic, the use of the properties of the products between periodic and anti-periodic vectors described in section 4.1 yields

$$
\begin{aligned}
\left\langle \boldsymbol{\Sigma} \cdot \boldsymbol{F}^T \right\rangle_0 &= \langle \boldsymbol{\Sigma} \rangle_0 \cdot \langle \boldsymbol{F} \rangle_0^T + \left\langle \boldsymbol{\Sigma} \cdot [\text{Grad}\boldsymbol{z}]^T \right\rangle_0 \\
&= \langle \boldsymbol{\Sigma} \rangle_0 \cdot \langle \boldsymbol{F} \rangle_0^T + \langle \boldsymbol{\Sigma} \,\tilde{\cdot}\, \text{Grad}\boldsymbol{z} + \text{Div}\boldsymbol{\Sigma} \otimes \boldsymbol{z} \rangle_0 \\
&= \langle \boldsymbol{\Sigma} \rangle_0 \cdot \langle \boldsymbol{F} \rangle_0^T + \left\langle [\text{Div}\,(\boldsymbol{z} \otimes \boldsymbol{\Sigma})]^T \right\rangle_0 \\
&= \langle \boldsymbol{\Sigma} \rangle_0 \cdot \langle \boldsymbol{F} \rangle_0^T + \langle \text{Div}\,(\boldsymbol{z} \otimes \boldsymbol{\Sigma}) \rangle_0^T \\
&= \langle \boldsymbol{\Sigma} \rangle_0 \cdot \langle \boldsymbol{F} \rangle_0^T + \lceil \boldsymbol{z} \otimes [\boldsymbol{\Sigma} \cdot \boldsymbol{N}] \rfloor_0^T = \langle \boldsymbol{\Sigma} \rangle_0 \cdot \langle \boldsymbol{F} \rangle_0^T ,
\end{aligned}
$$

$$
\begin{aligned}
\left\langle \boldsymbol{F} \cdot \boldsymbol{\Sigma}^T \right\rangle_0 &= \langle \boldsymbol{F} \rangle_0 \cdot \langle \boldsymbol{\Sigma} \rangle_0^T + \langle \text{Grad}\boldsymbol{z} \,\tilde{\cdot}\, \boldsymbol{\Sigma} \rangle_0 \\
&= \langle \boldsymbol{F} \rangle_0 \cdot \langle \boldsymbol{\Sigma} \rangle_0^T + \langle \text{Grad}\boldsymbol{z} \,\tilde{\cdot}\, \boldsymbol{\Sigma} + \boldsymbol{z} \otimes \text{Div}\boldsymbol{\Sigma} \rangle_0 \\
&= \langle \boldsymbol{F} \rangle_0 \cdot \langle \boldsymbol{\Sigma} \rangle_0^T + \langle \text{Div}\,(\boldsymbol{z} \otimes \boldsymbol{\Sigma}) \rangle_0 \\
&= \langle \boldsymbol{F} \rangle_0 \cdot \langle \boldsymbol{\Sigma} \rangle_0^T + \lceil \boldsymbol{z} \otimes [\boldsymbol{\Sigma} \cdot \boldsymbol{N}] \rfloor_0 = \langle \boldsymbol{F} \rangle_0 \cdot \langle \boldsymbol{\Sigma} \rangle_0^T .
\end{aligned}
$$

– The conservation of mass [2.16] and the properties of Table 1.3 lead to the relations

$$
\langle J \rangle_0 = \frac{V_t}{V_0}, \qquad \left\langle J^{-1} \right\rangle_t = \frac{V_0}{V_t}, \qquad \langle \rho \rangle_t = \left\langle \rho_0 J^{-1} \right\rangle_t = \frac{V_0}{V_t} \langle \rho_0 \rangle_0 .
$$

– For the boundary conditions of the theorem A1.1 and admissible fields, the following relation holds:

$$
\langle \boldsymbol{\sigma} \rangle_t = \frac{V_0}{V_t} \langle \boldsymbol{\Sigma} \rangle_0 \cdot \langle \boldsymbol{F} \rangle_0^T .
$$

Indeed, using the first identity in this section, the relation between Cauchy and Piola stress tensors, $\boldsymbol{\sigma} = J^{-1} \boldsymbol{\Sigma} \cdot \boldsymbol{F}^T$, and the properties of Table 1.3, we obtain

$$\langle \boldsymbol{\sigma} \rangle_t = \left\langle J^{-1} \boldsymbol{\Sigma} \cdot \boldsymbol{F}^T \right\rangle_t = \frac{V_0}{V_t} \left\langle \boldsymbol{\Sigma} \cdot \boldsymbol{F}^T \right\rangle_0 = \frac{V_0}{V_t} \langle \boldsymbol{\Sigma} \rangle_0 \cdot \langle \boldsymbol{F} \rangle_0^T .$$

– In a "brick" shaped body, periodic fields in undeformed configuration remain periodic in the deformed configuration. Extended discussion about this point can be found in the literature [COS 05, BET 12]. A "brick" shaped body is defined by three linearly independent vectors $\boldsymbol{r}_i$, $i = 1, 2, 3$, mutually orthogonal. Under these conditions, the boundary of the body consists of six faces such that for each face $\mathscr{F}$ we can find a unique corresponding face $\mathscr{F}_h$ that can be viewed as the translation of $\mathscr{F}$ by one of the six vectors $\pm \boldsymbol{r}_i$. For every point $\boldsymbol{X} \in \mathscr{F}$, a unique point $\boldsymbol{X}_h \in \mathscr{F}_h$ exists such that $\boldsymbol{X} - \boldsymbol{X}_h = \pm \boldsymbol{r}_i$. Under periodic motion, it holds that

$$\boldsymbol{u}(\boldsymbol{X}, t) = [\langle \boldsymbol{F} \rangle_0 - \boldsymbol{I}] \cdot \boldsymbol{X} + \boldsymbol{z}(\boldsymbol{X}, t),$$

with $\boldsymbol{z}(\boldsymbol{X}, t) = \boldsymbol{z}(\boldsymbol{X}_h, t)$. So, we obtain

$$\boldsymbol{u}(\boldsymbol{X}, t) - \boldsymbol{u}_h(\boldsymbol{X}_h, t) = [\langle \boldsymbol{F} \rangle_0 - \boldsymbol{I}] \cdot [\boldsymbol{X} - \boldsymbol{X}_h] = \pm [\langle \boldsymbol{F} \rangle_0 - \boldsymbol{I}] \cdot \boldsymbol{r}_i,$$

where the vector $\pm [\langle \boldsymbol{F} \rangle_0 - \boldsymbol{I}] \cdot \boldsymbol{r}_i$ is unique for the pair of homologous faces to which $\boldsymbol{X}$ and $\boldsymbol{X}_h$ belong. Thus, the two faces translate by a constant vector in the deformed configuration. The main consequence of the last equation is that fields that are periodic and anti-periodic over the body preserve their character throughout the deformation process even when viewed as fields defined over the deformed configuration.

Appendix 2

Periodic Homogenization in Large Deformations

Like in the small deformation processes, periodic homogenization under large deformations requires a description of the composite through the introduction of two scales. Generally, the thermomechanical problem can be described in the reference or the current configuration, as it is discussed in Chapter 2. In this appendix, the analysis focuses on the reference configuration.

A2.1. Thermomechanical processes

At the macroscale, the continuum body occupies the space $\overline{\mathcal{D}}_0$ with volume $\overline{V}_0$ and is bounded by the surface $\partial\overline{\mathcal{D}}_0$ with normal unit vector $\overline{\boldsymbol{N}}$. Each macroscopic point is assigned with a position vector $\overline{\boldsymbol{X}}$ in $\overline{\mathcal{D}}_0$. However, at the microscale level, there is a periodic unit cell that occupies the space $\mathcal{D}_0$ with volume V_0 and is bounded by the surface $\partial\mathcal{D}_0$ with normal unit vector $\boldsymbol{N}$. Each microscopic point is assigned to a position vector $\boldsymbol{X}$ in $\mathcal{D}_0$ (Figure A2.1). The two scales $\overline{\boldsymbol{X}}$ and $\boldsymbol{X}$ are connected by a characteristic length ϵ through the relation $\boldsymbol{X} = \overline{\boldsymbol{X}}/\epsilon$. The periodic homogenization theory provides accurate results only when the characteristic length is close to zero, i.e. when the microstructure is extremely small compared to the actual size of the composite.

When considering the composite in a global sense, the general coordinate system $\overline{\boldsymbol{X}}$ is used. The gradient operator Grad^{ϵ} is connected with the gradient operators of the macroscale ($\overline{\text{Grad}}$) and microscale (Grad) through the usual scale decomposition rule [TEM 12]

$$\text{Grad}^{\epsilon}\{\bullet\} = \overline{\text{Grad}}\{\bullet\} + \frac{1}{\epsilon}\text{Grad}\{\bullet\}. \qquad \text{[A2.1]}$$

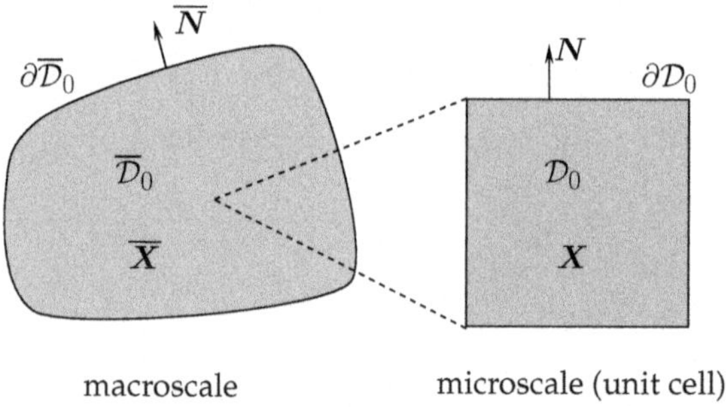

Figure A2.1. *Macroscale and microscale of composite at reference configuration*

Since all coordinate systems are Cartesian, the identity tensor $\boldsymbol{I}$ and the permutation tensor ϵ are common for all of them and thus the other operators (Div, Curl) follow a similar decomposition. All the variables that correspond to the general composite response are denoted with a superscript ϵ. Moreover, a bar above a symbol denotes a macroscopic quantity or variable, depending only on the vector $\overline{\boldsymbol{X}}$ and perhaps the time. Additionally, the mean operators $\langle\{\bullet\}\rangle_0$ and $\lceil\{\bullet\}\rfloor_0$ introduced in section A1.1 refer to volume and surface averaging respectively in the reference configuration.

A2.1.1. *Kinematics*

As described in Chapter 2, the deformation gradient $\boldsymbol{F}^{\epsilon}$ can be expressed as a function of the displacement vector $\boldsymbol{u}^{\epsilon}$ through the relation

$$\boldsymbol{F}^{\epsilon} = \boldsymbol{I} + \text{Grad}^{\epsilon}\boldsymbol{u}^{\epsilon}. \qquad \text{[A2.2]}$$

The periodicity of the composite microstructure allows us to assume the following asymptotic series expansion for the displacement [TEM 12],

$$\begin{aligned} \boldsymbol{u}^{\epsilon}(\overline{\boldsymbol{X}}, t) &= \boldsymbol{u}^{(0)}(\overline{\boldsymbol{X}}, \boldsymbol{X}, t) + \epsilon\, \boldsymbol{u}^{(1)}(\overline{\boldsymbol{X}}, \boldsymbol{X}, t) \\ &\quad + \epsilon^2 \boldsymbol{u}^{(2)}(\overline{\boldsymbol{X}}, \boldsymbol{X}, t) + ... \end{aligned} \quad \text{[A2.3]}$$

The main assumption in the asymptotic expansion homogenization framework is that all two-scale functions $\{\bullet\}^{(0)}$, $\{\bullet\}^{(1)}$ etc., are considered periodic in $\boldsymbol{X}$. Using equation [5.2] in Section [5.1] and taking into account [4.15] yields

$$\boldsymbol{F}^{\epsilon} = \frac{1}{\epsilon}\mathrm{Grad}\boldsymbol{u}^{(0)} + \boldsymbol{I} + \overline{\mathrm{Grad}}\boldsymbol{u}^{(0)} + \mathrm{Grad}\boldsymbol{u}^{(1)} + \epsilon....$$

For $\epsilon \to 0$, the deformation gradient is bounded only when the ϵ^{-1} term vanishes [TEM 12], leading to

$$\mathrm{Grad}\boldsymbol{u}^{(0)} = \boldsymbol{0} \quad \to \quad \boldsymbol{u}^{(0)} := \overline{\boldsymbol{u}}(\overline{\boldsymbol{X}}, t). \quad \text{[A2.4]}$$

Using the final result, we can write the deformation gradient in the general expanded form

$$\begin{aligned} \boldsymbol{F}^{\epsilon}(\overline{\boldsymbol{X}}, t) &= \boldsymbol{F}^{(0)}(\overline{\boldsymbol{X}}, \boldsymbol{X}, t) + \epsilon\, \boldsymbol{F}^{(1)}(\overline{\boldsymbol{X}}, \boldsymbol{X}, t) \\ &\quad + \epsilon^2 \boldsymbol{F}^{(2)}(\overline{\boldsymbol{X}}, \boldsymbol{X}, t) + ..., \end{aligned} \quad \text{[A2.5]}$$

with

$$\boldsymbol{F}^{(0)} = \boldsymbol{I} + \overline{\mathrm{Grad}}\overline{\boldsymbol{u}} + \mathrm{Grad}\boldsymbol{u}^{(1)}. \quad \text{[A2.6]}$$

Defining $\overline{\boldsymbol{F}} := \boldsymbol{I} + \overline{\mathrm{Grad}}\overline{\boldsymbol{u}}$ and taking into account the relation (see Section 4.1)

$$\left\langle \mathrm{Grad}\boldsymbol{u}^{(1)} \right\rangle_0 = \boldsymbol{0},$$

yields that

$$\overline{\boldsymbol{F}} = \left\langle \boldsymbol{F}^{(0)} \right\rangle_0 . \qquad \text{[A2.7]}$$

The tensor $\boldsymbol{F}^{(0)}$ is defined as the microscopic deformation gradient, while the tensor $\overline{\boldsymbol{F}}$, depending exclusively on $\overline{\boldsymbol{X}}$ and the time, is defined as the macroscopic deformation gradient.

A2.1.2. *Kinetics*

The Piola stress is assumed to have a similar asymptotic series expansion as the deformation gradient, i.e.

$$\begin{aligned}\boldsymbol{\Sigma}^{\epsilon}(\overline{\boldsymbol{X}}, t) &= \boldsymbol{\Sigma}^{(0)}(\overline{\boldsymbol{X}}, \boldsymbol{X}, t) + \epsilon\, \boldsymbol{\Sigma}^{(1)}(\overline{\boldsymbol{X}}, \boldsymbol{X}, t) \\ &\quad + \epsilon^2 \boldsymbol{\Sigma}^{(2)}(\overline{\boldsymbol{X}}, \boldsymbol{X}, t) + ... \end{aligned} \qquad \text{[A2.8]}$$

Similarly to the deformation gradient, the tensor $\boldsymbol{\Sigma}^{(0)}$ is defined as the microscopic Piola stress, while its weak limit, i.e. its average over the unit cell $\overline{\boldsymbol{\Sigma}} := \left\langle \boldsymbol{\Sigma}^{(0)} \right\rangle_0$, is defined as the macroscopic Piola stress. Obviously, $\overline{\boldsymbol{\Sigma}}$ does not depend on $\boldsymbol{X}$.

A2.1.3. *Conservation laws*

A2.1.3.1. *Conservation of mass*

The following asymptotic series expansion for density in the reference configuration is postulated:

$$\rho_0^{\epsilon}(\overline{\boldsymbol{X}}, t) = \rho_0^{(0)}(\overline{\boldsymbol{X}}, \boldsymbol{X}, t) + \epsilon\, \rho_0^{(1)}(\overline{\boldsymbol{X}}, \boldsymbol{X}, t) + \epsilon^2 \rho_0^{(2)}(\overline{\boldsymbol{X}}, \boldsymbol{X}, t) + ... \qquad \text{[A2.9]}$$

According to equation [2.17], in the reference configuration the conservation of mass is written as

$$\frac{D\rho_0^{\epsilon}}{Dt} = 0. \qquad \text{[A2.10]}$$

Inserting [A2.9] in [A2.10] yields

$$\frac{D\rho_0^{(0)}}{Dt} + \epsilon ... = 0. \qquad [A2.11]$$

Since the aim of homogenization is to study the system of equations when ϵ tends to zero, the following equation leads to the microscale conservation of mass:

$$\frac{D\rho_0^{(0)}}{Dt} = 0 \quad \Rightarrow \quad \rho_0^{(0)} := \rho_0^{(0)}(\overline{\boldsymbol{X}}, \boldsymbol{X}). \qquad [A2.12]$$

The macroscale conservation of mass is identified by obtaining the weak limit of equation [A2.12], i.e. averaging it over the unit cell:

$$\left\langle \frac{D\rho_0^{(0)}}{Dt} \right\rangle_0 = 0,$$

or, by defining $\bar{\rho}_0 := \left\langle \rho_0^{(0)} \right\rangle_0$,

$$\frac{D\bar{\rho}_0}{Dt} = 0 \quad \Rightarrow \quad \bar{\rho}_0 := \bar{\rho}_0(\overline{\boldsymbol{X}}). \qquad [A2.13]$$

From these results, it becomes clear that $\rho_0^{(0)}$ represents the microscopic density and $\bar{\rho}_0$ represents the macroscopic density.

A2.1.3.2. *Conservation of linear momentum*

According to equation [2.20], in the reference configuration the conservation of linear momentum is written as

$$\rho_0^\epsilon \frac{D\dot{\boldsymbol{u}}^\epsilon}{Dt} = \rho_0^\epsilon \boldsymbol{b}_0^\epsilon + \mathrm{Div}^\epsilon \boldsymbol{\Sigma}^\epsilon. \qquad [A2.14]$$

Considering an asymptotic series expansion for the body forces in the reference configuration

$$\boldsymbol{b}_0^\epsilon(\overline{\boldsymbol{X}}, t) = \boldsymbol{b}_0^{(0)}(\overline{\boldsymbol{X}}, \boldsymbol{X}, t) + \epsilon\, \boldsymbol{b}_0^{(1)}(\overline{\boldsymbol{X}}, \boldsymbol{X}, t) + \epsilon^2 \boldsymbol{b}_0^{(2)}(\overline{\boldsymbol{X}}, \boldsymbol{X}, t) + ..., [A2.15]$$

and with the help of equation [A2.1] and the expansions [A2.3], [A2.8] and [A2.9], we obtain

$$\rho_0^{(0)}\frac{D\dot{\bar{\boldsymbol{u}}}}{Dt}+\epsilon ... = \rho_0^{(0)}\boldsymbol{b}_0^{(0)}+\overline{\mathrm{Div}}\boldsymbol{\Sigma}^{(0)}+\mathrm{Div}\boldsymbol{\Sigma}^{(1)}+\frac{1}{\epsilon}\mathrm{Div}\boldsymbol{\Sigma}^{(0)}+\epsilon .... \quad \text{[A2.16]}$$

For $\epsilon \to 0$, the term ϵ^{-1} should vanish, leading to the conservation of linear momentum in the microscale:

$$\mathrm{Div}\boldsymbol{\Sigma}^{(0)}=\boldsymbol{0}. \quad \text{[A2.17]}$$

The conservation of linear momentum in the macroscale is obtained by averaging over the unit cell the ϵ^0 term of equation [A2.16]. Noting that $\left\langle \mathrm{Div}\boldsymbol{\Sigma}^{(1)}\right\rangle_0 = \boldsymbol{0}$ (see Section 4.1) and defining $\bar{\boldsymbol{b}}_0 := \left\langle \rho_0^{(0)}\boldsymbol{b}_0^{(0)}\right\rangle_0/\bar{\rho}_0$, the average of the ϵ^0 term takes the form

$$\bar{\rho}_0\frac{D\dot{\bar{\boldsymbol{u}}}}{Dt}=\bar{\rho}_0\bar{\boldsymbol{b}}_0+\overline{\mathrm{Div}}\bar{\boldsymbol{\Sigma}}. \quad \text{[A2.18]}$$

Similar to the small deformation case, the results of the asymptotic expansion approach are valid for the conservation of linear momentum only when the wavelength of the dynamic loading is much larger than the characteristic size of the microstructure.

NOTE.– The microscopic deformation gradient is expressed by equations [A2.6] and [A2.7]. Moreover, the microscopic Piola stress is statically admissible due to equation [A2.17]. Thus, according to the discussion in Section A1.2.1, $\boldsymbol{\Sigma}^{(0)}$ and $\boldsymbol{F}^{(0)}$ satisfy the Hill–Mandel power equivalence:

$$\left\langle \boldsymbol{\Sigma}^{(0)}:\dot{\boldsymbol{F}}^{(0)}\right\rangle_0=\left\langle \boldsymbol{\Sigma}^{(0)}\right\rangle_0:\left\langle \dot{\boldsymbol{F}}^{(0)}\right\rangle_0=\bar{\boldsymbol{\Sigma}}:\dot{\bar{\boldsymbol{F}}}. \quad \text{[A2.19]}$$

A2.1.3.3. *Conservation of angular momentum*

According to equation [2.23], in the reference configuration the conservation of angular momentum is written as

$$\boldsymbol{\Sigma}^{\epsilon}\cdot[\boldsymbol{F}^{\epsilon}]^T=\boldsymbol{F}^{\epsilon}\cdot[\boldsymbol{\Sigma}^{\epsilon}]^T. \quad \text{[A2.20]}$$

With the help of equations [A2.5] and [A2.8], the expanded form of the angular momentum is written as

$$\boldsymbol{\Sigma}^{(0)} \cdot \left[\boldsymbol{F}^{(0)}\right]^T + \epsilon ... = \boldsymbol{F}^{(0)} \cdot \left[\boldsymbol{\Sigma}^{(0)}\right]^T + \epsilon .... \qquad [A2.21]$$

Since $\epsilon \rightarrow 0$, the microscale conservation of angular momentum is obtained as

$$\boldsymbol{\Sigma}^{(0)} \cdot \left[\boldsymbol{F}^{(0)}\right]^T = \boldsymbol{F}^{(0)} \cdot \left[\boldsymbol{\Sigma}^{(0)}\right]^T, \qquad [A2.22]$$

while the volume average over the unit cell of the above expression provides the conservation of angular momentum in the macroscale. Taking into account the properties proven in Section A1.3, the final expression for the macroscale takes the form

$$\overline{\boldsymbol{\Sigma}} \cdot \overline{\boldsymbol{F}}^T = \overline{\boldsymbol{F}} \cdot \overline{\boldsymbol{\Sigma}}^T. \qquad [A2.23]$$

A2.1.3.4. *Conservation of energy*

According to equation [2.26], in the reference configuration the conservation of energy is written as

$$\rho_0^\epsilon \dot{e}_0^\epsilon = \boldsymbol{\Sigma}^\epsilon : \dot{\boldsymbol{F}}^\epsilon - \text{Div}\boldsymbol{q}_0^\epsilon + \rho_0^\epsilon \mathcal{R}_0^\epsilon. \qquad [A2.24]$$

Considering an asymptotic series expansion for the internal energy, the heat fluxes and the heat sources in the reference configuration

$$\begin{aligned} e_0^\epsilon(\overline{\boldsymbol{X}}, t) &= e_0^{(0)}(\overline{\boldsymbol{X}}, \boldsymbol{X}, t) + \epsilon\, e_0^{(1)}(\overline{\boldsymbol{X}}, \boldsymbol{X}, t) \\ &\quad + \epsilon^2 e_0^{(2)}(\overline{\boldsymbol{X}}, \boldsymbol{X}, t) + ..., \\ \boldsymbol{q}_0^\epsilon(\overline{\boldsymbol{X}}, t) &= \boldsymbol{q}_0^{(0)}(\overline{\boldsymbol{X}}, \boldsymbol{X}, t) + \epsilon\, \boldsymbol{q}_0^{(1)}(\overline{\boldsymbol{X}}, \boldsymbol{X}, t) \\ &\quad + \epsilon^2 \boldsymbol{q}_0^{(2)}(\overline{\boldsymbol{X}}, \boldsymbol{X}, t) + ..., \\ \mathcal{R}_0^\epsilon(\overline{\boldsymbol{X}}, t) &= \mathcal{R}_0^{(0)}(\overline{\boldsymbol{X}}, \boldsymbol{X}, t) + \epsilon\, \mathcal{R}_0^{(1)}(\overline{\boldsymbol{X}}, \boldsymbol{X}, t) \\ &\quad + \epsilon^2 \mathcal{R}_0^{(2)}(\overline{\boldsymbol{X}}, \boldsymbol{X}, t) + ..., \end{aligned} \qquad [A2.25]$$

and with the help of equation [A2.1] and the expansions [A2.5], [A2.8] and [A2.9], we obtain

$$\rho_0^{(0)}\dot{e}_0^{(0)} + \epsilon ... = \boldsymbol{\Sigma}^{(0)}:\dot{\boldsymbol{F}}^{(0)} - \overline{\text{Div}}\boldsymbol{q}_0^{(0)} - \text{Div}\boldsymbol{q}_0^{(1)} + \rho_0^{(0)}\mathcal{R}_0^{(0)} \\ -\frac{1}{\epsilon}\text{Div}\boldsymbol{q}_0^{(0)} + \epsilon .... \qquad [A2.26]$$

For $\epsilon \rightarrow 0$, the term ϵ^{-1} should vanish, leading to the conservation of energy in the microscale:

$$\text{Div}\boldsymbol{q}_0^{(0)} = \boldsymbol{0}. \qquad [A2.27]$$

The conservation of energy in the macroscale is obtained by volume averaging over the unit cell the ϵ^0 term of equation [A2.26]. Noting that $\left\langle \text{Div}\boldsymbol{q}_0^{(1)} \right\rangle_0 = \boldsymbol{0}$ (see section A1.1) and defining $\bar{e}_0 := \left\langle \rho_0^{(0)} e_0^{(0)} \right\rangle_0 / \bar{\rho}_0$, $\overline{\mathcal{R}}_0 := \left\langle \rho_0^{(0)}\mathcal{R}_0^{(0)} \right\rangle_0 / \bar{\rho}_0$ and $\bar{\boldsymbol{q}}_0 := \left\langle \boldsymbol{q}_0^{(0)} \right\rangle_0$, the average of the ϵ^0 term, with the help of the relation [5.18], takes the form

$$\bar{\rho}_0\dot{\bar{e}}_0 = \overline{\boldsymbol{\Sigma}}:\dot{\overline{\boldsymbol{F}}} - \overline{\text{Div}}\bar{\boldsymbol{q}}_0 + \bar{\rho}_0\overline{\mathcal{R}}_0. \qquad [A2.28]$$

A2.1.3.5. *Entropy inequality*

According to equation [2.29], in the reference configuration the entropy inequality is written as

$$\rho_0^\epsilon\theta^\epsilon\dot{\varsigma}_0^\epsilon - \frac{\boldsymbol{q}_0^\epsilon}{\theta^\epsilon}\cdot\boldsymbol{\nabla\theta}_0^\epsilon + \text{Div}\boldsymbol{q}_0^\epsilon - \rho_0^\epsilon\mathcal{R}_0^\epsilon \geq 0, \quad \boldsymbol{\nabla\theta}_0^\epsilon = \text{Grad}^\epsilon\theta^\epsilon.$$

Using the conservation of energy [A2.24] the inequality is expressed as

$$\theta^\epsilon\rho_0^\epsilon\dot{\varsigma}_0^\epsilon - \frac{1}{\theta^\epsilon}\boldsymbol{q}_0^\epsilon\cdot\boldsymbol{\nabla\theta}_0^\epsilon + \boldsymbol{\Sigma}^\epsilon:\dot{\boldsymbol{F}}^\epsilon - \rho_0^\epsilon\dot{e}_0^\epsilon \geq 0, \quad \boldsymbol{\nabla\theta}_0^\epsilon = \text{Grad}^\epsilon\theta^\epsilon. \qquad [A2.29]$$

The periodicity of the composite microstructure allows us to assume the following asymptotic series expansion for the absolute temperature and the entropy in the reference configuration:

$$\begin{aligned}\theta^{\epsilon}(\overline{\boldsymbol{X}},t) &= \theta^{(0)}(\overline{\boldsymbol{X}},\boldsymbol{X},t) + \epsilon\,\theta^{(1)}(\overline{\boldsymbol{X}},\boldsymbol{X},t) \\ &\quad + \epsilon^2\theta^{(2)}(\overline{\boldsymbol{X}},\boldsymbol{X},t) + ..., \\ \varsigma_0^{\epsilon}(\overline{\boldsymbol{X}},t) &= \varsigma_0^{(0)}(\overline{\boldsymbol{X}},\boldsymbol{X},t) + \epsilon\,\varsigma_0^{(1)}(\overline{\boldsymbol{X}},\boldsymbol{X},t) \\ &\quad + \epsilon^2\varsigma_0^{(2)}(\overline{\boldsymbol{X}},\boldsymbol{X},t) + ...\end{aligned} \qquad [A2.30]$$

Combining [A2.30]$_1$ with [A2.29]$_2$ and using [A2.1] yields

$$\boldsymbol{\nabla\theta}_0^{\epsilon} = \frac{1}{\epsilon}\text{Grad}\theta^{(0)} + \overline{\text{Grad}}\theta^{(0)} + \text{Grad}\theta^{(1)} + \epsilon....$$

For $\epsilon \to 0$, the temperature gradient is bounded only when the ϵ^{-1} term vanishes [TEM 12], leading to

$$\text{Grad}\theta^{(0)} = \mathbf{0} \quad \to \quad \theta^{(0)} := \bar{\theta}(\overline{\boldsymbol{X}},t). \qquad [A2.31]$$

Using the final result, we can write the temperature gradient in the reference configuration as

$$\begin{aligned}\boldsymbol{\nabla\theta}_0^{\epsilon}(\overline{\boldsymbol{X}},t) &= \boldsymbol{\nabla\theta}_0^{(0)}(\overline{\boldsymbol{X}},\boldsymbol{X},t) + \epsilon\,\boldsymbol{\nabla\theta}_0^{(1)}(\overline{\boldsymbol{X}},\boldsymbol{X},t) \\ &\quad + \epsilon^2\boldsymbol{\nabla\theta}_0^{(2)}(\overline{\boldsymbol{X}},\boldsymbol{X},t) + ...,\end{aligned} \qquad [A2.32]$$

with

$$\boldsymbol{\nabla\theta}_0^{(0)} = \overline{\text{Grad}}\bar{\theta} + \text{Grad}\theta^{(1)}. \qquad [A2.33]$$

Defining $\overline{\boldsymbol{\nabla\theta}}_0 := \overline{\text{Grad}}\bar{\theta}$ and recalling from section A1.1 that

$$\left\langle \text{Grad}\theta^{(1)} \right\rangle_0 = \mathbf{0},$$

we obtain that

$$\overline{\boldsymbol{\nabla}\boldsymbol{\theta}}_0 = \left\langle \boldsymbol{\nabla}\boldsymbol{\theta}_0^{(0)} \right\rangle_0 . \qquad [A2.34]$$

The equations [A2.33] and [A2.34] in conjunction with [A2.27] lead to the conclusion that the Hill–Mandel thermal power equivalence holds, i.e.

$$\left\langle \boldsymbol{q}_0^{(0)} \cdot \boldsymbol{\nabla}\boldsymbol{\theta}_0^{(0)} \right\rangle_0 = \left\langle \boldsymbol{q}_0^{(0)} \right\rangle_0 \cdot \left\langle \boldsymbol{\nabla}\boldsymbol{\theta}_0^{(0)} \right\rangle_0 = \bar{\boldsymbol{q}}_0 \cdot \overline{\boldsymbol{\nabla}\boldsymbol{\theta}}_0, \qquad [A2.35]$$

(see the discussion in Section 4.2.2).

The expanded form of equation [A2.29] is written as

$$\bar{\theta}\rho_0^{(0)}\dot{\varsigma}_0^{(0)} - \frac{1}{\bar{\theta}+\epsilon\ldots}\boldsymbol{q}_0^{(0)} \cdot \boldsymbol{\nabla}\boldsymbol{\theta}_0^{(0)} + \boldsymbol{\Sigma}^{(0)} : \dot{\boldsymbol{F}}^{(0)} - \rho_0^{(0)}\dot{e}_0^{(0)} + \epsilon\ldots \geq 0. \qquad [A2.36]$$

For $\epsilon \to 0$, the entropy inequality in the microscale is obtained:

$$\bar{\theta}\rho_0^{(0)}\dot{\varsigma}_0^{(0)} - \frac{1}{\bar{\theta}}\boldsymbol{q}_0^{(0)} \cdot \boldsymbol{\nabla}\boldsymbol{\theta}_0^{(0)} + \boldsymbol{\Sigma}^{(0)} : \dot{\boldsymbol{F}}^{(0)} - \rho_0^{(0)}\dot{e}_0^{(0)} \geq 0. \qquad [A2.37]$$

Defining $\bar{\varsigma}_0 := \left\langle \rho_0^{(0)}\varsigma_0^{(0)} \right\rangle / \bar{\rho}_0$, the macroscale entropy inequality takes the form

$$\bar{\theta}\bar{\rho}_0\dot{\bar{\varsigma}}_0 - \frac{1}{\bar{\theta}}\bar{\boldsymbol{q}}_0 \cdot \overline{\boldsymbol{\nabla}\boldsymbol{\theta}}_0 + \bar{\boldsymbol{\Sigma}} : \dot{\bar{\boldsymbol{F}}} - \bar{\rho}_0\dot{\bar{e}}_0 \geq 0, \qquad [A2.38]$$

where equations [A2.19] and [A2.35] have been utilized.

Before closing this section, it is worth noting that we could equivalently develop a similar framework in the current configuration by utilizing the Cauchy stress $\boldsymbol{\sigma}$ and the inverse of deformation gradient $\boldsymbol{F}^{-1}$. These two ways of homogenization have been shown to be consistent with each other, providing meaningful average quantities [COS 05, BET 12]. However, caution is required with the time derivatives, since the spatial coordinates depend on time.

Bibliography

[ABO 03] ABOUDI J., PINDERA M.-J., ARNOLD S.M., "Higher-order theory for periodic multiphase materials with inelastic phases", *International Journal of Plasticity*, vol. 19, no. 6, pp. 805–847, 2003.

[ABO 04] ABOUDI J., "Micromechanics-based thermoviscoelastic constitutive equations for rubber-like matrix composites at finite strains", *International Journal of Solids and Structures*, vol. 41, pp. 5611–5629, 2004.

[ACC 13] ACCURATUS, "Silicon Carbide, SiC Ceramic Properties", available at: http://accuratus. com/silicar.html, 2013.

[AGG 04] AGGELIS D.G., TSINOPOULOS S.V., POLYZOS D., "An iterative effective medium approximation (IEMA) for wave dispersion and attenuation predictions in particulate composites, suspensions and emulsions", *The Journal of the Acoustical Society of America*, vol. 116, pp. 3443–3452, 2004.

[ALL 92] ALLAIRE G., "Homogenization and two-scale convergence", *SIAM Journal on Mathematical Analysis*, vol. 23, pp. 1482–1518, 1992.

[ALL 04] ALLAIRE G., JOUVE F., TOADER A.-M., "Structural optimization using sensitivity analysis and a level-set method", *Journal of Computational Physics*, vol. 194, no. 1, pp. 363–393, 2004.

[ASA 07] ASADA T., OHNO N., "Fully implicit formulation of elastoplastic homogenization problem for two-scale analysis", *International Journal of Solids and Structures*, vol. 44, nos. 22–23, pp. 7261–7275, 2007.

[BAN 12] BANIASSADI M., AHZI S., GARMESTANI H. *et al.*, "New approximate solution for N-point correlation functions for heterogeneous materials", *Journal of the Mechanics and Physics of Solids*, vol. 60, pp. 104–119, 2012.

[BEN 78] BENSOUSSAN A., LIONS J., PAPANICOLAOU G., *Asymptotic Methods for Periodic Structures*, North Holland, Amsterdam, 1978.

[BEN 89] BENVENISTE Y., DVORAK G.J., CHEN T., "Stress fields in composites with coated inclusions", *Mechanics of Materials*, vol. 7, pp. 305–317, 1989.

[BEN 90] BENVENISTE Y., DVORAK G.J., "On a Correspondence Between Mechanical and Thermal Effects in Two-Phase Composites", in WENG G.J., TAYA M., ABE H. (eds.), *Toshio Mura Anniversary Volume: Micro-mechanics and Inhomogeneity*, Springer, Berlin, 1990.

[BEN 91] BENVENISTE Y., DVORAK G.J., CHEN T., "On diagonal and elastic symmetry of the approximate effective stiffness tensor oh heterogeneous media", *Journal of the Mechanics and Physics of Solids*, vol. 39, no. 7, pp. 927–946, 1991.

[BEN 01] BENVENISTE Y., MILOH T., "Imperfect soft and stiff interfaces in two-dimensional elasticity", *Mechanics of Materials*, vol. 33, no. 6, pp. 309–323, 2001.

[BEN 14] BENAARBIA A., CHRYSOCHOOS A., ROBERT G., "Kinetics of stored and dissipated energies associated with cyclic loadings of dry polyamide 6.6 specimens", *Polymer Testing*, vol. 34, pp. 155–167, 2014.

[BEN 15] BENAARBIA A., CHRYSOCHOOS A., ROBERT G., "Thermomechanical behavior of PA6.6 composites subjected to low cycle fatigue", *Composites Part B*, vol. 76, pp. 52–64, 2015.

[BER 07] BERTHEL B., Mesures thermographiques de champs de dissipation accompagnant la fatigue à grand nombre de cycles des aciers, PhD Thesis, University of Montpellier II - Sciences et Techniques du Languedoc, 2007.

[BET 12] BETTAIEB M.B., DÉBORDES O. *et al.*, "Averaging properties for periodic homogenization and large deformation", *Journal for Multiscale Computational Engineering*, vol. 10, no. 3, pp. 281–293, 2012.

[BOD 01] BODOVILLÉ G., "The implicit standard material theory for modelling the nonassociative behaviour of metals", *Archive of Applied Mechanics*, vol. 71, pp. 426–435, 2001.

[BRA 88] BRAHIM-OTSMANE S., FRANCFORT G.A., MURAT F., "Homogenization in thermoelasticity", in KOHN R.V., MILTON G.W. (eds.), *Random Media and Composites*, S.I.A.M. Press, Philadelphia, 1988.

[BRA 92] BRAHIM-OTSMANE S., FRANCFORT G.A., MURAT F., "Correctors for the homogenization of the wave and heat equations", *Journal de Mathématiques Pures et Appliquées*, vol. 71, pp. 197–231, 1992.

[BRA 09] BRAVO-CASTILLERO J., RODRÍGUEZ-RAMOS R., MECHKOUR H. *et al.*, "Homogenization and effective properties of periodic thermomagnetoelectroelastic composites", *Journal of Mechanics of Materials and Structures*, vol. 4, no. 5, pp. 819–836, 2009.

[CAV 09] CAVALCANTE M.A.A., MARQUES S.P.C., PINDERA M.J., "Transient thermomechanical analysis of a layered cylinder by the parametric finite-volume theory", *Journal of Thermal Stresses*, vol. 32, pp. 112–134, 2009.

[CAV 11] CAVALCANTE M.A.A., KHATAM H., PINDERA M.J., "Homogenization of elastic-plastic periodic materials by FVDAM and FEM approaches - An assessment", *Composites Part B- Engineering*, vol. 42, pp. 1713–1730, 2011.

[CAV 16] CAVALCANTE M.A.A., PINDERA M.J., "Generalized FVDAM theory for elastic-plastic periodic materials", *International Journal of Plasticity*, vol. 77, pp. 90–117, 2016.

[CHA 05] CHABOCHE J., KANOUTE P., ROSS A., "On the capabilities of mean field approaches for the description of plasticity in metal matrix composites", *International Journal of Plasticity*, vol. 21, pp. 1409–1434, 2005.

[CHA 09] CHATZIGEORGIOU G., CHARALAMBAKIS N., MURAT F., "Homogenization of a pressurized tube made of elastoplastic materials with discontinuous properties", *International Journal of Solids and Structures*, vol. 46, no. 21, pp. 3902–3913, 2009.

[CHA 10a] CHARALAMBAKIS N., "Homogenization Techniques and Micromechanics. A Survey and Perspectives", *Applied Mechanics Reviews*, vol. 63, no. 3, p. 030803, 2010.

[CHA 10b] CHARALAMBAKIS N., MURAT F., "Stability by homogenization of thermoviscoplastic problems", *Mathematical Models and Methods in Applied Sciences*, vol. 20, pp. 1591–1616, 2010.

[CHA 10c] CHARALAMBAKIS N., MURAT F., "Two stable by homogenization models in simple shearing of rate-dependent non-homogeneous materials", *Quarterly of Applied Mathematics*, vol. 68, pp. 395–419, 2010.

[CHA 11] CHATZIGEORGIOU G., EFENDIEV Y., LAGOUDAS D.C., "Homogenization of aligned "fuzzy fiber" composites", *International Journal of Solids and Structures*, vol. 48, no. 19, pp. 2668–2680, 2011.

[CHA 12a] CHATZIGEORGIOU G., EFENDIEV Y., CHARALAMBAKIS N. *et al.*, "Effective Thermoelastic Properties of Composites with Periodicity in Cylindrical Coordinates", *International Journal of Solids and Structures*, vol. 49, no. 18, pp. 2590–2603, 2012.

[CHA 12b] CHATZIGEORGIOU G., SEIDEL G.D., LAGOUDAS D.C., "Effective mechanical properties of aligned 'fuzzy fiber' composites", *Composites Part B: Engineering*, vol. 43, pp. 2577–2593, 2012.

[CHA 14] CHATZIGEORGIOU G., JAVILI A., STEINMANN P., "Unified magnetomechanical homogenization framework with application to magnetorheological elastomers", *Mathematics and Mechanics of Solids*, vol. 19, no. 2, pp. 194–212, 2014.

[CHA 15] CHATZIGEORGIOU G., CHEMISKY Y., MERAGHNI F., "Computational micro to macro transitions for shape memory alloy composites using periodic homogenization", *Smart Materials and Structures*, vol. 24, p. 035009, 2015.

[CHA 16] CHATZIGEORGIOU G., CHARALAMBAKIS N., CHEMISKY Y. *et al.*, "Periodic homogenization for fully coupled thermomechanical modeling of dissipative generalized standard materials", *International Journal of Plasticity*, vol. 81, pp. 18–39, 2016.

[CHE 95] CHERKAOUI M., SABAR H., BERVEILLER M., "Elastic composites with coated reinforcements: a micromechanical approach for nonhomothetic topology", *International Journal of Engineering Science*, vol. 33, no. 6, pp. 829–843, 1995.

[CHE 01] CHEN W., FISH J., "A Dispersive Model for Wave Propagation in Periodic Heterogeneous Media Based on Homogenization With Multiple Spatial and Temporal Scales", *Journal of Applied Mechanics*, vol. 68, pp. 153–161, 2001.

[CHE 11] CHEMISKY Y., DUVAL A., PATOOR E. *et al.*, "Constitutive model for shape memory alloys including phase transformation, martensitic reorientation and twins accommodation", *Mechanics of Materials*, vol. 43, no. 7, pp. 361–376, 2011.

[CHR 79] CHRISTENSEN R.M., LO K.H., "Solutions for effective shear properties in three phase sphere and cylinder models", *Journal of the Mechanics and Physics of Solids*, vol. 27, pp. 315–330, 1979.

[CHR 89] CHRYSOCHOOS A., CHEZEAUX J.-C., CAUMON H., "Analyse thermomécanique des lois de comportement par thermographie infrarouge", *Revue de Physique Appliquée*, vol. 24, pp. 215–225, 1989.

[CHR 10] CHRYSOCHOOS A., HUON V., JOURDAN F. *et al.*, "Use of Full-Field DIC & IRT Measurements for the Thermomechanical Analysis of Material Behavior", *Strain*, vol. 46, no. 1, pp. 117–130, 2010.

[CIA 05] CIARLET P.G., "An Introduction to Differential Geometry with Applications to Elasticity", *Journal of Elasticity*, vols. 78–79, nos. 1–3, pp. 1–215, 2005.

[COL 67] COLEMAN B.D., GURTIN M.E., "Thermodynamics with Internal State Variables", *The Journal of Chemical Physics*, vol. 47, no. 2, pp. 597–613, 1967.

[COS 05] COSTANZO F., GRAY G.L., ANDIA P.C., "On the definitions of effective stress and deformation gradient for use in MD: Hill's macro-homogeneity and the virial theorem", *International Journal of Engineering Science*, vol. 43, no. 7, pp. 533–555, 2005.

[DES 01] DESRUMAUX F., MERAGHNI F., BENZEGGAGH M.L., "Generalised Mori-Tanaka Scheme to Model Anisotropic Damage Using Numerical Eshelby Tensor", *Journal of Composite Materials*, vol. 35, no. 7, pp. 603–624, 2001.

[DES 16] DESPRINGRE N., CHEMISKY Y., BONNAY K. *et al.*, "Micromechanical modeling of damage and load transfer in particulate composites with partially debonded interface", *Composite Structures*, vol. 155, pp. 77–88, 2016.

[DHA 15] DHALA S., RAY M.C., "Micromechanics of piezoelectric fuzzy fiber-reinforced composite", *Mechanics of Materials*, vol. 81, pp. 1–17, 2015.

[DOG 03] DOGHRI I., OUAAR A., "Homogenization of two-phase elasto-plastic composite materials and structures: Study of tangent operators, cyclic plasticity and numerical algorithms", *International Journal of Solids and Structures*, vol. 40, pp. 1681–1712, 2003.

[DUN 93] DUNN M.L., TAYA M., "Micromechanics predictions of the effective electroelastic moduli of piezoelectric composites", *International Journal of Solids and Structures*, vol. 30, no. 2, pp. 161–175, 1993.

[ENE 83] ENE H.I., "On linear thermoelasticity of composite materials", *International Journal of Engineering Science*, vol. 21, no. 5, pp. 443–448, 1983.

[ESH 57] ESHELBY J.D., "The Determination of the Elastic Field of an Ellipsoidal Inclusion, and Related Problems", *Proceedings of the Royal Society of London. Series A, Mathematical and Physical Sciences*, vol. 241, no. 1226, pp. 376–396, 1957.

[FER 37] FERMI E., *Thermodynamics*, Prentice-Hall, New York, 1937.

[FIS 97] FISH J., SHEK K., PANDHEERADI M. *et al.*, "Computational plasticity for composite structures based on mathematical homogenization: Theory and practice", *Computer Methods in Applied Mechanics and Engineering*, vol. 148, pp. 53–73, 1997.

[FIS 01] FISHER F.T., BRINSON L.C., "Viscoelastic interphases in polymer-matrix composites: theoretical models and finite-element analysis", *Composites Science and Technology*, vol. 61, pp. 731–748, 2001.

[FRA 86a] FRANCFORT G., SUQUET P., "Homogenization and mechanical dissipation in thermoviscoelasticity", *Archive for Rational Mechanics and Analysis*, vol. 96, no. 3, pp. 265–293, 1986.

[FRA 86b] FRANCFORT G., MURAT F., "Homogenization and optimal bounds in linear elasticity", *Archive for Rational Mechanics and Analysis*, vol. 94, no. 4, pp. 307–334, 1986.

[FRA 12] FRANCFORT G., GIACOMINI A., "Small strain heterogeneous elasto-plasticity revisited", *Communications on Pure and Applied Mathematics*, vol. 65, pp. 1185–1241, 2012.

[FRA 14] FRANCFORT G., GIACOMINI A., "On periodic homogenization in perfect elasto-plasticity", *Journal of European Mathematical Society*, vol. 16, pp. 409–461, 2014.

[FRA 15] FRANCFORT G., GIACOMINI A., "The role of a vanishing interfacial layer in perfect elasto-plasticity", *Chinese Annals of Mathematics. Series B*, vol. 36B, no. 5, pp. 813–828, 2015.

[FRE 02] FREMOND M., *Non-Smooth Thermomechanics*, Springer, New York, 2002.

[GAV 90] GAVAZZI A.C., LAGOUDAS D.C., "On the Numerical Evaluation of Eshelby's Tensor and its Application to Elastoplastic Fibrous Composites", *Computational Mechanics*, vol. 7, pp. 13–19, 1990.

[GEE 10] GEERS M.G.D., KOUZNETSOVA V.G., BREKELMANS W.A.M., "Multi-scale computational homogenization: Trends and challenges", *Journal of Computational and Applied Mathematics*, vol. 234, no. 7, pp. 2175–2182, 2010.

[GER 73] GERMAIN P., *Cours de mécanique des milieux continus, Tome I: Théorie Générale*, Masson, Paris, 1973.

[GER 82] GERMAIN P., "Sur certaines définitions liées à l'énergie en mécanique des solides", *International Journal of Engineering Science*, vol. 20, no. 2, pp. 245–259, 1982.

[GER 83] GERMAIN P., NGUYEN Q.S., SUQUET P., "Continuum thermodynamics", *Journal of Applied Mechanics*, vol. 50, pp. 1010–1020, 1983.

[GON 04] GONZALEZ C., SEGURADO J., LLORCA J., "Numerical simulation of elasto-plastic deformation of composites: evolution of stress microfields and implications for homogenization models", *Journal of the Mechanics and Physics of Solids*, vol. 52, pp. 1573–1593, 2004.

[GUE 90] GUEDES J.M., KIKUCHI N., "Preprocessing and posprocessing for materials based on the homogenization method with adaptive finite element methods", *Computer Methods in Applied Mechanics and Engineering*, vol. 83, pp. 143–198, 1990.

[GUI 01] GUINOVART-DÍAZ R., BRAVO-CASTILLERO J., RODRÍGUEZ-RAMOS R. *et al.*, "Closed-form expressions for the effective coefficients of fibre-reinforced composite with transversely isotropic constituents. I: Elastic and hexagonal symmetry", *Journal of the Mechanics and Physics of Solids*, vol. 49, no. 7, pp. 1445–1462, 2001.

[HAL 75] HALPHEN B., NGUYEN Q.S., "Sur les matériaux standards généralisés", *Journal de Mécanique*, vol. 14, no. 1, pp. 39–63, 1975.

[HAR 10] HARTL D.J., CHATZIGEORGIOU G., LAGOUDAS D.C., "Three-dimensional modeling and numerical analysis of rate-dependent irrecoverable deformation in shape memory alloys", *International Journal of Plasticity*, vol. 26, no. 10, pp. 1485–1507, 2010.

[HAS 63] HASHIN Z., SHTRIKMAN S., "A variational approach to the theory of the elastic behaviour of multiphase materials", *Journal of the Mechanics and Physics of Solids*, vol. 11, no. 2, pp. 127–140, 1963.

[HAS 64] HASHIN Z., ROSEN B.W., "The elastic moduli of fiber-reinforced materials", *Journal of Applied Mechanics*, vol. 31, pp. 223–232, 1964.

[HAS 83] HASHIN Z., "Analysis of Composite Materials: A Survey", *Journal of Applied Mechanics*, vol. 50, pp. 481–505, 1983.

[HAS 98] HASSANI B., HINTON E., "A review of homogenization and topology optimization I - homogenization theory for media with periodic structure", *Computers and Structures*, vol. 69, pp. 707–717, 1998.

[HAT 86] HATTA H., TAYA M., "Equivalent inclusion method for steady state heat conduction in composites", *International Journal of Engineering Science*, vol. 24, no. 7, pp. 1159–1172, 1986.

[HER 07] HERZOG H., JACQUET E., "From a shape memory alloys model implementation to a composite behavior", *Computational Materials Science*, vol. 39, pp. 365–375, 2007.

[HIL 63] HILL R., "Elastic properties of reinforced solids: Some theoretical principles", *Journal of the Mechanics and Physics of Solids*, vol. 11, pp. 357–372, 1963.

[HIL 65] HILL R., "Continuum micro-mechanics of elastoplastic polycrystals", *Journal of the Mechanics and Physics of Solids*, vol. 13, pp. 89–101, 1965.

[HIL 67] HILL R., "The essential structure of constitutive laws for metal composites and polycrystals", *Journal of the Mechanics and Physics of Solids*, vol. 15, no. 2, pp. 79–95, 1967.

[HIL 72a] HILL R., "On Constitutive Macro-Variables for Heterogeneous Solids at Finite Strain", *Proceedings of the Royal Society of London A*, vol. 326, no. 1565, pp. 131–147, 1972.

[HIL 72b] HILL R., RICE J., "Constitutive analysis of elastic-plastic crystals at arbitrary strain", *Journal of the Mechanics and Physics of Solids*, vol. 20, no. 6, pp. 401–413, 1972.

[HÖH 03] HÖHNE G.W.H., HEMMINGER W., FLAMMERSHEIM H.-J., *Differential Scanning Calorimetry*, 2nd ed., Springer-Verlag, New York, 2003.

[HOR 93] HORI M., NEMAT-NASSER S., "Double-inclusion model and overall moduli of multi-phase composites", *Mechanics of Materials*, vol. 14, pp. 189–206, 1993.

[HU 00a] HU G.K., HUANG G.L., "Influence of residual stress on the elastic-plastic deformation of composites with two- or three-dimensional randomly oriented inclusions", *Acta Mechanica*, vol. 141, pp. 193–200, 2000.

[HU 00b] HU G.K., WENG G.J., "The connections between the double-inclusion model and the Ponte Castaneda-Willis, Mori-Tanaka, and Kuster-Toksoz models", *Mechanics of Materials*, vol. 32, pp. 495–503, 2000.

[HUI 14] HUI T., OSKAY C., "A high order homogenization model for transient dynamics of heterogeneous media including micro-inertia effects", *Computer Methods in Applied Mechanics and Engineering*, vol. 273, pp. 181–203, 2014.

[HUT 76] HUTCHINSON J.W., "Bounds and self-consistent estimates for creep of polycrystalline materials", *Proceedings of Royal Society of London A*, vol. 348, pp. 101–127, 1976.

[JAV 13] JAVILI A., CHATZIGEORGIOU G., STEINMANN P., "Computational homogenization in magneto-mechanics", *International Journal of Solids and Structures*, vol. 50, pp. 4197–4216, 2013.

[JEN 09] JENDLI Z., MERAGHNI F., FITOUSSI J. *et al.*, "Multi-scales modelling of dynamic behaviour for discontinuous fibre SMC composites", *Composites Science and Technology*, vol. 69, no. 1, pp. 97–103, 2009.

[KAL 97] KALAMKAROV A.L., KOLPAKOV A.G., *Analysis, Design and Optimization of Composite Structures*, Wiley, West Sussex, 1997.

[KAN 09] KANOUTÉ P., BOSO D.P., CHABOCHE J.L. *et al.*, "Multiscale methods for composites: a review", *Archives of Computational Methods in Engineering*, vol. 16, pp. 31–75, 2009.

[KEL 13] KELLY P., Mechanics lecture notes, available at: http://homepages.engineering.auckland.ac.nz/~pkel015/SolidMechanics Books, 2013.

[KHA 95] KHAN A.S., HUANG S., *Continuum Theory of Plasticity*, John Wiley & Sons, New York, 1995.

[KHA 09] KHATAM H., PINDERA M.J., "Parametric finite-volume micromechanics of periodic materials with elastoplastic phases", *International Journal of Plasticity*, vol. 25, pp. 1386–1411, 2009.

[KRU 11] KRUCH S., CHABOCHE J.L., "Multi-scale analysis in elasto-viscoplasticity coupled with damage", *International Journal of Plasticity*, vol. 27, pp. 2026–2039, 2011.

[KUN 14] KUNDALWAL S.I., RAY M.C., "Effect of carbon nanotube waviness on the effective thermoelastic properties of a novel continuous fuzzy fiber reinforced composite", *Composites Part B: Engineering*, vol. 57, pp. 199–209, 2014.

[LAG 91] LAGOUDAS D.C., GAVAZZI A.C., NIGAM H., "Elastoplastic behavior of metal matrix composittes based on incremental plasticity and the Mori-Tanaka averaging scheme", *Computational Mechanics*, vol. 8, pp. 193–203, 1991.

[LAG 08] LAGOUDAS D.C., *Shape Memory Alloys: Modeling and Engineering Applications*, Springer, New York, 2008.

[LAG 12] LAGOUDAS D.C., HARTL D., CHEMISKY Y. *et al.*, "Constitutive model for the numerical analysis of phase transformation in polycrystalline shape memory alloys", *International Journal of Plasticity*, vols. 32–33, pp. 155–183, 2012.

[LAH 13] LAHELLEC N., SUQUET P., "Effective response and field statistics in elasto-plastic and elasto-viscoplastic composites under radial and non-radial loadings", *International Journal of Plasticity*, vol. 42, pp. 1–30, 2013.

[LAI 10] LAI W.M., RUBIN D., KREMPL E., *Introduction to Continuum Mechanics*, 4th ed., Elsevier, Burlington, 2010.

[LEM 02] LEMAITRE J., CHABOCHE J.L., *Mechanics of Solid Materials*, Cambridge University Press, Cambridge, 2002.

[LEQ 08] LE QUANG H., HE Q.-C., ZHENG Q.-S., "Some general properties of Eshelby's tensor fields in transport phenomena and anti-plane elasticity", *International Journal of Solids and Structures*, vol. 45, pp. 3845–3857, 2008.

[LES 11] LESTER B.T., CHEMISKY Y., LAGOUDAS D.C., "Transformation characteristics of shape memory alloy composites", *Smart Materials and Structures*, vol. 20, no. 9, p. 094002, 2011.

[LEV 67] LEVIN V.M., "On the Coefficients of Thermal Expansion of Heterogeneous Materials (in Russian)", *Mekhanika Tverdogo Tela*, vol. 1, pp. 88–94, 1967.

[LOV 06] LOVE B., BATRA R.C., "Determination of effective thermomechanical parameters of a mixture of two elastothermoviscoplastic constituents", *International Journal of Plasticity*, vol. 22, pp. 1026–1061, 2006.

[LUB 72] LUBLINER J., "On the thermodynamic foundations of non-linear solid mechanics", *International Journal of Non-Linear Mechanics*, vol. 7, no. 3, pp. 237–254, 1972.

[MAG 03] MAGHOUS S., CREUS G.J., "Periodic homogenization in thermoviscoelasticity: case of multilayered media with ageing", *International Journal of Solids and Structures*, vol. 40, pp. 851–870, 2003.

[MAL 69] MALVERN L.E., *Introduction to the Mechanics of a Continuous Medium*, Prentice Hall, Hertfordshire, 1969.

[MAU 92] MAUGIN G.A., *The Thermomechanics of Plasticity and Fracture*, Cambridge University Press, Cambridge, 1992.

[MAU 99] MAUGIN G.A., *The Thermomechanics of Nonlinear Irreversible Behaviors – An Introduction*, World Scientific, Singapore, 1999.

[MER 95] MERAGHNI F., BENZEGGAGH M.L., "Micromechanical modelling of matrix degradation in randomly oriented discontinuous-fibre composites", *Composites Science and Technology*, vol. 55, no. 2, pp. 171–186, 1995.

[MER 02] MERAGHNI F., DESRUMAUX F., BENZEGGAGH M.L., "Implementation of a constitutive micromechanical model for damage analysis in glass mat reinforced composite structures", *Composites Science and Technology*, vol. 62, no. 16, pp. 2087–2097, 2002.

[MER 09] MERCIER S., MOLINARI A., "Homogenization of elastic-viscoplastic heterogeneous materials: Self-consistent and Mori-Tanaka schemes", *International Journal of Plasticity*, vol. 25, pp. 1024–1048, 2009.

[MER 12] MERCIER S., MOLINARI A., BERBENNI S. *et al.*, "Comparison of different homogenization approaches for elastic-viscoplastic materials", *Modelling and Simulation in Materials Science and Engineering*, vol. 20, no. 2, p. 024004, 2012.

[MIC 99] MICHEL J.C., MOULINEC H., SUQUET P., "Effective properties of composite materials with periodic microstructure: a computational approach", *Computer Methods in Applied Mechanics and Engineering*, vol. 172, pp. 109–143, 1999.

[MOR 70] MOREAU J.J., "Sur les lois de frottement, de plasticité et de viscosité", *Comptes Rendus de l'Académie des Sciences, Series A*, vol. 271, pp. 608–611, 1970.

[MOR 73] MORI T., TANAKA K., "Average stress in matrix and average elastic energy of materials with misfitting inclusions", *Acta Metallurgica*, vol. 21, no. 5, pp. 571–574, 1973.

[MOR 05] MOREAU S., CHRYSOCHOOS A., MURACCIOLE J.-M. *et al.*, "Analysis of thermoelastic effects accompanying the deformation of PMMA and PC polymers", *Comptes Rendus Mécanique*, vol. 333, pp. 648–653, 2005.

[MOU 98] MOULINEC H., SUQUET P., "A numerical method for computing the overall response of nonlinear composites with complex microstructure", *Computer methods in applied mechanics and englneering*, vol. 157, pp. 69–94, 1998.

[MUR 78] MURAT F., "Compacité par compensation", *Annali della Scuola Normale Superiore di Pisa*, vol. 5, pp. 489–507, 1978.

[MUR 87] MURA T., "Micromechanics of Defects in Solids", in NEMAT-NASSER S., ORAVAS G.A.E. (eds.), *Mechanics of Elastic and Inelastic Solids*, 2nd ed., Kluwer Academic Publishers, Dordrecht, 1987.

[MUR 97] MURAT F., TARTAR L., "H-convergence, in Topics in the mathematical modelling of composite materials", in CHERKAEV A., KOHN R.V. (eds.), *Progress in Nonlinear Differential Equations and their Applications*, Birkhäuser, Boston, 1997.

[NEM 99a] NEMAT-NASSER S., "Averaging theorems in finite deformation plasticity", *Mechanics of Materials*, vol. 31, pp. 493–523, 1999.

[NEM 99b] NEMAT-NASSER S., HORI M., *Micromechanics: Overall Properties of Heterogeneous Materials*, 2nd ed., North-Holland, Amsterdam, 1999.

[NEM 11] NEMAT-NASSER S., WILLIS J.R., SRIVASTAVA A. *et al.*, "Homogenization of periodic elastic composites and locally resonant sonic materials", *Physical Review B*, vol. 83, p. 104103, 2011.

[NGU 88] NGUYEN Q.S., "Mechanical modelling of anelasticity", *Revue de Physique Appliquée*, vol. 23, pp. 325–330, 1988.

[NGU 89] NGUETSENG G., "A general convergence result for a functional related to the theory of homogenization", *SIAM Journal of Mathematical Analysis*, vol. 20, pp. 608–623, 1989.

[ONS 31] ONSAGER L., "Reciprocal relations in irreversible processes I", *Physical Review*, vol. 37, pp. 405–426, 1931.

[ORT 86] ORTIZ M., SIMO J.C., "An analysis of a new class of integration algorithms for elastoplastic constitutive relations", *International Journal for Numerical Methods in Engineering*, vol. 23, pp. 353–366, 1986.

[PAL 92] PALEY M., ABOUDI J., "Micromechanical analysis of composites by the generalized cells model", *Mechanics of Materials*, vol. 14, pp. 127–139, 1992.

[PIN 09] PINDERA M.J., KHATAM H., DRAGO A.S. *et al.*, "Micromechanics of spatially uniform heterogeneous media: A critical review and emerging approaches", *Composites Part B: Engineering*, vol. 40, no. 5, pp. 349–378, 2009.

[PON 91] PONTE-CASTAÑEDA P., "The effective mechanical properties of nonlinear isotropic composites", *Journal of the Mechanics and Physics of Solids*, vol. 39, pp. 45–71, 1991.

[PON 95] PONTE-CASTAÑEDA P., WILLIS J.R., "The effect of spatial distribution on the effective behavior of composite materials and cracked media", *Journal of the Mechanics and Physics of Solids*, vol. 43, no. 12, pp. 1919–1951, 1995.

[PON 97] PONTE-CASTAÑEDA P., SUQUET P., "Nonlinear composites", *Advances in Applied Mechanics*, vol. 34, pp. 171–302, 1997.

[QID 00] QIDWAI M.A., LAGOUDAS D.C., "Numerical Implementation of a Shape Memory Alloy Thermomechanical Constitutive Model Using Return Mapping Algorithms", *International Journal for Numerical Methods in Engineering*, vol. 47, pp. 1123–1168, 2000.

[QU 06] QU J., CHERKAOUI M., *Fundamentals of Micromechanics of Solids*, Wiley, New Jersey, 2006.

[RAN 13] RANC N., CHRYSOCHOOS A., "Calorimetric consequences of thermal softening in Johnson-Cook's model", *Mechanics of Materials*, vol. 65, pp. 44–55, 2013.

[REM 16] REMOND Y., AHZI S., BANIASSADI M. *et al.*, *Applied RVE Reconstruction and Homogenization of Heterogeneous Materials*, ISTE Ltd., John Wiley & Sons, London, 2016.

[ROC 70] ROCKAFELLAR T., *Convex Analysis*, Princeton, New Jersey, 1970.

[ROD 01] RODRÍGUEZ-RAMOS R., SABINA F.J., GUINOVART-DÍAZ R. *et al.*, "Closed-form expressions for the effective coefficients of a fiber-reinforced composite with transversely isotropic constituents - I. Elastic and square symmetry", *Mechanics and Materials*, vol. 33, no. 4, pp. 223–235, 2001.

[ROS 70] ROSEN B.W., HASHIN Z., "Effective thermal expansion coefficients and specific heats of composite materials", *International Journal of Engineering Science*, vol. 8, pp. 157–173, 1970.

[ROS 00] ROSAKIS P., ROSAKIS A.J., RAVICHANDRAN G. *et al.*, "A thermodynamic internal variable model for the partition of plastic work into heat and stored energy in metals", *Journal of the Mechanics and Physics of Solids*, vol. 48, pp. 581–607, 2000.

[SAN 78] SANCHEZ-PALENCIA E., *Non-Homogeneous Media and Vibration Theory*, Springer-Verlag, Berlin, 1978.

[SAX 02] DE SAXCÉ G., BOUSSHINE L., *Implicit Standard Materials*, Springer-Verlag, New York, 2002.

[SEG 02] SEGURADO J., LLORCA J., "A numerical approximation to the elastic properties of sphere-reinforced composites", *Journal of the Mechanics and Physics of Solids*, vol. 50, pp. 2107–2121, 2002.

[SEN 12] SENGUPTA A., PAPADOPOULOS P., TAYLOR R.L., "A multiscale finite element method for modeling fully coupled thermomechanical problems in solids", *International Journal for Numerical Methods in Engineering*, vol. 91, pp. 1386–1405, 2012.

[SIM 92] SIMO J.C., MIEHE C., "Associative coupled thermoplasticity at finite strains: Formulation, numerical analysis and implementation", *Computer Methods in Applied Mechanics and Engineering*, vol. 98, pp. 41–104, 1992.

[SIM 98] SIMO J.C., HUGHES T.J.R., *Computational Inelasticity*, Springer-Verlag, New York, 1998.

[STE 08] STEINMANN P., "On boundary potential energies in deformational and configurational mechanics", *Journal of the Mechanics and Physics of Solids*, vol. 56, no. 3, pp. 772–800, 2008.

[STE 15] STEINMANN P., *Geometrical Foundations of Continuum Mechanics: An Application to First- and Second-Order Elasticity and Elasto-Plasticity*, Springer, 2015.

[STR 11] STRÁNSKÝ J., VOREL J., ZEMAN J. *et al.*, "Mori-Tanaka Based Estimates of Effective Thermal Conductivity of Various Engineering Materials", *Micromachines*, vol. 2, pp. 129–149, 2011.

[SUN 04] SUN L.Z., JU J.W., "Elastoplastic Modeling of Metal Matrix Composites Containing Randomly Located and Oriented Spheroidal Particles", *Journal of Applied Mechanics*, vol. 71, pp. 774–785, 2004.

[SUQ 87] SUQUET P.M., *Elements of Homogenization for Inelastic Solid Mechanics*, Springer, Berlin, 1987.

[SUQ 12] SUQUET P., "Four exact relations for the effective relaxation function of linear viscoelastic composites", *Comptes Rendus Mécanique*, vol. 340, pp. 387–399, 2012.

[TAK 85] TAKAO Y., TAYA M., "Thermal Expansion Coefficients and Thermal Stresses in an Aligned Short Fiber Composite With Application to a Short Carbon Fiber/Aluminum", *Journal of Applied Mechanics*, vol. 52, pp. 806–810, 1985.

[TAR 77] TARTAR L., Homogénéisation et compacité par compensation, Cours Peccot, Collège de France, 1977.

[TAR 78] TARTAR L., *Nonlinear Constitutive Relations and Homogenization*, North-Holand, Amsterdam, 1978.

[TAR 79] TARTAR L., "Compensated compactness and applications to partial differential equations", *Nonlinear Analysis and Mechanics, Heriot Watt Symposium IV*, pp. 136–212, Pitman, San Francisco, USA, 1979.

[TEK 10] TEKOĞLU C., PARDOEN T., "A micromechanics based damage model for composite materials", *International Journal of Plasticity*, vol. 26, pp. 549–569, 2010.

[TEM 12] TEMIZER I., "On the asymptotic expansion treatment of two-scale finite thermoelasticity", *International Journal of Engineering Science*, vol. 53, pp. 74–84, 2012.

[TER 01] TERADA K., KIKUCHI N., "A class of general algorithms for multi-scale analyses of heterogeneous media", *Computer Methods in Applied Mechanics and Engineering*, vol. 190, pp. 5427–5464, 2001.

[TSA 12] TSALIS D., CHATZIGEORGIOU G., CHARALAMBAKIS N., "Homogenization of structures with generalized periodicity", *Composites Part B: Engineering*, vol. 43, pp. 2495–2512, 2012.

[TSA 13a] TSALIS D., BAXEVANIS T., CHATZIGEORGIOU G. *et al.*, "Homogenization of elastoplastic composites with generalized periodicity in the microstructure", *International Journal of Plasticity*, vol. 51, pp. 161–187, 2013.

[TSA 13b] TSALIS D., CHATZIGEORGIOU G., CHARALAMBAKIS N., "Admissible deformation fields for the homogenization of elastoplastic materials with generalized periodicity", *Mechanics Research Communications*, vol. 53, pp. 43–46, 2013.

[TSA 15] TSALIS D., CHATZIGEORGIOU G., TSAKMAKIS C. *et al.*, "Dissipation inequality-based periodic homogenization of wavy materials", *Composites Part B*, vol. 76, pp. 89–104, 2015.

[TSU 12] TSUKROV I., DRACH B., GROSS T.S., "Effective stiffness and thermal expansion coefficients of unidirectional composites with fibers surrounded by cylindrical orthotropic matrix layers", *International Journal of Engineering Science*, vol. 58, pp. 129–143, 2012.

[WAC 83] WACK B., TERRIEZ J.-M., GUELIN P., "A Hereditary Type, Discrete Memory, Constitutive Equation with Applications to Simple Geometries", *Acta Mechanica*, vol. 50, pp. 9–37, 1983.

[WAK 91] WAKASHIMA K., TSUKAMOTO H., "Mean-field micromechanics model and its application to the analysis of thermomechanical behaviour of composite materials", *Materials Science and Engineering: A*, vol. 146, nos. 1–2, pp. 291–316, 1991.

[WAL 81] WALPOLE L.J., "Elastic Behavior of Composite Materials: Theoretical Foundations", *Advances in Applied Mechanics*, vol. 21, pp. 169–242, 1981.

[WAN 16a] WANG G., PINDERA M.J., "Locally-exact homogenization of unidirectional composites with coated or hollow reinforcement", *Materials and Design*, vol. 93, pp. 514–528, 2016.

[WAN 16b] WANG G., PINDERA M.J., "Locally exact homogenization of unidirectional composites with cylindrically orthotropic fibers", *Journal of Applied Mechanics*, vol. 83, p. 071010, 2016.

[WIK 13] WIKIPEDIA, Curvilinear coordinates, available at: http://en.wikipedia.org/wiki/Curvilinear_coordinates, 2013.

[WON 99] WONG C.P., BOLLAMPALLY R.S., "Thermal Conductivity, Elastic Modulus, and Coefficient of Thermal Expansion of Polymer Composites Filled with Ceramic Particles for Electronic Packaging", *Journal of Applied Polymer Science*, vol. 74, pp. 3396–3403, 1999.

[YAN 94] YANG R.-B., MAL A.K., "Multiple scattering of elastic waves in a fiber-reinforced composite", *Journal of the Mechanics and Physics of Solids*, vol. 42, no. 12, pp. 1945–1968, 1994.

[YU 02] YU Q., FISH J., "Multiscale asymptotic homogenization for multiphysics problems with multiple spatial and temporal scales: a coupled thermo-viscoelastic example problem", *International Journal of Solids and Structures*, vol. 39, no. 26, pp. 6429–6452, 2002.

[YVO 09] YVONNET J., GONZALEZ D., HE Q.-C., "Numerically explicit potentials for the homogenization of nonlinear elastic heterogeneous materials", *Computer Methods in Applied Mechanics and Engineering*, vol. 198, nos. 33–36, pp. 2723–2737, 2009.

[ZIE 63] ZIEGLER H., "Some extremum principles in irreversible thermodynamics with application to continuum mechanics", in SNEDDON I.R. *et al.* (eds.), *Progress in Solid Mechanics*, vol. 4, North Holland, New York, 1963.